# Optical Properties of Narrow-Gap Low-Dimensional Structures

# NATO ASI Series

## Advanced Science Institutes Series

*A series presenting the results of activities sponsored by the NATO Science Committee,
which aims at the dissemination of advanced scientific and technological knowledge,
with a view to strengthening links between scientific communities.*

The series is published by an international board of publishers in conjunction with the
NATO Scientific Affairs Division

| | | |
|---|---|---|
| **A** | **Life Sciences** | Plenum Publishing Corporation |
| **B** | **Physics** | New York and London |
| | | |
| **C** | **Mathematical** | D. Reidel Publishing Company |
| | **and Physical Sciences** | Dordrecht, Boston, and Lancaster |
| | | |
| **D** | **Behavioral and Social Sciences** | Martinus Nijhoff Publishers |
| **E** | **Engineering and** | The Hague, Boston, Dordrecht, and Lancaster |
| | **Materials Sciences** | |
| | | |
| **F** | **Computer and Systems Sciences** | Springer-Verlag |
| **G** | **Ecological Sciences** | Berlin, Heidelberg, New York, London, |
| **H** | **Cell Biology** | Paris, and Tokyo |

*Recent Volumes in this Series*

*Series B: Physics*

# Optical Properties of Narrow-Gap Low-Dimensional Structures

Edited by

## C. M. Sotomayor Torres

University of St. Andrews
St. Andrews, Scotland

## J. C. Portal

CNRS–INSA
Toulouse, France
and CNRS–SNCI
Grenoble, France

## J. C. Maan

Max-Planck-Institut für Festkörperforschung
Grenoble, France

and

## R. A. Stradling

Imperial College of Science and Technology
London, England

Plenum Press
New York and London
Published in cooperation with NATO Scientific Affairs Division

Proceedings of a NATO Advanced Research Workshop on
Optical Properties of Narrow-Gap Low-Dimensional Structures,
held July 29–August 1, 1986,
at St. Andrews, Scotland

Library of Congress Cataloging in Publication Data

NATO Advanced Research Workshop on Optical Properties of Narrow-Gap
   Low-Dimensional Structures (1986 Saint Andrews, Fife)
   Optical properties of narrow-gap low-dimensional structures.

   (NATO ASI series. Series B, Physics; v. 152)
   "Proceedings of a NATO Advanced Research Workshop on Optical Properties of Narrow-Gap Low-Dimensional Structures, held July 29–August 1, 1986,
at St. Andrews, Scotland"—T.p. verso.
   "Published in cooperation with NATO Scientific Affairs Division."
   Includes bibliographies and indexes.
   1. Narrow gap semiconductors—Optical properties—Congresses. 2. One-
dimensional conductors—Optical properties—Congresses. I. Sotomayor To-
rres, C.M. II. North Atlantic Treaty Organization. Scientific Division. III. Title.
IV. Series.
QC611.8.N35N38   1986              537.6′22                    87-12355
ISBN-13: 978-1-4612-9047-6      e-ISBN-13: 978-1-4613-1879-8
DOI: 10.1007/ 978-1-4613-1879-8

PREFACE


    This volume contains the Proceedings of the NATO Advanced Research
Workshop on "Optical Properties of Narrow-Gap Low-Dimensional
Structures", held from July 29th to August 1st, 1986, in St. Andrews,
Scotland, under the auspices of the NATO International Scientific
Exchange Program.

    The workshop was not limited to optical properties of narrow-gap
semiconductor structures (Part III). Sessions on, for example, the
growth methods and characterization of III-V, II-VI, and IV-VI
materials, discussed in Part II, were an integral part of the workshop.
Considering the small masses of the carriers in narrow-gap low-
dimensional structures (LDS), in Part I the enhanced band mixing and
magnetic field effects are explored in the context of the envelope
function approximation. Optical nonlinearities and energy relaxation
phenomena applied to the well-known systems of HgCdTe and GaAs/GaAlAs,
respectively, are reviewed with comments on their extension to narrow-
gap LDS. The relevance of optical observations in quantum transport
studies is illustrated in Part IV. A review of devices based on
epitaxial narrow-gap materials defines a frame of reference for future
ones based on two-dimensional narrow-gap semiconductors; in addition, an
analysis of the physics of quantum well lasers provides a guide to
relevant parameters for narrow-gap laser devices for the infrared (Part
V). The roles and potentials of special techniques are explored in Part
VI, with emphasis on hydrostatic pressure techniques, since this has a
pronounced effect in small-mass, narrow-gap, non-parabolic structures.
Poster contributions were displayed throughout this NATO workshop, and
their results were incorporated in the relevant sessions. An informal
session on band offsets was held one evening, with many short
contributions on recent results, followed by a lively discussion.

    The Organizing Committee would like to express its sincere thanks
to Eric Thirkell, John Speed, Ian Ferguson, and Morag Watt for their
assistance with the smooth running of the workshop. It would also like
to thank the following companies for financial assistance for
entertainment: Hughes Microelectronics (Glenrothes), Barr and Stroud
(Glasgow), and Ferranti Defense Systems Ltd (Edinburgh). Their
contribution in stationery is also acknowledged. The hospitality of the
Department of Physics of St. Andrews University is cordially
acknowledged.

    Finally, we would like to express our appreciation of the
assistance provided by Madeleine Carter (Plenum Publishing Corporation),
Karen Lumsden, and Morag Watt in the preparation of this volume.

Autumn 1986
                                        C.M. Sotomayor Torres
                                        J.C. Portal
                                        J.C. Mann
                                        R.A. Stradling

v

CONTENTS

# ELECTRONIC ENERGY LEVELS IN NARROW-GAP LOW-DIMENSIONAL STRUCTURES

G. Bastard, J.A. Brum and J.M. Berroir

Groupe de Physique des Solides de
l'Ecole Normale Superieure
24 rue Lhomond,
F-75005 Paris (France)

## I  INTRODUCTION

The last few years have witnessed an increasing effort to better describe the electronic structure of semiconductor heterostructures (quantum wells, superlattices...). Here, we restrict our considerations to the envelope description of the subband structure in heterolayers[1-5]. Such a description has proved to be versatile and reliable for the electronic states which are energetically close to the hosts' band extrema. Meanwhile, efforts in growth techniques have considerably increased the number of heterolayers. In particular, the quality of heterostructures involving narrow bandgap materials (eg. Ga(In)As, Hg(Cd)Te) has been significantly improved. These materials are important for the infra-red detection.

The band structure of bulk narrow gap materials is in the vicinity of the $\Gamma$ point well described by the Kane model[6]. In this model the $\mathbf{k.p}$ interaction between the closely spaced $\Gamma_6$, $\Gamma_7$, $\Gamma_8$ bands is exactly diagonalised while the effect of remote bands is taken into account only up to the second order in $\mathbf{k}$. Let us denote by $\mathbf{z}$ the quantization axis of the total angular momentum $\mathbf{J}$ of the carrier. Then, $m_J = \pm 3/2$ correspond to the $\Gamma_8$ heavy hole band while $m_J = \pm 1/2$ correspond to the light particle bands, ie. the $\Gamma_6$, $\Gamma_7$ and the light $\Gamma_8$ bands. This classification holds only if the carrier wavevector $\mathbf{k}$ is parallel to $\mathbf{J}$. It is always possible in bulk materials to find a basis where $\mathbf{k}$ and $\mathbf{J}$ are lined up. In A-B heterostructures the existence of a preferential axis, the growth ($\mathbf{z}$) axis, and of a band edge profile, which is z-dependent, implies that the splitting of electronic states into decoupled heavy hole and light particle states is possible only if the carrier wavevectors $\mathbf{k}_A$ and $\mathbf{k}_B$ are parallel to z; in other words if the in-plane wavevector $\mathbf{k}_\perp$, which is conserved at the interfaces, is equal to zero. If $\mathbf{k}_\perp \neq \mathbf{0}$ the light and heavy particle states become hybridized and the in-plane subband structure becomes complicated. We shall return to this point in section III.

## II  HETEROLAYER ELECTRONIC STATES AT $\mathbf{k}_\perp = \mathbf{0}$

If $\mathbf{k}_\perp = \mathbf{0}$, one may describe the heavy hole and light particle states independently. The heavy hole states are without surprise. If we

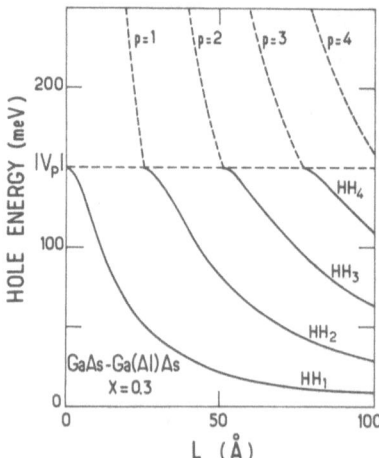
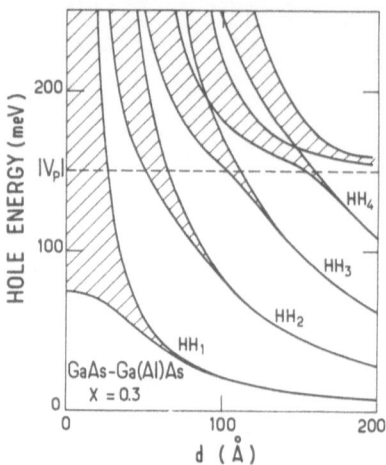

Fig. 1. The heavy hole confinement energies in GaAs-Ga(Al)As, x=0.3 single quantum wells are plotted versus the GaAs slab thickness L. The solid lines correspond to bound states and the dashed lines to virtual bound states.

Fig. 2. The heavy hole superlattice bands in GaAs-Ga(Al)As, x=0.3 superlattices are plotted versus the super-lattice period d. The hatched areas correspond to allowed superlattice states $L_A=L_B$.

denote by $V_p$ the algebraic energy shift of the $\Gamma_8$ edge when going from the A to the B materials and if $V_p<0$ the A material is a potential well for holes whose motion becomes size-quantized for hole energies such that $0<\epsilon_h<|V_p|$ (see fig.1). There are $1+\text{Int}(2m_{hh}|V_p|L_A^2/\hbar^2\pi^2)^{1/2}$ such bound levels, where $m_{hh}$ is the heavy hole mass along the growth axis in the A material, $L_A$ is the A slab thickness and $\text{Int}(x)$ denotes the integer part of x. If the heterostructure consists of a superlattice instead of a single quantum well, the bound states of isolated wells hybridize and give rise to the superlattice minibands (fig.2) whose widths decrease almost exponentially with the barrier thickness $L_B$.

The allowed light particle states of a A-B superlattice are the solutions of:

$$\cos(qd)=\cos(k_AL_A)\cos(k_BL_B)-1/2(\xi+1/\xi)\sin(k_AL_A)\sin(k_BL_B) \qquad (1)$$

with:

$$\epsilon(\epsilon+\epsilon_A)(\epsilon+\epsilon_A+\Delta_A)=\hbar^2k_A^2P^2(\epsilon+\epsilon_A+2\Delta_A/3) \qquad (2)$$

$$(\epsilon-V_S)(\epsilon-V_S+\epsilon_B)(\epsilon-V_S+\epsilon_B+\Delta_B)=\hbar^2k_B^2P^2(\epsilon-V_S+2\Delta_B/3) \qquad (3)$$

$$\xi=k_A\nu_B/k_B\nu_A \qquad (4)$$

$$\xi=k_A/k_B[2/(\epsilon+\epsilon_A)+1/(\epsilon+\epsilon_A+\Delta_A)]/[2/(\epsilon-V_S+\epsilon_B)+1/(\epsilon-V_S+\epsilon_B+\Delta_B)] \qquad (5)$$

$$d=L_A+L_B \qquad (6)$$

In eqs(1-4) the energy zero has been taken at the $\Gamma_6$ edge of the A material. q is the superlattice wavevector along the growth axis, $V_S$ the algebraic energy shift of the $\Gamma_6$ edge when going from the A to the B layer; $\varepsilon_A$ ($\varepsilon_B$) is the $\Gamma_6$-$\Gamma_8$ bandgap energy of the A(B) material and $\Delta_A$ ($\Delta_B$) the spin orbit energies. Finally, P is the Kane matrix element:

$$P = -i|<S|p_x|X>|^2/m_o \qquad (7)$$

To obtain eqs(1-4) we have dropped the higher bands as well as the free electron kinetic energy contributions. The hosts band non-parabolicity shows up in eqs(2,3).

In fact eq(3) can be rewritten in the form of $\varepsilon=\hbar^2k_A{}^2/2\mu_A(\varepsilon)$ where $\mu_A(\varepsilon)$ is an energy-dependent effective mass which increases when the carrier kinetic energy (as measured from the $\Gamma_6$ edge) increases. Roughly, the $\Gamma_6$ mass increases in relative proportion like $1+\varepsilon/\varepsilon_A$. This is very significant in narrow gap materials (where the fraction $\varepsilon/\varepsilon_A$ can be larger than one) and markedly affects the carrier confinement energies in these heterostructures.

An extreme case is obtained when $\varepsilon_A = 0$, which happens in $Hg_{0.84}Cd_{0.16}Te$ alloys at low temperatures. If $\varepsilon_A = 0$, ie. if the $\Gamma_6$ and $\Gamma_8$ bands are accidentally degenerate, the bulk light particle energy spectra are linear upon $k$, which alters the thickness dependence of the bound state confinement energies in quantum wells in fact, these confinement energies decrease like $L_A^{-1}$ instead of decreasing like $L_A^{-2}$ as found in quantum wells whose hosts display quadratic dispersion relations. This point is illustrated in fig.(3) where we have shown the thickness dependence of the bound states in $Hd_{0.84}Cd_{0.16}Te$ $Hg_{0.76}Cd_{0.24}Te$ single quantum wells ($\varepsilon_A = 0$, $\varepsilon_B = 0.152$ eV, $\Delta_A = \Delta_B = 1$ eV and $V_S = 148$ meV, which amounts to assuming that the valence offset between $Hg_{1-x}Cd_xTe$ and $Hg_{1-y}Cd_yTe$ varies like (x-y) times 40 meV[7]).

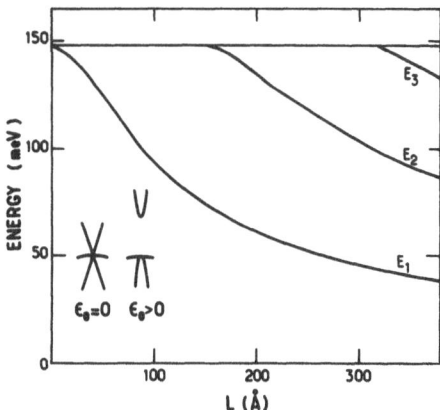

Fig. 3. The electron confinement energies in $Hg_{0.84}Cd_{0.16}Te$-$Hg_{0.76}Cd_{0.24}Te$ single quantum wells are plotted versus the $Hg_{0.84}Cd_{0.16}Te$ slab thickness L. Notice the slow decrease of $E_n$ with L. It results from the extreme band non-parabolicity of the $\Gamma_6$ band.

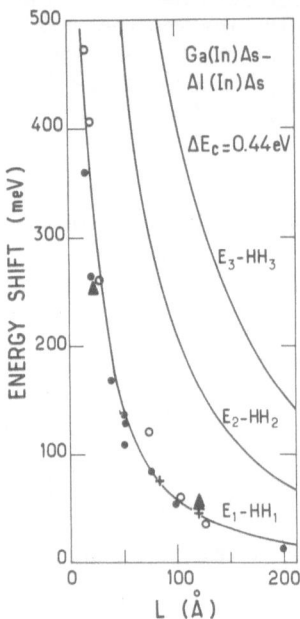

Fig. 4. The sum of the confinement energies of
the nth heavy hole $HH_n$ and nth electron
bound state $E_n$, n = 1,2,2... is plotted
versus the $Ga_{0.47}In_{0.53}As$ slab thickness
L in $Ga_{0.47}In_{0.53}As$-$Al_{0.48}In_{0.52}As$ single
quantum wells. T = 300 K. $V_S$ = 0.44 eV.
$E_1$-$HH_1$ also represents the energy shift
of the band-to-band recombination line in
$Ga_{0.47}In_{0.53}As$ quantum wells over that
observed in bulk $Ga_{0.47}In_{0.53}As$. The
symbols are taken from reference [8].

Other heterostructures which are potentially interesting for the
1.3µm, 1.5µm-operating optoelectronic devices are the $Ga_{0.47}In_{0.53}As$-
$Al_{0.48}In_{0.52}As$ quantum wells and superlattices which are lattice-matched
to InP. There have recently been several photoluminescence[8] and
absorption[9] measurements performed on these systems. From the fitting
of excitonic absorption peaks at room temperature, a conduction band
offset $V_S$ of 0.44 eV was deduced[9]. No allowance for band non-
parabolicity was however made in the energy level calculations.
Actually, a conduction band offset of 0.44 eV explains very well the
room temperature photoluminescence data[8] (see fig.(4)) as well as some
77 K absorption experiments[10] (see fig.(5)). The theoretical values
appearing in figs. (4,5) have been obtained with a full account of the
band non-parabolicity and by neglecting the small exciton binding energy

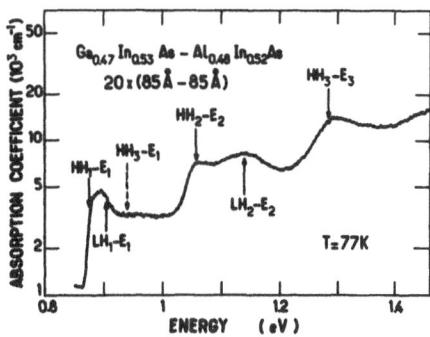

Fig. 5. Energy dependence of the absorption coefficient in $Ga_{0.47}In_{0.53}As$-$Al_{0.48}In_{0.52}As$ multiple quantum wells ($L_A = L_B = 85$ Å). T = 77 K. Courtesy J.Y. Marzin and L. Goldstein. The arrows represent the calculated transition energies at $k_\perp = 0$ for $Ga_{0.47}In_{0.53}As$-$Al_{0.48}In_{0.52}As$ single quantum wells with $V_S = 0.44$ eV, L = 85 Å.

(5-8 meV). The parameters entering in the calculations are: $\epsilon_A = 0.81$eV (0.75eV) at 77 K (300 K), $\epsilon_B = 1.47$ eV, $\Delta_A = \Delta_B = 0.36$ eV, $V_S = 0.44$ eV, $m_{\Gamma_6} = 0.041$ $m_0$ (0.038 $m_0$) at T = 77 K (300 K), $m_{hh} = 0.465$ $m_0$. It may be remarked that without band non-parabolicity the $E_3$ level would be found unbound in the 0.44 eV deep conduction well if $L_A = 85$ Å, while the $E_3$-HH absorption peak is clearly seen in the spectra. However, we have checked that the energy levels are very weakly dependent upon $V_S$. Thus, a more definite conclusion about $V_S$ must await for more optical data eg. obtained in specially designed heterostructures (pseudo-parabolic wells[11] or Separate Confinement Heterostructures[12]).

As shown in fig.(1) (dashed lines) and discussed elsewhere[13], the quantum well bound states do not disappear abruptly when their confinement energies have come to exceed the top of the confining well. They survive as virtual bound states whose properties (eg. the piling up of the wavefunction inside the quantum well) are similar to those of the true bound states, except that a particle, once trapped in the well, finally escapes towards infinity. There exist heterolayers such as GaSb-InAs-GaSb where the band line-up[14] ($V_S = 0.96$ eV, $V_P = 0.56$ eV between InAs and GaSb) is such that the lowest energy states "bound" in InAs are actually virtually bound. A light hole in GaSb (predominantly $\Gamma_8$-like) has to be hybridized with an InAs electron (predominantly $\Gamma_6$-like). In a parabolic, ie. decoupled, description of the hosts' bands, the transmission of a particle across the heterostructure would be zero. In the multi-band envelope function framework one easily calculates the transmission coefficient $T(\epsilon)$ of a GaSb light hole across the InAs layer[15]. Recognizing that the carrier is in propagating states in both kinds of layers if $0 < \epsilon < \Lambda$, where $\epsilon^{InAs} = 0$ and $\epsilon^{GaSb} = \Lambda$, one finds:

$$T(\epsilon) = [1 + 1/4(\zeta - 1/\zeta)^2 \sin^2(k_A L_A)]^{-1} \tag{8}$$

$$\zeta = k_A v_B(\epsilon)/k_B v_A(\epsilon) = [\epsilon(\epsilon + \epsilon_A)/(\epsilon - V_S)(\epsilon - V_S + \epsilon_B)]^{1/2}(\epsilon - V_S + \epsilon_B)/(\epsilon + \epsilon_A) \tag{9}$$

In eqs.(8,9) the two-band model has been used ($\Delta_A = \Delta_B = \infty$). A(B) stands for InAs (GaSb) and $k_A$, $k_B$ are the carrier wavevectors in the InAs and GaSb layers respectively. Eq.(9) can be rewritten in a more transparent form:

$$\zeta = (k_A/k_B \times m_B/m_A) \times \varepsilon_A/(\varepsilon + \varepsilon_A) \times (\Lambda - \varepsilon)/\varepsilon_B \qquad (10)$$

where $m_A > 0$ ($m_B < 0$) is the $\Gamma_6$ ($\Gamma_8$) band edge effective mass in the InAs (GaSb) layer respectively. The first part of the right-hand side of eq.(10) is the result that would be obtained when analyzing the transmission coefficient between two materials displaying band edges of the same symmetry but different effective masses. The second factor merely corrects the InAs band edge mass to account for the $\Gamma_6$ band non-parabolicity. Finally, the third term can be viewed as a genuine band-mixing effect. It is an admixture coefficient [$(\Lambda - \varepsilon)/\varepsilon_B$] of $\Gamma_6$-related states into a predominantly $\Gamma_8$ light hole wavefunction. Notice that if we let $\varepsilon_A$, $\varepsilon_B$ diverge while keeping $m_A$, $m_B$ fixed (which mimics parabolic bands), $\zeta$ as well as $T(\varepsilon)$ vanish. Fig.(6) shows the transmission coefficient of a GaSb light hole across InAs layers of different thicknesses. T vanishes at $\varepsilon = \Lambda$. Above that energy there is no impinging hole and only regular bound states, essentially localized in InAs, may exist (if $\varepsilon < \Lambda + \varepsilon_B$). $T(\varepsilon = 0)$ does not vanish. For $\varepsilon < 0$ the transmission coefficient is finite, the InAs layer behaving like a barrier. In the energy segment $[0, \Lambda]$ the transmission coefficient exhibits resonances any time that:

$$k_A L_A = p\pi \; ; \; p = 1, 2 \ldots \qquad (11)$$

These transmission resonances are not very narrow. Typically, a carrier spends $10^{-13}$ s oscillating back and forth in the InAs layer while at resonance. This amounts only to four oscillations before escaping to infinity. Despite the apparent lack of bound levels for the z motion, the GaSb-InAs-GaSb double heterostructures exhibit clearly defined Hall plateaux[16]. Thus, these heterolayers markedly contrast with eg. the GaAs-Ga(Al)As ones, where Hall plateaux appear to be associated with a pronounced size-quantization along the growth axis[17].

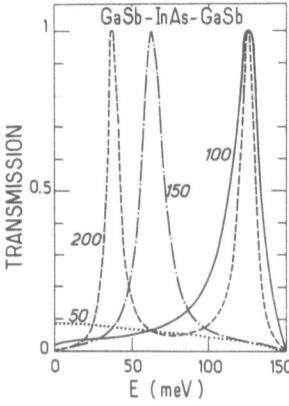

Fig. 6. The transmission coefficient of a carrier across a GaSb-InAs-GaSb double hetero-structure is plotted versus the energy E for four different InAs slab thicknesses.

In contrast with the situation where $k_\perp = 0$, there does not seem to exist any analytical solution to the problem of the heterostructure energy levels at finite $k_\perp$. This is due to the $k_\perp$-induced coupling between the heavy hole and light particle states. Actually, the problem would be analytically tractable if the remote band parameters were set equal to zero. Such a simplification is reasonable only for $\Gamma_6$-related subbands (see section IV for $\Gamma_6$ Landau levels) but completely fails to reproduce, even qualitatively, the $\Gamma_8$ valence subbands. This is not so surprising since the heavy hole curvature is entirely fixed by the remote band parameters. In the most general case, one ends up with a 8x8 coupled second order differential system which, for a given $k$, governs the z-dependent envelope functions. Following Altarelli et al's works[18], there have been a substantial amount of theoretical investigations[19-21] of the in-plane subband dispersions in both undoped and doped heterolayers. Here, for simplicity, we restrict our considerations to undoped quantum wells and superlattices. Let us first discuss the $\Gamma_8$-related valence subbands of heterostructures built out of relatively wide gap materials, eg. GaAs-Ga(Al)As. Thus, we adopt a parabolic description of the $\Gamma_8$ host states and therefore use the Luttinger valence hamiltonian[22], suitably generalized to include the band-edge profile $V_p(z)$. The generic shape of the 4x4 system is:

$$
H = \begin{bmatrix}
H_{hh} & & & \\
& H_{lh} & C(k_\perp) & \\
C^*(k_\perp) & H_{lh} & \\
& & & H_{hh}
\end{bmatrix}
\tag{12}
$$

where $H_{hh}$ and $H_{lh}$ stand for the diagonal heavy and light hole (along the growth axis) contributions and $C(k_\perp)$ is a coupling term which vanishes if $k_\perp = 0$. $H_{hh}$ and $H_{lh}$ exhibit the mass reversal effect. Namely, the longitudinal mass appearing in $H_{hh}(H_{lh})$ is heavy (light) while the in-plane is heavier for $H_{lh}$ than for $H_{hh}$. This means that if $C(k_\perp)$ were always negligible the (decoupled) solutions of $H_{hh}$ and $H_{lh}$ would cross

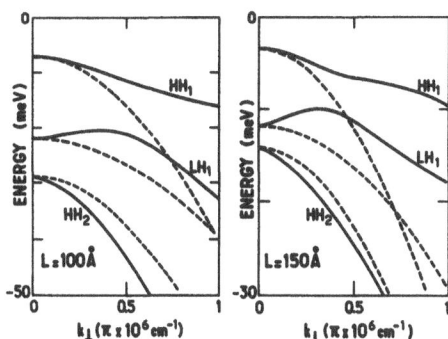

Fig. 7. The in-plane dispersion of $\Gamma_8$-related subbands in GaAs-Ga$_{0.7}$Al$_{0.3}$As single quantum wells is shown for two GaAs slab thicknesses L. The dashed lines correspond to subband dispersion that would result if HH$_n$ and LH$_n$ were uncoupled.

(see the dashed lines of fig.(7)). Such crossings are actually replaced
by anti-crossings since $C(k_\perp)$ is non-vanishing (solid lines in fig.(7)).
The valence subbands are strongly non-parabolic in $k_\perp$. This non-
parabolicity may even be such that electron-like segments appear in the
in-plane dispersions of several subbands. The results shown in fig.(7)
were obtained for rectangular GaAs-Ga(Al)As quantum wells in the axial
approximation[3] which renders the dispersion relations isotropic in the
layer plane. Notice that each level is twice degenerate (Kramers
degeneracy) owing to the symmetric band edge profile and the exclusion
of the inversion assymmetry splitting which exists in zinc-blende
materials. The non-parabolicity also manifests itself by a pronounced
mixing of the $k_\perp = 0$ -eigenstates at finite $k_\perp$. In fact, the
denominations $HH_n, LH_m$ are justified only at vanishing $k_\perp$ while at $k_\perp \neq 0$
the eigensolutions have a mixed heavy and light hole character. This is
more easily appreciated if one calculates the expectation value of $J_z^2$
at finite $k_\perp$. The results of such a calculation is presented in fig.(8)
for the same structures as shown in fig.(7).

The band mixing effects are all the more pronounced when the host
materials are non-parabolic, eg. InAs-GaSb[3] or HgTe-CdTe heterolayers.
The fig.(9) shows the in-plane dispersion relations of a 100Å-100Å InAs-
GaSb superlattice at $q = 0$. If $k_\perp = 0$, the misaligned band edges of
InAs and GaSb make the $E_1$ subband (whose wavefunctions are essentially
localized within the InAs layers) to lie below the $HH_1$ and $HH_2$ subbands
whose wavefunctions are heavily localized in the GaSb layers. In the
absence of coupling between light particle and heavy hole states, $E_1$
would display a positive curvature upon $k_\perp$ and would cross the heavy

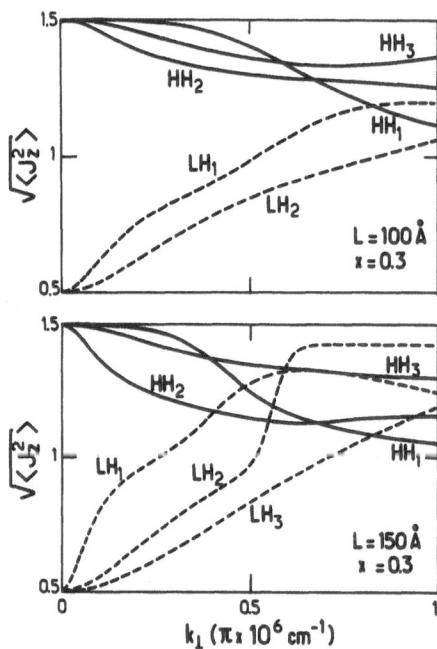

Fig. 8. The quantity $\sqrt{\langle J_z^2 \rangle}$ is plotted versus
the in-plane wavevector $k_\perp$ in GaAs-
$Ga_{0.7}Al_{0.3}As$ single quantum wells (L =
100Å and L = 150Å respectively). If
the $HH_n$ and $LH_n$ subbands were uncoupled
$\sqrt{\langle J_z^2 \rangle}$ would be equal to 3/2 ($HH_n$) and
$1/2$ ($LH_n$) at any $k_\perp$.

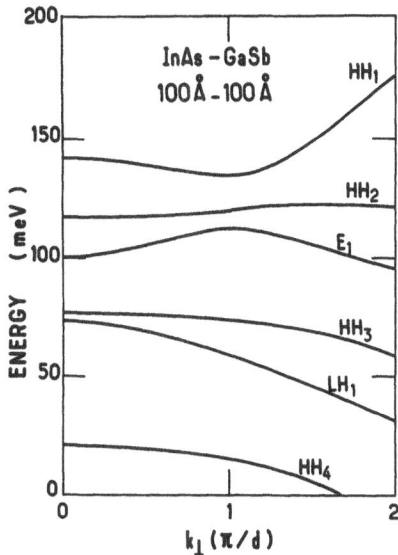

Fig. 9.   In-plane dispersion relations of
a 100 Å - 100 Å InAs-GaSb superlattice assuming flat band
conditions.  q=0.  Λ=0.15 eV.

hole-like $HH_1$ and $HH_2$ subbands.  This would result in the formation of a
semimetallic  phase  as electrons would leave the GaSb layers  and  flow
into  the  InAs ones until the Fermi energy remains constant  throughout
the heterolayer.    The $k_\perp$-induced mixing between the two kinds of states
suppress the crossings and give rise to small but non-zero hybridization
gaps between $HH_1$, $HH_2$ and $E_1$.  Thus the Fermi level constancy as well as
the  charge  neutrality  can  both  be  achieved  without  implying  the
existence of free carriers.    The conclusion of this discussion is  that
perfect  InAs-GaSb  heterolayers  should always  be  semiconductors  and
therefore  become  insulating at low temperatures whereas a  substantial
amount of charges have always been evidenced in these heterolayers ($L_{InAs}$ >
100Å).   It  should however be stressed that the residual dopings in InAs
and  GaSb  layers  lead  to the existence of  a  significant  number  of
extrinsic carriers.

     There  has  recently  been a considerable body of work  devoted  to
HgTe-CdTe heterolayers.  These materials are potentially interesting for
the far-infrared detection.  Early magneto-optical data were interpreted
in terms of a small valence band offset ($\Lambda = \epsilon^{HgTe} - \epsilon^{CdTe} = 40$  meV).    New
optical  and  magneto-optical data support such a small  positive  value
within  the  limits  [0-100 meV] but  recent  X-ray  photo-emission
spectroscopy  (XPS)  measurements[23] lead  to a much  larger  value  ($\Lambda$ =
0.35 meV).   The electronic properties of HgTe-CdTe heterolayers will be
thoroughly  discussed  by  McGill[24] and  Altarelli[25] at  this  workshop.
Here,  we  limit ourselves to giving a flavour of  the  band  structure
intricacies  by  presenting  in fig.(10) the  dispersion  relations  of  a
100Å - 36Å HgTe-CdTe superlattice[7].   The subband labeled I  corresponds
at $k_\perp = 0$ to an interface state[15,26,27].  It is a genuine feature of the
HgTe-CdTe  materials  which arises from the sign reversal of the  carrier
effective  mass  across the interfaces.  In fact,  HgTe is a symmetry-
induced  zero gap semiconductor whose $\Gamma_8$ conduction and $\Gamma_8$ valence bands

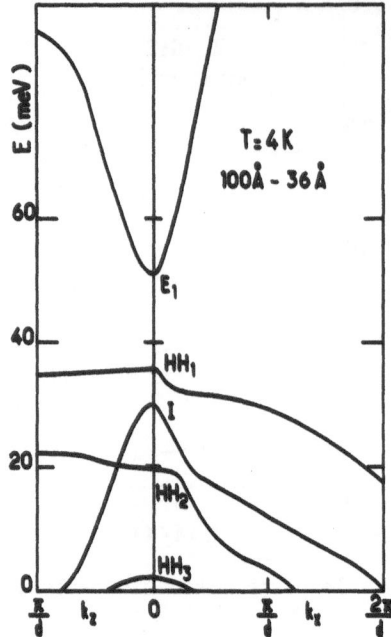

Fig. 10. In-plane dispersion relation (right panel) and along the axis (left panel) of a 100 Å - 36 Å HgTe-CdTe superlattice. Λ = 40 meV.

are degenerate at the zone centre while CdTe is a conventional direct gap semiconductor. If we now focus our attention on the light particle states, the carrier either behaves as an electron in HgTe or as a light hole in CdTe. This leads to a piling of the ground state envelope function near the interfaces (fig.(11)). The heavy hole levels $HH_n$ are essentially localized in the HgTe layers. At $k_\perp \neq 0$, I and $HH_n$ hybridize to give rise to a complex subband pattern. The details of the subband structure are sensitive to Λ, stresses, interdiffusion etc... Thus, more work is required to fully understand the subband structure of these heterolayers.

## IV $\Gamma_6$ LANDAU LEVELS IN RECTANGULAR QUANTUM WELLS

If we neglect the coupling between the $\Gamma_6$, $\Gamma_7$ and $\Gamma_8$ states and the remote bands as well as the free electron terms, the Landau level ladders of $\Gamma_6$-related subbands can be obtained in closed forms. By using the projection[2] technique one may eliminate all the envelope functions at the benefits of the two ($f_1$ and $f_2$) which are associated with the S↓ and S↑ edges respectively. The effective, energy-dependent, hamiltonian which acts on these two functions is thus a 2x2 matrix which can be written:

$$H_{\Gamma_6} = \begin{bmatrix} H_{11} & H_{12} \\ H_{21} & H_{22} \end{bmatrix} \tag{13}$$

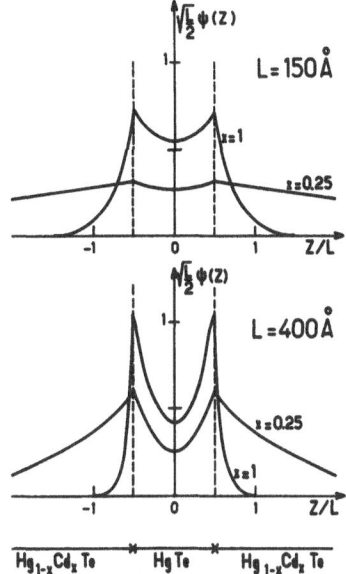

Fig. 11. An example of interface state
wavefunctions in $Hg_{1-x}Cd_xTe$-
$HgTe$-$Hg_{1-x}Cd_xTe$ double heterostructures.

where

$$H_{11}=V_S(z)+2P^2/3\pi_+\beta(z)\pi_-+P^2/3\pi_z\beta(z)\pi_z+\hbar^2P^2/\lambda^2[\epsilon+\epsilon_A-V_P(z)]^{-1} \qquad (14)$$

$$H_{22}=V_S(z)+2P^2/3\pi_-\beta(z)\pi_++P^2/3\pi_z\beta(z)\pi_z+\hbar^2P^2/\lambda^2[\epsilon+\epsilon_A-V_P(z)]^{-1} \qquad (15)$$

$$H_{12}=P^2\sqrt{2}/3\pi_+[\pi_z,g(z)] \qquad (16)$$

$$g(z)=1/[\epsilon+\epsilon_A-V_P(z)]-1/[\epsilon+\epsilon_A+\Delta_A-V_G(z)] \qquad (17)$$

$$\beta(z)=2/[\epsilon+\epsilon_A-V_P(z)]+1/[\epsilon+\epsilon_A+\Delta_A-V_G(z)] \qquad (18)$$

In eqs.(14-16) P has been defined in eq.(7), $\Pi = p + eA/c$ where $A$
is the vector potential associated with the magnetic field $B$, $\lambda$ is the
magnetic length ($\lambda^2=\hbar c/eB$) and $\pi_\pm = (\pi_x\pm i\pi_y)/\sqrt{2}$.

If $B$ // z the eigenvalues of $H_{\Gamma_6}$ can be written in the form:

$$\psi(\mathbf{r})= \begin{cases} \varphi_{n+1}(x+\lambda^2k_y)\exp(ik_yy)\chi_1(z) \\ \\ \varphi_n(x+\lambda^2k_y)\exp(ik_yy)\chi_2(z) \end{cases} \qquad (19)$$

where $\chi_1$, $\chi_2$ have opposite parities with respect to the centre of the quantum well. In eq.(19) n = -1, 0... and $\varphi_n$ is the nth harmonic oscillator wavefunction. It is interesting to point out that $H_{12}$ vanishes in the absence of spin-orbit coupling, of a magnetic field (or in-plane wavevector $k_\perp$ if B = 0) or in bulk materials (flat band everywhere). Moreover, $H_{12}$ is induced by the band non-parabolicity as witnessed by the energy denominators in eq.(16). In addition, $H_{12}$ manifests itself only in the boundary conditions set by $\chi_1$ and $\chi_2$ (as it is zero in both kinds of layers). These conditions are obtained by requiring the continuity of $\chi_1$ and $\chi_2$ and that of a linear combination of $\chi_1$, $\chi_2$, $d\chi_1/dz$, $d\chi_2/dz$ which is itself deduced from the integration of $H_{\Gamma_6}$ across the interfaces. Let $k_1$ and $k_2$ be the real wavevectors characterizing $\chi_1$ and $\chi_2$ in the A layer and $k_1'$ and $k_2'$ the wavevectors of the evanescent waves in the B (barrier) layer. The eigenvalues of $H_{\Gamma_6}$ are the roots of the two equations;

$$2(n+1)/\lambda^2(g_A-g_B)^2C_1S_2+(\beta_Ak_2C_2+\beta_Bk_2'S_2)(\beta_Ak_1S_1-\beta_Bk_1'C_1)=0 \qquad (20)$$

$$2(n+1)/\lambda^2(g_A-g_B)^2S_1C_2+(\beta_Ak_2S_2-\beta_Bk_2'C_2)(\beta_Ak_1C_1+\beta_Bk_1'S_1)=0 \qquad (21)$$

where:

$$C_1=\cos(k_1L_A/2) \;\; ; \;\; C_2=\cos(k_2L_A/2) \qquad (22)$$

$$S_1=\sin(k_1L_A/2) \;\; ; \;\; S_2=\sin(k_2L_A/2) \qquad (23)$$

and $g_A(g_B)$, $\beta_A$ ($\beta_B$) are the g(z) and $\beta$(z) values appropriate to the A(B) layers. The first set of solutions correspond to even $\chi_1$'s and odd $\chi_2$'s while the solutions of eq.(21) correspond to odd $\chi_1$'s and even $\chi_2$'s. In the case of imaginary $k_1$ or $k_2$ (or both) the eqs.(20,21) should appropriately be changed. One may immediately anticipate from the form of the eigenfunctions that electric dipole spin-flip transitions (spin resonance $\Delta n=0$, combined resonances $\Delta n=\pm 1$) will be allowed. This is not specific to the heterostructures but also exists in bulk materials[28].

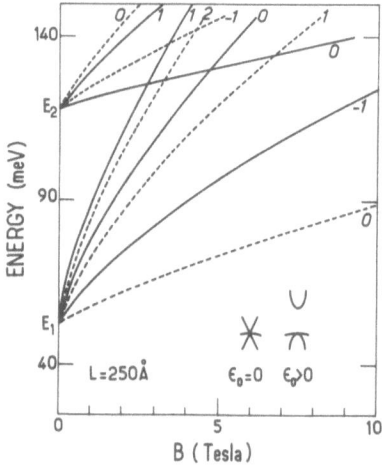

Fig. 12. Calculated magnetic field dependence of the $\Gamma_6$-related Landau levels in a $Hg_{0.84}Cd_{0.16}Te$-$Hg_{0.76}Cd_{0.24}Te$ single quantum well. L=250 Å. $V_S$=148.3 meV.

The effects should however be larger in heterostructures since, even at B=0, there already exists a band mixing due to the finite $k_1$, $k_2$. Band mixing effects similar to the ones considered here for rectangular quantum wells were already analyzed by Lassnig[29] in modulation-doped heterostructures. We show in fig.(12) the calculated $\Gamma_6$-related Landau levels of a 200 Å-thick $Hg_{0.84}Cd_{0.16}Te$-$Hg_{0.76}Cd_{0.24}Te$ quantum well. The band parameters are the same as those used in fig.(3). One may notice the strongly non linear behaviour of the Landau levels upon the magnetic field strength. This is not surprising since the well-acting material is such that $\epsilon_A=0$.

For a more complete treatment of the heterostructure Landau levels, including those of the valence subbands, the reader is referred to Altarelli's lectures[25] at this workshop.

ACKNOWLEDGEMENTS

We are pleased to thank J.Y. Marzin and L. Goldstein for communicating to us their experimental data on Ga(In)As-Al(In)As prior to publication. Discussions with C. Delalande, Y. Guldner, J. Orgonasi, P. Voisin and M. Voos were most helpful. One of us (J.A.B.) expresses his gratitude to CNPq (Brazil) for financial support. The Groupe de Physique des Solides de l'Ecole Normale Superieure is Laboratoire Associe au CNRS (LA 17). This work has been supported by the GRECO "Experimentations numeriques".

REFERENCES

1.  S. White and L.J. Sham, Phys.Rev.Lett. 47:879 (1981).
2.  G. Bastard, Phys.Rev. B24:5693 (1981) and B25:7584 (1982).
    See also G. Bastard and J.A. Brum, IEEE Journ. of Quant. Electr. (1986) in press.
3.  M. Altarelli, Phys.Rev. B28:842 (1983).
    See also M. Altarelli in Proceedings of the Les Houches Winterschool "Semiconductor Superlattices and Heterojunctions" Springer Verlag (1986) in press.
4.  M.F.H. Schuurmans and G.W.'t Hooft, Phys.Rev. B31:8041 (1985).
5.  D.L. Smith and C. Mailhiot, Phys.Rev. B (1986) in press.
6.  E.O. Kane, J.Phys.Chem.Solids 1:249 (1957).
7.  Y. Guldner, G. Bastard, J.P. Vieren, M. Voos, J.P. Faurie and A. Million, Phys.Rev.Lett. 51:907 (1983).
    See also J.M. Berroir, Y. Guldner, J.P. Vieren, M. Voos and J.P. Faurie, Phys.Rev. B (1986) in press.
8.  W. Stolz, K. Fujiwara, L. Tapfer, H. Oppolzer and K. Ploog in "GaAs and related compounds" Biarritz 1984. Inst.Phys.Conf.Ser. 74:139 (1985) edited by B. de Cremoux. Adam Hilger (1985 Bristol) and references cited therein.
9.  J.S. Weiner, D.S. Chemla, D.A.B. Miller, T.H. Wood, D. Sivco and A.Y. Cho, Appl.Phys.Lett. 46:619 (1985).
10. J.Y. Marzin, L. Goldstein (1985), unpublished results.
11. R.C. Miller, A.C. Gossard, D.A. Kleinman and O. Munteanu, Phys.Rev. B29:3740 (1984).
12. M.H. Meynadier, C. Delalande, G. Bastard, M. Voos, F. Alexandre and J.L. Lievin, Phys.Rev. B31:5539 (1985).
13. See eg. D. Bohm "Quantum Theory" (Prentice-Hall, New York 1951).
14. See eg. L. Esaki in "Narrow Gap Semiconductors - Physics and Applications" edited by W. Zawadzki. Lecture Notes in Physics vol. 133 Springer Verlag, Berlin (1980).

15. G. Bastard, Surf.Sci. 170:426 (1986).
16. S. Washburn, R.A. Webb, E.E. Mendez, L.L. Chang and L. Esaki, Phys. Rev. B31:1198 (1985).
17. See however H.L. Stormer, J.P. Eisenstein, A.C. Gossard, W. Wiegman and K. Baldwin, Phys.Rev.Lett. 56:85 (1986).
18. U. Ekenberg and M. Altarelli, Phys.Rev. B30:3369 (1984).
    See also A. Fasolino and M. Altarelli, Surf.Sci. 142:322 (1984).
19. E. Bangert and G. Landwehr, Superl. and Microstr. 1:363 (1985).
20. D.A. Broido and L.J. Sham, Phys.Rev. B31:888 (1985).
    See also S.R. Eric Yang, D.A. Broido and L.J. Sham, Phys.Rev. B32:6630 (1985).
21. T. Ando, J.Phys.Soc.Japan 54:1528 (1985).
22. J.M. Luttinger, Phys.Rev. 102:1030 (1956).
23. S.P. Kowalczyk, J.T. Cheung, E.A. Drant and R.W. Grant, Phys.Rev. Lett. 56:1605 (1986).
24. T.C. McGill, this volume.
25. M. Altarelli, this volume.
26. Y.C. Chang, J.N. Schulman, G. Bastard, Y. Guldner and M. Voos, Phys.Rev. B31:2557 (1985).
27. Y.R. Lin Liu and L.J. Sham, Phys.Rev. B32:5561 (1985).
28. P. Kacman and W. Zawadzki, Phys.Stat.Sol. (b) 47:629 (1971).
29. R. Lassnig, Phys.Rev. B31:8076 (1985).

MAGNETIC FIELD EFFECTS ON THE ELECTRONIC STATES OF NARROW-

GAP LOW-DIMENSIONAL STRUCTURES

M. Altarelli

Max-Planck-Institut fuer Festkoerperforschung
Hochfeld-Magnetlabor
BP 166 X, F-38042 Grenoble, Frankreich

ABSTRACT

The electronic structure of small-gap semiconductors is
characterized by important band mixing and non-parabolicity at
energies of interest for the interpretation of all experiments.
In heterostructures, the coupling of bands with different
character produces a strongly non-parabolic dispersion of the
subbands and, when a magnetic field is present, a complicated
Landau level pattern. This will be exemplified by results of
calculations performed with the envelope-function method for
InAs-GaSb and HgTe-CdTe heterostructures· Current problems in
the interpretation of magneto-optical and Quantum Hall experiments are discussed.

INTRODUCTION

This paper is devoted to a description of the energy
levels of narrow-gap semiconductor heterostructures in an external magnetic field, as obtained by the envelope-function
method. This is a complicated problem by definition, because
narrow-gap materials are those for which the fundamental
energy gap $E_g$ is not much larger than the energies of interest
for experiments: the confinement energies of subbands, the
Fermi energy with respect to the bottom of the subband, the
cyclotron energy at high fields, etc. This means that experiments on subband spectroscopy probe a non-parabolic region of
the band structure, which is conveniently described by considering a set of coupled bands in k.p perturbation theory.
The envelope-function method produces then a set of coupled
differential equations, with boundary conditions, which are to
be solved to obtain the energy levels. The resort to numerical
solutions is unavoidable in most cases.

In spite of these complications, and often of inprecise
knowledge of important input parameters, the envelope-function
calculations always give good qualitative insight into electronic states and sometimes even quantitative agreement. This
is a non-trivial achievement, in view of the extreme compli-

cation of the dependence of the Landau levels on the applied field, characterized by a strong non-linearity, by anticrossing and mixing of states from different bands. We shall illustrate the state of the art in the comparison of theory and experiment by considering two systems: InAs-GaSb and CdTe-HgTe heterostructures.

This paper is organized as follows. We first recall the basic results of k.p theory and of the envelope-function method for coupled bands, as exemplified by the six-band model used to describe narrow-gap materials with large spin-orbit splitting, and applied to the InAs-GaSb system. The following section is devoted to the calculation of Landau levels and to the interpretation of magneto-optical and quantum Hall experiments in InAs-GaSb superlattices and quantum wells. A short discussion of Landau levels in CdTe-HgTe superlattices completes the paper.

## THE MODEL AND ITS APPLICATION TO InAs-GaSb HETEROSTRUCTURES

In the k.p formalism of band theory[1] the k-dependence of the energy bands in the neighbourhood of the $\Gamma$ point ($\vec{k}=0$) for a bulk semiconductor, is described in terms of the eigenvalues at $k=0$, $E_n(0)$, and of the matrix elements of the momentum operator between the corresponding eigenfunctions $u_n$. The method is of practical value if only a small number of bands need to be considered to get satisfactory accuracy in the energy region of interest. This is generally implemented by separating the energy bands in two groups: n bands which are important, and whose k.p coupling is retained explicitly, and all others, whose influence on the former n bands is only evaluated in second order perturbation theory. One then writes:

$$H_{lm}(\vec{k}) = E_l(0)\ \delta_{lm} + \Sigma^3_{\alpha=1}\ P^\alpha_{lm}\ k_\alpha + \Sigma^3_{\alpha,\beta=1}\ D^{\alpha,\beta}_{lm}\ k_\alpha\ k_\beta \qquad (1)$$

where $l,m=1,2,\ldots,n$ and $\alpha,\beta$ run over the x,y and z directions. Given a k-vector, the n band energies $E_l(\vec{k})$ are given by the eigenvalues of the nxn matrix $H_{lm}(\vec{k})$. The direct k.p coupling between the n bands is thus retained in the terms $P^\alpha_{lm} k_\alpha$, where the matrix $P^\alpha$ is given by:

$$P^\alpha_{lm} = \frac{\hbar}{m_0}\ <\ u_l|\ P^\alpha\ |u_m\ > \qquad (2)$$

($m_0$ is the free electron mass).

The k-quadratic terms proportional to the matrix $D^{\alpha,\beta}$, on the other hand, represent the indirect $\vec{k}\cdot\vec{p}$ coupling between two of the n bands via the other bands (n+1 to $\infty$) not included in the set:

$$D^{\alpha,\beta}_{lm} = \frac{\hbar^2}{2m_0}\ \delta_{\alpha\beta} + \frac{\hbar^2}{m_0^2}\ \Sigma^\infty_{j=n+1}\ \frac{<u_l|P^\alpha|u_j><u_j|P^\beta|u_m>}{E_l(0)\ -\ E_j(0)} \qquad (3)$$

In practice, the $P_{lm}$ and $D_{lm}$ matrices have very few independent elements and they are determined empirically from the fitting of bulk experiments·

For the description of the semiconductors of interest here, a six-band model is adopted, including the $\Gamma_5$ s-like conduction band minimum, two-fold degenerate (including spin), and the $\Gamma_8$ $P_{3/2}$-like valence band maximum, four-fold degenerate and comprising the heavy- and light-hole bands. The $\Gamma_7$ $P_{1/2}$-like split-off valence band is not included explicitly, because of the large value of the spin-orbit splitting. It produces, however, an important contribution, via Eq. (3), to the effective mass and effective g-factor of the conduction band·

If an electron is moving in a bulk semiconductor described by Eq· (1) and subject to a weak, slowly varying potential $U(\vec{r})$, its energy levels are determined by many-band effective-mass or envelope-function equations:

$$\sum_{m=1}^{n} \{H_{lm} (-i \vec{\nabla}) + U(\vec{r}) \delta_{lm}\} F_m (\vec{r}) = E F_l (\vec{r}) \quad (l=1,2,\ldots n) \quad (4)$$

The F functions are slowly varying on the unit cell scale and the complete wavefunction of the system is to the lowest order of approximation given by:

$$\psi (\vec{r}) = \sum_{l=1}^{n} F_l (\vec{r}) u_l (\vec{r}) \tag{5}$$

In order to extend this type of effective-mass description (introduced originally to describe, e.g., acceptor impurities) to heterostructures, we need a set of boundary conditions to match the solution in material A with that in material B· This is in general not possible without some additional information on the Bloch functions $u^A, u^B$ of the two semiconductors· Fortunately, we are interested in structures composed of 111-V and 11-Vl compounds, which share a large amount of structural and chemical features. It appears then that the rather drastic assumption [2,3]

$$u_l^A (\vec{r}) \approx u_l^B (\vec{r}) \qquad (l=1,2,\ldots,n) \tag{6}$$

produces reasonable results, provided both materials have a direct gap at the $\Gamma$ point and the n band edges in A correspond in symmetry character and chemical nature to those in B· Once Eq· (6) is assumed, a plausible set of boundary conditions can be written, which involve only the envelope function $F^A_l$, $F^B_l$, without any further reference to the Bloch functions[4].1

They require (i) the continuity of the $F_l$ and (ii) the continuity of the expressions:

$$\sum_{m=1}^{n} \sum_{\alpha=x,y} \{(D_{lm}^{z\alpha} + D_{lm}^{\alpha z}) k_\alpha - 2i D_{lm}^{zz} \frac{\partial}{\partial z}\} F_m \quad (l=1,2,\ldots n) \quad (7)$$

As implied by Eq· (7), $k_x$ and $k_y$ are good quantum numbers, and Eqs· (4)-(7) yield the energy levels as functions of $k = (k_x, k_y)$, i·e· the k -parallel dispersion of the subbands·

In InAs-GaSb superlattices this dispersion is especially

interesting because of the peculiar band line-up of these two
materials [5]. Indeed the electronic properties of the superlattices indicate that the top of the GaSb valence band is
0·15 eV higher than the bottom of the InAs conduction band.
Therefore, for layer thickness in excess of ~ 8nm, there are
GaSb hole-like subbands. However, the mixing of hole-like and
electron-like states prevents in most cases the formation of a
true semimetal by opening small hybridization gaps [4], of the
order of one meV. The effects of hybridization are therefore
masked in most experiments by small amounts of disorder, carrier
imbalance etc. However, they are brought out more prominently
in experiments involving high magnetic fields and hydrostatic
pressure, as discussed by Maan in these Proceedings. An
interesting consequence of the mixing of InAs electron-like
and GaSb hole-like states is the large subband width in the
growth direction, in the energy range where they overlap. In
fact electron states are not exponentially attenuated in the
GaSb "barriers", but rather go through them as propagating
light-holes. Thus widths of ~15 meV are encountered for 12 nm
InAs-8nm GaSb superlattices. An even more dramatic effect has
been emphasized by Bastard [6]. An InAs quantum well between
GaSb barriers does not confine electrons at energies below the
GaSb valence band maximum. Electrons are trapped in quasi-bound
resonant states for ~$10^{-13}$ s only, in ~10 nm wells, before
escaping into the continuum of GaSb valence states. We shall
discuss the electronic states of the single quantum wells, which
differ in significant respects from those of superlattices,
after the description of the Landau level calculations.

## LANDAU LEVELS IN InAs-GaSb HETEROSTRUCTURES

The inclusion of a perpendicular magnetic field in the
many-band envelope-function formalism follows the lines of the
classical work of Luttinger on the cyclotron resonance of holes
in semiconductors. The field $\vec{B} = (0, 0, B)$ is described by the
vector potential $\vec{A}$. (It is convenient to choose a gauge with
$A_z = 0$). In the $\vec{k} \cdot \vec{p}$ bulk Hamiltonian, Eq·(1), $\vec{k}$ is to be replaced by $\vec{k}' = \vec{k} + (e/c)\vec{A}$. Then it is easy to see that the x
and y components of this new operator do not commute, but instead ($\hbar = 1$)

$$\{k'_x, k'_y\} = -i(e/c)B \tag{8}$$

As a consequence of Eq· (8), we can define operators a, a+

$$a = \sqrt{\frac{c}{2eB}} (k'_x - ik'_y) \tag{9}$$

$$a^+ = \sqrt{\frac{c}{2eB}} (k'_x + ik'_y)$$

with commutator

$$\{a, a^+\} = 1 \tag{9'}$$

so that all terms in $k_x$ or $k_y$ in the Hamiltonian can be expressed in terms of these harmonic oscillator raising and

lowering operators. Besides these "orbital" terms in the Hamiltonian, new diagonal terms also arise, representing the direct coupling of the electron and hole spins to the field. They
introduce additional parameters, representing the g-factor for
electrons and holes. In the 4x4 submatrix representing the valence bands, the new terms can be written [7]

$$\frac{e}{c} \kappa J_z B + \frac{e}{c} q J_z^3 B \qquad (10)$$

where $J_z$ is the spin $3/2$ matrik and $\kappa$ and $q$ are material parameters. Actually, q turns out to be very small and the second
term of (10) is usually neglected.

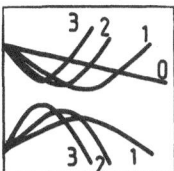

Fig. 1. Energy levels vs.
magnetic field for
a two-band model
with k.p inter-
action

    Before proceeding to a discussion of the Landau level
spectrum for InAs-GaSb superlattices[8], it is instructive to
inspect the solutions for a simple model of two coupled bands
with opposite curvature, to gain insight into the magnetic
field effect on coupled, non-parabolic subbands. Consider indeed the subband structure given for k in the (x,y) plane by
the following matrix:

$$H = \begin{matrix} \Delta/2 - \alpha k^2 & P(k_x + ik_y) \\ P(k_x - ik_y) & -\Delta/2 + \beta k^2 \end{matrix} \qquad (11)$$

in which a hole-like subband is an energy $\Delta$ above an electron
like one at k=0, the two being coupled by a k.p term with matrix
element P. Adding a field B in the z direction, one obtains
solutions in terms of harmonic oscillator eigenfunctions of
the form $(C_{1n}, C_{2n-1})$ where $C_2=0$ for n=0 and $C_1, C_2 \neq 0$ for
n=1,2,.... The eigenvalues as a function of the field are sk-
etched in Fig.1. The subband coupling induces strong deviations
from linearity for all Landau levels except n=0, which is purely hole-like.

    This simple model provides good qualitative insight into

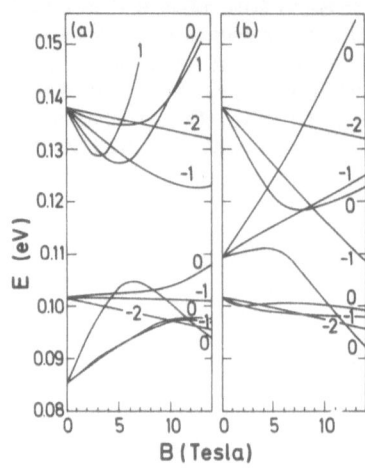

Fig. 2. Landau levels of an
InAs-GaSb superlattice
with period d=12+8 nm;
(a) at the superlattice
Brillouin zone center;
(b) at the zone edge·
The Landau quantum num-
bers -2,-1,0,... are
indicated·

the Landau level spectrum of InAs-GaSb superlattices (Fig. 2),
computed within the 6x6 band model. In this case too the six-
component envelope wavefunction is expressed in closed form
via the harmonic oscillator eigenstates, if the bulk materials
are assumed to be isotropic in the plane normal to the growth
direction. Notice the $n=-2$ level, which is linear in the field
and is a purely hole-like state, and the strong deviation from
linearity which characterizes all other levels. In terms of
these energy levels it is possible to interpret magneto-optical
experiments on superlattices[9,10] with reasonable success.[8,11]
In order to compare theory and experiment it is necessary to
position the Fermi level as a function of field and of the carrier concentration in the sample· The Fermi level position is
easily determined for high fields, when all Landau levels have
a clear hole-like or electron-like character. In an intrinsic
sample, the Fermi level lies then in between the highest, hole-
like level and the lowest electron-like one. As the field decreases, this gap shrinks to zero and the levels begin to cross.
It is then not always easy to follow its path, especially when
(see Fig. 2) in the superlattice geometry, the subband width
in the z direction is comparable or larger than the separation
of the Landau levels, so that the situation becomes in effect
3-dimensional. Because of the large bandwidths encountered in
InAs-GaSb, as discussed in the previous section, and of the
relatively large field at which levels uncouple, one must
often perform the calculations at fields much higher than experimentally accessible, to position the Fermi level correctly.

If the sample is not intrinsic, the position of the Fermi
level is accordingly modified. The interpretation of magneto-

optical experiments on InAs-GaSb superlattices indicates a nearly intrinsic situation, with at most a slight electron concentration in some samples. The signature of the intrinsic or nearly intrinsic character is in the many low energy transitions which are observed throughout the low field regions. This conclusion is also in agreement with an analysis of the low-field Hall and magnetoresistance results.[12]

The situation is quite different if instead of superlattices, single InAs quantum wells between GaSb barriers are considered. Here concentrations of electrons as large as $8 \times 10^{11} cm^{-2}$ are reported.[13-15] The presence of a large number of defects in the quantum wells (but, for some reason, not in the superlattices) appears unlikely, in view of the exceedingly high mobilities ($> 10^5$ $cm^2$/Vs). According to a recent proposal,[16] the difference from the superlattice case is that the top GaSb layer is only 20 nm thick[13,14] and, if one assumes that the Fermi level is pinned somewhere in the GaSb gap at the interface with vacuum, the whole InAs well lies in the depletion region, characterized by strong band bending. Thus the electrons that flood the quantum well come from the surface 20 nm away (Fig. 3) and have therefore the mobility behaviour of a modulation doped system. The position of the Fermi level, deep into the thick GaSb layer on the side opposite the free surface, must coincide with the GaSb native acceptor level, which is 34.5 meV above the valence band.[17] Therefore the states at the Fermi energy are always two-dimensional in character in spite of the fact that those degenerate with the valence band continuum of GaSb are not.[6] It can be argued,[16] however, that this is sufficient to explain the observation of the quantum Hall effect,[13-14] a quantum Hall plateau being observed whenever the Fermi level is in between two-dimensional Landau levels.

LANDAU LEVELS IN CdTe-HgTe SUPERLATTICES

A system of great interest is the CdTe-HgTe superlattice, as a consequence of the zero-gap character of HgTe and of the band line-up, which puts the $\Gamma_8$ HgTe edge above that of CdTe. The precise value $\Delta$ of this band line-up is controversial, the estimate $\Delta = 0.04$ eV from magneto-optics[18] being confirmed by optical measurements,[19] but contradicted by photoemission results,[20] which suggest a value $\Delta = 0.35$ eV.

The electronic structure of these superlattices is discussed in these Proceedings by McGill. Here we only notice that, given the positive sign of the band offset, $\Gamma_8$ heavy holes are confined in the HgTe well, together with $\Gamma_8$ electron-like levels. There are also unusual interface states arising in the gap between $\Gamma_8$ light holes in CdTe and $\Gamma_8$ electrons in HgTe.[21-22] Such interface states, associated with bands of the same symmetry but with opposite mass could also originate from the $\Gamma_6$ bands (see Fig. 4), however, given the large gap between the two, they would not be accessible to an envelope-function treatment. If these $\Gamma_6$-related interface states exist and are in the energy region of the relevant $\Gamma_8$-related states, considerable doubt on the possibility of using the effective-mass method at all should arise.

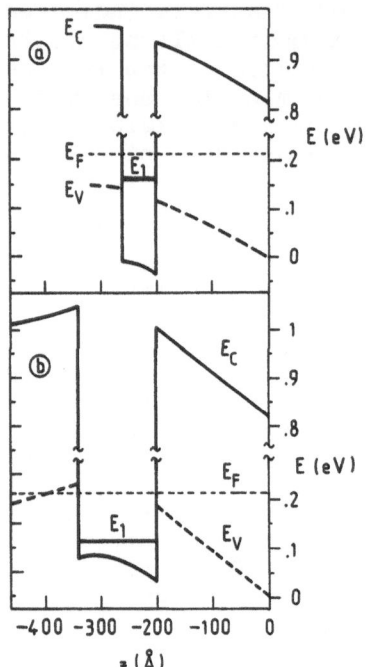

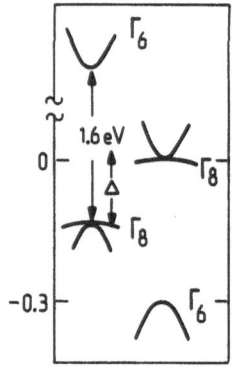

Fig. 3. (a) Band structure profile for a 6 nm InAs quantum well. $E_1$ is the position of the occupied electron subband, $E_C$ and $E_V$ denote the conduction and valence band edges. The Fermi level $E_F$ is pinned at z=0, the GaSb-vacuum interface. (b) same as (a) for a 15 nm InAs well.

Fig. 4. Schematic band line-up diagram of CdTe and HgTe. The zero of energy is at the HgTe $\Gamma_8$-edge. $\Delta$ denotes the band offset.

From this simple analysis of the energy level scheme, it seems that the determination of the band offset via infrared magneto-optics is difficult because, unlike in InAs-GaSb, the important transitions take place between states mostly located in the HgTe wells. Complications such as strain, doping, and the dispersion in the growth direction[23] make the positioning of the Fermi level complicated. Here we report, just to give a flavour for the behaviour of the Landau levels, a calculation for a superlattice with 3.6 nm CdTe layers and 10 nm HgTe, at zero superlattice wavevector, neglecting strain effects. Fig. 5 is obtained assuming $\Delta$=0.04 eV, while Fig. 6 is for $\Delta$=0.35 eV. In both cases, a complicated analysis, involving the $k_z$-dependence, is needed to position the Fermi level properly.

It is clear that more experimental and theoretical work is

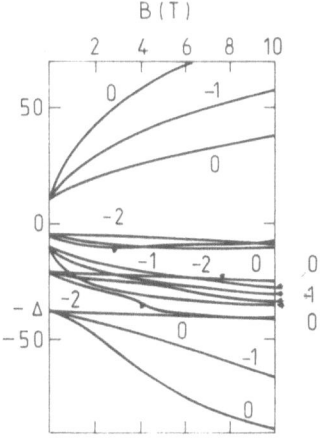
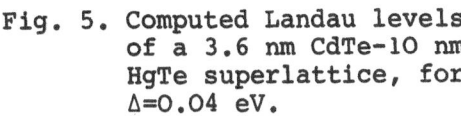
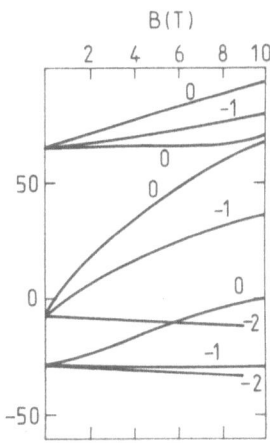

Fig. 5. Computed Landau levels of a 3.6 nm CdTe-10 nm HgTe superlattice, for Δ=0.04 eV.

Fig. 6. Same as Fig. 5, for Δ=0.35 eV.

needed to sort out the remarkable electronic properties of this system.

ACKNOWLEDGMENTS

It is a pleasure to thank A. Fasolino for many contributions to the results presented here and J.M. Berroir for a useful correspondence on the CdTe-HgTe system. Numerical calculations were performed with the support of the Centre de Calcul Vectoriel pour la Recherche, Palaiseau, France.

REFERENCES

1.  E.O.Kane, in: "Semiconductors and Semimetals", R.K. Willardson and A.C.Beer eds., Academic, New York (1966) vol. 1, p. 75
2.  G.Bastard, Phys.Rev.B. 24, 5693 (1981)
3.  S.R.White and L.J.Sham, Phys.Rev.Letters 47, 879 (1981)
4.  M.Altarelli, Phys.Rev.B 28, 842 (1983)
5.  See e.g. L.L.Chang, in: "Semiconductor Superlattices and Heterojunctions", G.Allan, G.Bastard, N.Boccara, M.Lannoo and M.Voos, Springer, Berlin (1986) and references therein
6.  G.Bastard, Surf.Sci. 170, 426 (1986)
7.  J.M.Luttinger, Phys.Rev. 102, 1030 (1956)
8.  A.Fasolino and M.Altarelli, Surf.Sci. 142, 322 (1984)
9.  J.C.Maan, Y.Guldner,J.P.Vieren, P.Voisin, M.Voos, L.L.Chang and L.Esaki, Solid State Commun. 39, 683 (1981)
10. P.Voisin, Thesis, Université Paris Sud, unpublished

11. A.Fasolino and M.Altarelli, in: "Two-Dimensional Systems, Heterojunctions and Superlattices", G.Bauer, F.Kuchar and H.Heinrich, eds. Springer, Berlin (1984) p.176
12. M.Altarelli, J.of Lumin. $\underline{30}$, 472 (1985)
13. E.E.Mendez, L.L.Chang, C.-A.Chang, L.F.Alexander and L.Esaki, Surf.Sci. $\underline{142}$, 215 (1984)
14. E.E.Mendez, L.Esaki and L.L.Chang, Phys.Rev.Lett. $\underline{55}$, 2216 (1985)
15. H.Munekata, E·E.Mendez, Y.Iye and L.Esaki, Proceedings of the MSS-11 Conference, Kyoto, Surf.Sci., in print (1986)
16. M.Altarelli and J.C.Maan, unpublished
17. O.Madelung, ed. "Landolt-Boernstein Numerical Data and Functional Relationships in Science and Technology", vol. 17, sec. 2.11
18. See e.g. Y.Guldner, in: "Semiconductor Superlattices and Heterojunctions", G.Allan, G.Bastard, N.Boccara, M.Lannoo and M.Voos, eds., Springer, Berlin (1986) and references therein
19. D.J.Olego, J.P.Faurie and P.M.Raccah, Phys.Rev.Lett. $\underline{55}$, 328 (1985)
20. S.P.Kowalczyk, J.T.Cheung, E.A.Kraut and R.W.Grant, Phys.Rev.Lett. $\underline{56}$, 1605 (1986)
21. Y.-C.Chang, J.N.Schulman, G.Bastard, Y.Guldner and M.Voos, Phys.Rev.B 31, 2557 (1985)
22. Y.R.Lin-Liu and L.J.Sham, Phys.Rev.B $\underline{32}$, 5561 (1985)
23. J.N Schulman and Y.-C·Chiang, Proceedings of the MSS-11 Conference, Kyoto, Surface Sci., in print (1986)

# GROWTH AND PROPERTIES OF Hg-BASED SUPERLATTICES

Jean-Pierre Faurie

Department of Physics
University of Illinois at Chicago
Chicago, Illinois 60680

ABSTRACT

This paper reports on recent developments concerning the growth and characterization of $Hg_{1-x}Cd_xTe$-CdTe superlattices and other Hg based superlattices such as HgTe-ZnTe, $Hg_{1-x}Zn_xTe$-CdTe and $Hg_{1-x}Mn_xTe$-CdTe. These superlattices have been grown in order to investigate the type III-type I transition in these superlattices. Thus a special attention has been given to the study of magneto transport properties.

## INTRODUCTION

Hg based superlattices have received a great deal of attention over the last several years as potential materials for far infrared detectors. Since 1979 when HgTe-CdTe superlattice (SL) system was first proposed as a new material for application in infrared optoelectronic devices[1] significant theoretical and experimental attention has been given to the study of this new superlattice system. The interest in HgTe-CdTe SL is due to the fact that it is a new structure involving a II-VI semiconductor and a II-VI semimetal and that it appears to have great potential as a material for infrared detectors.

In the classification proposed for hetero-interfaces[2] the HgTe-CdTe SL appears to belong to a new class of superlattices called Type III. This is due to the inverted band structure ($\Gamma_6$ and $\Gamma_8$) in the zero gap semiconductor HgTe as compared to that of CdTe, which is a normal semi-conductor. Thus the $\Gamma_8$ light-hole band in CdTe becomes the conduction band in HgTe. When bulk states made of atomic orbitals of the same

25

symmetry but with effective masses of opposite signs are used, the matching up of bulk states belonging to these bands has as a consequence the existence of a quasi-interface state which could contribute significantly to optical and transport properties.[3]

Most of the studies have focused primarily on the determination of the superlattice bandgap as a function of layer thicknesses and as a function of temperature. Also the description of the electronic and optical properties at energies close to the fundamental gap has received much attention.[4,5,6]

The growth of this novel superlattice was first reported in 1982[7] and has subsequently been reported by several other groups.[8,9,10,11] Some differences have been observed in the past between theoretical predictions and experimental determinations of the SL bandgaps. Since then, the theory has been refined, the control of the layer thicknesses has been improved and the understanding of the interpretation of the experimental data used to determine the bandgap has deepened.

Does this mean that this SL system is now a well understood system? No, because many questions remain open. For example, the value of the interdiffusion at the HgTe-CdTe interface, the value of the valence band offset $\Lambda = \Gamma_{8HgTe} - \Gamma_{8CdTe}$, the role of the strain,[12,13] the role and the nature of the interface state,[3] and the existence of high hole mobilities in p-type SLs.[14]

In this paper I will report on the growth and characterization of $Hg_{1-x}Cd_xTe$-CdTe SLs with a special emphasis on the transport properties. Since the study of the Type III-Type I transition appeared to be very interesting[15] we have extended this investigation towards three others SL systems i.e. $Hg_{1-x}Zn_xTe$-CdTe, $Hg_{1-x}Zn_xTe$-ZnTe, and $Hg_{1-x}Mn_xTe$-CdTe. These SL systems compared to $Hg_{1-x}Cd_xTe$-CdTe have additional property. $Hg_{1-x}Zn_xTe$-CdTe and $Hg_{1-x}Zn_xTe$-ZnTe are strained layer SLs (SLSL) whereas $Hg_{1-x}Mn_xTe$-CdTe is a diluted magnetic semiconductor (DMS) SL.

GROWTH

HgTe-CdTe superlattices were grown for the first time on a CdTe $(\overline{111})$B substrate in a Riber 1000 MBE system. In our laboratory at the University of Illinois the growth experiments are currently carried out in a Riber 2300 MBE machine using three different effusion cells containing CdTe, for the growth of CdTe, Te and Hg for the growth of HgTe. We have shown that on a CdTe substrate, the substrate temperature must be above 180° in order to grow high quality superlattice crystals.[7] At this temperature, the condensation coefficient for mercury is close to $10^{-3}$. This requires a high mercury flux during the growth of HgTe.[16] Nevertheless the background pressure during the growth is in the high $10^{-7}$ torr range. Most of the time the Hg cell is left open during the growth of the CdTe layers. Thus a competition occurs between Hg and Cd. As a result we have found that the CdTe layers contain a few percent of Hg (up to 5%) and are in fact (Cd,Hg)Te layers. This is not supposed to affect even slightly the calculations. Thus we will neglect this effect here. HgTe-CdTe superlattices have also been grown on $Cd_{0.96}Zn_{0.04}Te$ $(\overline{111})$Te substrates and on GaAs(100) substrates.[17] On GaAs(100), both (100)SL//(100)GaAs and (111)SL//(100)GaAs epitaxial relationships have been obtained. The orientation can be controlled by the preheating temperature as previously reported.[18]

For CdTe(111) grown on GaAs(100) we have recently reported that according to selective etching, X-ray photoelectron spectroscopy and electron diffraction investigations, the orientation of the CdTe film is the $(\overline{111})$Te face.[19] We have grown on both CdTe$(\overline{111})$//GaAs(100) and CdTe(100)/-/GaAs(100) substrates and have experienced a difference in the mercury condensation coefficient. This has already been reported for the growth of $Hg_{1-x}Cd_xTe$ films on substrates of different crystallographic orientations.[20] It turns out that growing on a (100) orientation requires about 4.4 times more mercury than growing on a $(\overline{111})$Te orientation. But in the (100) orientation no microtwinning due to the formation of antiphase boundaries are observed which makes the growth more easy to control than in the $(\overline{111})$B orientation.

In order to obtain high quality superlattices we use typical growth rates of $3-As^{-1}$ for HgTe and $1As^{-1}$ for CdTe. This represents the best compromise between the low growth rate required for high crystal quality, especially for CdTe which should be grown at a higher temperature than 180°C, and the duration of the growth, which should be as short as possible in order to save mercury and to limit the interdiffusion process which cannot be completely neglected between these interfaces.[21]

Compared to the growth of HgTe-CdTe SLs that of $Hg_{1-x}Cd_xTe$-CdTe presents an additional difficulty since we have to control the ternary alloy $Hg_{1-x}Cd_xTe$ instead of the binary HgTe. Furthermore, since our goal is the study of the Type III – Type I transition the composition (x) should be very well controlled. In order to have the necessary flexibility for the composition x, a Cd cell plus a CdTe cell or two CdTe cells are required. The growth of $Hg_{1-x}Cd_xTe$ by MBE has already been discussed in numerous papers.[22]

Concerning HgTe and ZnTe, there is a 6.5% difference between their lattice parameters (6.46A and 6.09A, respectively, at 300K). The existence of such a strain could make it difficult to grow this alternate microstructure. But in the same way as we have grown CdTe-ZnTe SLs with a comparable strain[23] HgTe-ZnTe superlattices have been successfully grown.[24]

The growth of HgTe-ZnTe SLs have been achieved at 185°C on a $Cd_{0.5}Zn_{0.5}Te(100)$ buffer layer previously deposited on a GaAs(100) substrate. This is done since such a buffer layer is expected to have a lattice parameter lying between those of HgTe and ZnTe.

The ternary alloy $Hg_{1-x}Zn_xTe$ has already been grown by MBE[25] and the growth of superlattices involving this alloy is currently undertaken in our laboratory.

MnTe is not stable in the zinc-blende structure thus the growth of MnTe-CdTe SLs has not been yet attempted. Nevertheless, the ternary alloy $Hg_{1-x}Mn_xTe$ exists in the zinc-blende structure for manganese concentration x up to 0.30. Since the Type III - Type I transition should occur at 77K for x about 7% the growth of $Hg_{1-x}Mn_xTe$-CdTe SLs has been carried out.[26] High quality $Hg_{1-x}Mn_xTe$ alloy has been obtained by MBE[27] at 185°C thus the growth of the superlattices has been achieved on CdTe($\overline{111}$)-/GaAs(100) substrates using three effusion cells containing Hg, Mn and Te for the growth of the alloy and a CdTe cell for the growth of CdTe.

The proof that this novel superlattice system has successfully been grown is attested to by X-ray diffraction, as illustrated in Figure 1. In addition to the Bragg peaks one can see the existence of satellite peaks due to the new periodicity.

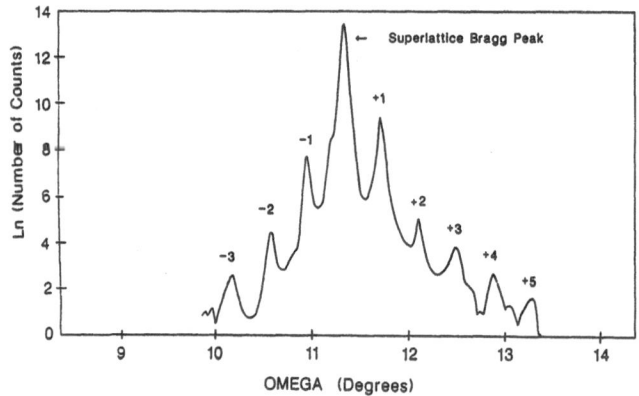

Fig. 1. Room temperature X-ray diffraction profile about the (111) reflection of a $Hg_{0.87}Mn_{0.13}Te$-CdTe superlattice with 150 periods of 112 A each (HgMnTe: 66 A, CdTe 46 A).

The band structure of HgTe-CdTe superlattices can be calculated by using the LCAO or the envelope function models which give very similar results. An important parameter, which determines most of the HgTe-CdTe SL's properties, is the valence band discontinuity $\Lambda$ between HgTe and CdTe. The value of $\Lambda$ is presently disputed.

From the phenomenological common anion rule[28] and the LCAO approach of Harrison[29] one can deduce that $\Lambda$ is small i.e. < 0.1eV. But recent theoretical results, based on the role of interface dipoles do not support the common anion rule and predict a much larger value $\Lambda = 0.5eV$ [30].

The first experimental determination of $\Lambda$ was obtained from far-infrared magneto-optical experiments at T = 1.6K on a superlattice consisting of 100 periods of HgTe (180A) - CdTe (44A). The best agreement between experiment and theory (done in the envelope function approximation) was obtained for $\Lambda = 40$ meV.[3]

Since then, additional magneto-absorption experiments have been performed on several other SLs and it has been constantly found that a small positive offset $\Lambda$ within the limits (0-100meV) provides the best fit.[32]

Resonant Raman Scattering was applied recently to investigate electronic properties of HgTe - CdTe SLs. From these experiments, it has been shown that the $\Gamma_7$ holes are confined in the CdTe layers which implies an upper limit of 120 meV for $\Lambda$.[33]

Recently, $\Lambda$ was also measured by X-ray photoemission spectroscopy (XPS) and a much larger value $\Lambda = 0.35eV$ was obtained.[34] It is important to point out that magneto-optical data at 2K as well as the infrared transmission measurements that we have performed at 300K cannot be interpreted by using such a large valence band offset either in the envelope function model or in the LCAO approach. In fact, most of the investigated SLs are calculated to be semimetallic at 4K for $\Lambda = 0.35eV$ which is not compatible with the magneto-optical data.[35]

In order to clarify whether there is a discrepancy between the optical and XPS data, we have performed very careful XPS measurements under well controlled conditions.[36] Both CdTe-HgTe ($\overline{111}$) B and HgTe-CdTe ($\overline{111}$) B heterojunctions have been grown "in situ" at the same temperature, 190°C, known to give no interdiffusion. The samples are transfered directly from the MBE chamber to a SSX-100 spectrometer without passing through the atmosphere. Therefore no contamination occurs as shown by the absense of C1s and O1s peaks (detection limit < 0.1 monolayer). The thickness of CdTe or HgTe thin overlayer on the counterpart compound, determined from the XPS peak areas, varies from 5A to 40A. The intensity of the substrate and overlayer peaks varies linearly with thickness indicating that the interface is abrupt.

The principle of the valence band discontinuity determination is derived from ref. 34 with however different parameters. The valence band maxima are obtained by linear extrapolation of the leading edge and the Cd 4d core levels are the resolved in Cd $4d_{5/2}$ and Cd $4d_{3/2}$ levels. The results obtained on a series of 15 measurements on independent samples show a valence band discontinuity $\Lambda = 0.34 \pm 0.06eV$ confirming the result of ref. 34. Note that this value is found to be independent of the Fermi level position which shifts by 0.15eV with overlayer thickness. However, when one considers different core levels such as Te $4d_{5/2}$ Cd $3d_{5/2}$ and Hg $4f_{7/2}$ one can observe that differences between core levels are not fixed quantities independent of the coverage.[36] Such observations are in contradiction with the principle of the valence band discontinuity determination. In addition, the choice of core levels other than Cd $4d_{5/2}$ and Hg $5d_{5/2}$ gives $\Lambda$ values which can differ from 0.34 eV.

Nevertheless, we have to conclude that direct XPS determination gives definitively a large value for $\Lambda$. It is important to point out that XPS measurements are carried out at 300K whereas magnetoptical and RRS experiments are performed at 2K and 10K respectively. Thus temperature dependent photoemission experiments should be carried out in order to find out if the discrepancy between the different experimental measurements is due to the fact that $\Lambda$ is temperature dependent in this peculiar heterojunction.

TRANSPORT PROPERTIES

One of the most interesting unanswered questions of HgTe-CdTe super-lattices is the mobility enhancement in the p-type structures. Hole mobilities have been reported as high as 30,000 $cm^2$/V.sec, but all are above 1,000 $cm^2$/V.sec.[14] The mobility of bulk p-type $Hg_{1-x}Cd_xTe$ is usually less than 500 $cm^2$/V.sec. Mixing of light and heavy holes has been suggested for the enhancement of the hole mobilities.[14] Several theoretical investigations have been carried out to study this problem. The band structure calculation has been refined using a multi-band tight binding model [13] and the effect of the lattice mismatch between the HgTe and CdTe has been investigated.[12,13] These studies conclude that the light holes should not contribute to the in-plane transport properties.

In order to investigate this interesting problem we have grown a related superlattice system i.e., $Hg_{1-x}Cd_xTe$-CdTe. HgTe-CdTe is called a Type III superlattice because of the inverted band structure of HgTe. In

this new system at T = 77K when x is smaller than 0.14 it is a Type III
SL. Whereas, when x is larger than 0.14 it is a Type I SL, similar to
GaAs-AlGaAs SL, since HgCdTe is now a semiconductor with both electrons
and holes confined in the smaller bandgap material. In Table I the Hall
characterization of six p-type SL samples is reported. When x changes
from 0.08 to 0.16 a drop of about 1 order of magnitude is observed for
the hole mobility.[15]

Table 1. Characteristics of $Hg_{1-x}Cd_xTe$–CdTe superlattices grown at 190°C on CdTe(111)/GaAs(100)
substrates. The Hall mobilities were
measured at 30 K except for sample No.
18124 which was measured at 10 K. $D_1$ =
$Hg_{1-x}Cd_xTe$ layer thickness; $D_2$ = CdTe layer
thickness; n = numbers of periods; x = cadmium
composition in $Hg_{1-x}Cd_xTe$ layers.

| Sample | x | $D_1$ (Å) | $D_2$ (Å) | n | $\mu_H$ ($cm^2V^{-1}s^{-1}$) |
|--------|------|-----------|-----------|-----|------------------------------|
| 18124  | 0    | 70        | 45        | 70  | p-2.5X10$^3$                 |
| 20539  | 0.01 | 82        | 34        | 120 | p-1.8X10$^3$                 |
| 20842  | 0.08 | 70        | 32        | 100 | p-2.5X10$^3$                 |
| 20943  | 0.16 | 70        | 40        | 100 | p-3.5X10$^2$                 |
| 18929  | 0.23 | 48        | 22        | 90  | p-1.3X10$^2$                 |
| 18728  | 0.27 | 69        | 22        | 100 | p-5X10                       |

Interestingly, the mobility enhancement ceases when the $Hg_{1-x}Cd_xTe$ in the HgCdTe-CdTe superlattices changes from a semimetal to a
semiconductor. This strongly suggests that the mobility enhancement only
occurs for the Type III superlattices and not in Type I superlattices in
the HgCdTe-CdTe system. The interfacial strain and the valence band
offset in all these samples should be the same. One of the differences
between the Type III and Type I superlattices is the existence of interface
states in the Type III superlattices but not in the Type I superlattices.
It is possible that the drastic difference in mobility in these superlattices
is related to these interfactial states. Unlike the GaAs – $Al_xGa_{1-x}As$
system this mobility enhancement is not due to modulation doping nor is
alloy scattering a factor.

In order to determine what carriers are responsible for transport properties we have performed Shubnikov-De Haas experiments. We have determined the effective mass of $Hg_{1-x}Cd_xTe$-CdTe superlattices from the temperature dependence of the amplitude of the Shubnikov-De Haas oscillations and find that the dominating carrier at low temperatures is the heavy hole.[37]

Our results also indicate that carriers are in the HgTe or $Hg_{1-x}Cd_xTe$ layers of the superlatttices. The lowest value for the effective mass of CdTe is about $0.7m_e$ (38) which is larger than we observed. Furthermore, if the carriers are in the CdTe layer, the effective mass of the carriers should be the same for both HgTe-CdTe and $Hg_{0.92}Cd_{0.08}Te$-CdTe superlattices. Our results, determined under identical conditions, are $0.30m_e$ and $0.36m_e$, respectively, for the two superlattices which is consistent with the band-structure calculation that the heavy-hole effective mass increases as x decreases in $Hg_{1-x}Cd_xTe$.

Figure 2 shows the $\rho_{xx}$ and $\rho_{xy}$ of a $Hg_{0.92}Cd_{0.08}Te$-CdTe superlattice. This sample has 100 periods of $Hg_{0.92}Cd_{0.08}Te$ (70 A) and CdTe (40 A). The Quantized Hall Effect (QHE) is observed in $\rho_{xy}$. Such an effect has been observed in n-type superlattices.[39,40] This represents the first p-type superlattice that shows QHE. Assuming the minima in $\rho_{xx}$ at 5.5 T and 11 T to be the Landau level, index i = 2 and i = 1, the two-dimensional hole density is $2.65 \times 10^{11}$ per sq.cm. The value of the plateau ( R (in ohm) = 25,812 / (i x n), where n is the number of layers contacted) indicates nine layers of the superlattice are contacted. With the strong oscillations above 5 T, it is surprising that no oscillation is detected below 5 tesla. Preliminary results of the n-type heterojunctions show the expected gradual increase in the amplitude of the Shubnikov-De Haas oscillations and the sudden onset of the quantum oscillations in the p-type structures is a peculiar property of the system.

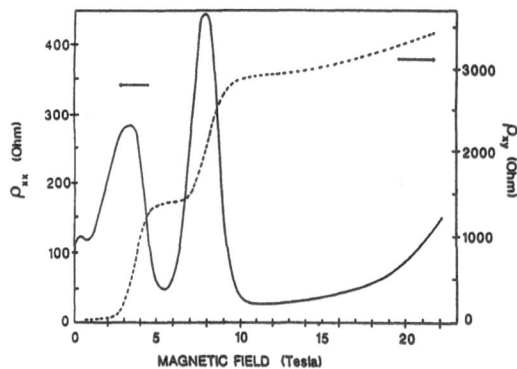

Fig. 2:   Quantized Hall Effect of $Hg_{0.92}Cd_{0.08}Te$-CdTe Superlattice at 0.5K.

The fact that $\rho_{xx}$ is not going down to zero when $\rho_{xy}$ reaches a plateau value is attributed to a dispersion in the carrier concentration among the interfaces participating in the QHE. Indeed, QHE carried out recently on a single HgTe-pHg$_{1-x}$Cd$_x$Te heterojunction does show that $\rho_{xx}$ can be equal to zero.[41]

The HgTe-ZnTe superlattice system has not yet been investigated from a theoretical point of view. Compared to the HgTe-CdTe SLs this novel SL system is expected to have the same potential of being able to produce an infrared material whose bandgap is controlled by HgTe and ZnTe layer thicknesses. But in addition to that, with a 6.5% difference between HgTe and ZnTe lattice parameters, the strain is supposed to offer another parameter to tailor the SL band structure. This has already been discussed for other superlattices.[42]

Table 2 presents three of these superlattices. It can be seen that they exhibit a p-type character with a hole mobility which is larger than those reported for HgZnTe alloys.[25,43] The existence of satellite peaks on the X-ray diffraction pattern confirms that these materials are superlattices. Infrared transmission data taken at 300K shows that the cutoff wavelength for these materials is in the 6-8 μm range compared to 2 μm for the corresponding alloy.[24] In Table 2 is also reported the Hall mobility of the first Hg$_{1-x}$Zn$_x$Te-CdTe SL ever grown. The hole mobility value is comparable with the highest reported for a HgTe-ZnTe and two times higher than the best hole mobility for a HgTe-CdTe SL grown on GaAs substrate (see Table 1). This result confirms that alloy scattering is not playing a major role in the enhancement of hole mobility and shows that investigation of Type III – Type I transition in this SL system looks very promising.

Table 2. Hall characterization of HgTe-ZnTe and Hg$_{1-x}$Zn$_x$Te-CdTe SLs at 30K with B = 0.3T, $d_1$ is HgTe or Hg$_{1-x}$Zn$_x$Te layer thickness, $d_2$ is ZnTe or CdTe layer thickness.

| Sample | Orientation | $d_1$(A) | $d_1$(A) | T(K) for $R_H$ = 0 | $\mu_H$(cm$^2$V$^{-1}$s$^{-1}$) |
|--------|-------------|----------|----------|------------------|----------------------------------|
| 17219 HgTe/ZnTe | (100) | 33 | 18 | 160 | p-2.6x10$^3$ |
| 18023 HgTe/ZnTe | (100) | 40 | 23 | 140 | p-5.0x10$^3$ |
| 18427 HgTe/ZnTe | (100) | 20 | 20 | 160 | p-2.0x10$^3$ |
| 47685 Hg$_{.94}$Zn$_{.06}$Te/ CdTe | $(\overline{111})$ | 40 | 20 | 140 | p-4.7x10$^3$ |

Several $Hg_{1-x}Mn_xTe-CdTe$ SLs have been grown with x ranging from 0.02 to 0.12. Unfortunately, none of these superlattices are p-type thus the Type III - Type I transition for holes has not been investigated. Concerning the mobility of electrons such a transition is not expected to have an effect since the interface states involved in the transition have a light hole character. As a matter of fact we have not observed a sudden change in the electron mobility for SLs when x is about 0.07.[44]

CONCLUSION

In this paper we have reported on very recent developments concerning the growth and characterization of $Hg_{1-x}Cd_xTe-CdTe$ SLs and related Hg based superlattice systems.

These SLs are now currently grown on CdTe, CdZnTe or GaAs substrates. The success of the epitaxial growth on the latter substrate represents an important opening due to the high crystal quality of GaAs, its availability in large area and its interest for electronic devices. The only concern with GaAs is its large mismatch with HgTe and CdTe which could generate, even after growth of a buffer layer, some residual strain in the super-lattices.

Concerning the valence band discontinuity $\Lambda$ if a small value i.e. less than 100 meV is determined from low temperature experiments such as magneto-optics and Resonant Raman scattering a much larger value of 0.34 eV has been calculated from X.P.S. experiments performed at 300K. None of these experiments is a direct measurement of $\Lambda$. In addition, if $\Lambda$ is equal to 0.34 eV, both theories (envelop function and tight binding models) must be drastically revised since their predictions for the superlattice energy bandgaps are in good agreement with experimental bandgaps when $\Lambda$, which is an adjustable parameter in these calculations, is less than 100 meV.

Such a discrepancy between experimental determination is not presently understood.

$Hg_{1-x}Cd_xTe-CdTe$ superlattices of both Type III and Type I have been grown and characterized in terms of transport properties. These superlattices are p-type. Their Hall characterization, along with magneto transport experiments seem to indicate that high hole mobilities observed in p-type HgTe-CdTe superlattices are due to some type of relationship between the two-dimensional heavy hole gas and the interface state existing in Type III superlattices.

Quantized Hall effect has also been observed in these p-type superlattices.

## ACKNOWLEDGEMENTS

I would like to thank many participants in the Microphysics Laboratory where these superlattices have been grown and their transport properties carried out. This work was entirely supported by Defense Advanced Research Projects Agency under contract No. MDA-903-85K-0030.

## REFERENCES

1. J. N. Schulman and T. C. McGill, Appl. Phys. Lett. 34, 663 (1979).

2. L. Esaki, Proceedings of the 17th International Conference on the Physics of Semiconductors, edited by J. D. Chadi and W. A. Harris on (Springer-Verlag, New York, Inc.), 473 (1985).

3. Y. C. Chang, J. N. Schulman, G. Bastard, Y. Guildner and M. Voos, Phys. Rev. B31, 2557 (1985).

4. G. Bastard, Phys. Rev. B25, 7584 (1982).

5. D. L. Smith, T. C. McGill and J.N. Schulman, Appl. Phys. Lett. 43, 180 (1983).

6. Y. Guldner, G. Bastard and M. Voos, J. Appl. Phys. 57, 1403 (1985).

7. J. P. Faurie, A. Million and J. Piaguet, Appl. Phys. Lett. 41, 713 (1982).

8. J. T. Cheung, J. Bajaj and M. Khoshnevisan, Proceedings of Infrared Information Symposia, Detector Specialty, Boulder (1983).

9. P. P. Chow and D. Johnson, J. Vac. Sci. Technol. A3, 67 (1985).

10. K. A. Harris, S. Hwang, D. K. Blanks, J. W. Cook Jr., and J. F. Schetzina, J. Voc. Sci. Technical A4, 2061 (1986).

11. D. J. Leopold, M. L. Wroge, J. M. Ballingall, B. J. Morris, D. J. Peterman and J. G. Broerman, 1985 U.S. Workshop on the Physics and Chemistry of Mercury Cadmium Telluride – San Diego.

12. J. N. Schulman and Y. C. Chang, Phys. Rev. B33, 2594 (1986).

13. G. Y. Wu and T. C. McGill, Apl. Phys. Lett. 47, 634 (1985).

14. J. P. Faurie, M. Boukerche, S. Sivananthan, J. Reno and C. Hsu, Superlattices and Microstructures 1, 237 (1985).

15. J. Reno, I. K. Sou, P. S. Wijewarnasuriya and J. P. Faurie, Appl. Phys. Lett. 48, 1069 (1986).

16. J. P. Faurie, A. Million, R. Boch and J. L. Tissot, J. Vac. Sci. Technol. A1, 1593 (1983).

17. J. P. Faurie, J. Reno and M. Boukerche, J. of Cryst. Growth 72, 11 (1985).

18. J. P. Faurie, C. Hsu, S. Sivananthan and X. Chu – Surface Science 168, 473 (1986).

19. C. Hsu, X. Chu, S. Sivananthan and J. P. Faurie, Appl. Phys. Lett. 48, 908 (1986).

20. S. Sivananthan, J. Reno, X. Chu and J. P. Faurie, J. Appl. Phys. 60, 1359 (1986).

21. D. K. Arch, J. L. Staudenmann and J. P. Faurie, Appl. Phys. Lett. 48, 1588 (1986).

22. J. P. Faurie, M. Boukerche, J. Reno, S. Sivananthan and C. Hsu, J. Vac. Sci. Technol. A3, 55 (1985) (and references therein).

23. G. Monfroy, S. Sivananthan, X. Chu, J. L. Staudenmann and J. P. Faurie, Appl. Phys. Lett. 49, 152 (1986).

24. J. P. Faurie, S. Sivananthan and X. Chu, Appl. Phys. Lett. 48, 785 (1986).

25. S. Sivananthan, X. Chu, M. Boukerche and J. P. Faurie, Appl. Phys. Lett. 47, 1291 (1985).

26. J. P. Faurie - MRC Meeting. Superlattices and heterostructures. LaJolla, July 1986 (unpublished results).

27. J. Reno, I. K. Sou, P. S. Wijewarnasuriya and J. P. Faurie, Appl. Phys. Lett. 47, 1168 (1985).

28. J. O. McCaldin, T. C. McGill and C. A. Mead, Phys. Rev. Lett. 36, 56 (1976).

29. W. Harrison, J. Vac. Sci. Techn. 14, 1016 (1977).

30. J. Tersoff, Phys. Rev. Lett. 56, 2755 (1986).

31. Y. Guldner, G. Bastard, J. P. Vieren, M. Voos, J. P. Faurie and A. Million, Phys. Rev. Lett. 51, 907 (1983).

32. J. M. Berroir, Y. Guldner, J. P. Vieren, M. Voos and J. P. Faurie, Phys. Rev. B 34, (1986).

33. D. J. Olego, P. M. Raccah and J. P. Faurie, Phys. Rev. Lett. 55, 328 (1985).

34. S. P. Kowalczyk, J. T. Cheung, E. A. Kraut and R. W. Grant, Phys. Rev. Lett. 56, 1605 (1986).

35. J. Reno, I. K. Sou, J. P. Faurie, J. M. Berroir, Y. Guldner and J. P. Vieren, Appl. Phys. Lett. 49, 106 (1986).

36. C. Hsu, Tran Minh Duc, J. P. Faurie (to be published).

37. K.C. Woo, S. Rafol and J. P. Faurie, Phys. Rev. B 34, October 15 (1986).

38. D. Kranzer, J. Phys. C6, 2977 (1973).

39. H. L. Stormer, J. P. Eisenstein, A. C. Gossard, W. Wiegmann and K. Baldwin, phys. Rev. Lett. 56, 85 (1986).

40. J. T. Cheung, G. Nizawa, J. Moyle, N. P. Ong, T. Vreland and B. Paine, J. Vac. Sci. Technol., July-August 1986.

41. J. P. Faurie, I. K. Sou, P. S. Wijewarnasuriya, S. Rafol and K. C. Woo, Phys. Rev. B 34, October 15 (1986).

42. G. C. Osborn, J. Vac. Sci. Technol. B1, 379 (1983) and references there in.

43. A. Sher, D. Eger, A. Zemel, H. Feldstein and A. Raizman, J. Vac. Sci. Techol. A4, 2024 (1986).

44. X. Chu, S. Sivananthan and J. P. Faurie (to be published).

MOCVD-GROWTH, CHARACTERIZATION AND APPLICATION OF

III-V SEMICONDUCTOR STRAINED HETEROSTRUCTURES

M. Razeghi, P. Maurel, F. Omnes and E. Thörngren

Thomson-CSF
Domaine de Corbeville, B.P. 10
91401 Orsay, France.

ABSTRACT

High quality InP and related compounds strained heterostructures have been grown on alternative substrates by the low pressure metalorganic chemical vapor deposition growth technique. Photoluminescence, SIMS and Auger measurements showed the high quality optical and electrical proper- ties of these layers. Buried ridge structure lasers emitting at 1.3 μm have been fabricated from the GaInAsP-InP double heterojunction grown on a GaAs substrate. MESFETs of GaInAs-InP heterojunctions using GaInP for a Schottky contact have been made.

INTRODUCTION

During the past few years it has been demonstrated[1-31] that the low pressure metalorganic chemical vapor deposition growth technique (LP-MOCVD) is well adapted for the growth of a variety of III-V semiconductor binary, ternary and quaternary heterojunctions, multiquantum wells (MQW) and super- lattices on lattice matched or alternative substrates for optoelectronic or microwave device applications. In this paper we describe the growth and characterization of strained heterostructures of InP, InAs and GaAs on various substrates ; the fabrication of a GaInAsP-InP buried ridge structure (BRS) laser emitting at 1.3 μm grown by two step (LP-MOCVD) on a GaAs sub- strate ; and finally we report the first growth and fabrication of $Ga_{0.49}In_{0.51}P$ /$Ga_{0.47}In_{0.53}As$/InP MESFET.

GROWTH PROCEDURE

The growth apparatus has been described in detail in ref.[17]. Growth was carried out at 76 torr. The optimum conditions for the low pressure growth of these layers as determined during these investigations are presen- ted in table I. Pretreatment of the substrates was found to be critical. The pretreatment procedure for InP and InAs was given in ref.[9].

Smooth single crystal films exhibiting mirror like surfaces have been obtained, even in the presence of a large layer-substrate lattice parameter mismatch. These layers tend, however, to be heavily dislocated, and their electrical and optical characteristics, especially those related to minority carrier properties such as diffusion length and lifetime, are generally inferior to the typical lattice-match system. These effects tend to be especially severe for thin layers, but we found that they can be partly eliminated by the use of thick buffer layers or special grading or superlattice techniques[27].

Table I. Optimized growth parameters

| (1.3 μm) | | InP | GaAs | InAs | GaP | GaInAs | GaInP | GaInAsP |
|---|---|---|---|---|---|---|---|---|
| Growth temperature | °C | 550 | 550 | 550 | 550 | 550 | 550 | 630 |
| Total flow rate ($N_2+H_2$) | ℓ/min | 6 | 6 | 6 | 6 | 6 | 6 | 7 |
| $N_2$/TEI bubbler flow | $cm^3$/min | 200 | – | 200 | – | 200 | 200 | 350 |
| $H_2$/TEG bubbler flow | $cm^3$/min | – | 120 | – | 120 | 120 | 120 | 60 |
| $PH_3$ flow | $cm^3$/min | 300 | – | – | 300 | – | 300 | 530 |
| $AsH_3$ flow | $cm^3$/min | – | 90 | 90 | – | 90 | – | 21 |
| Growth rate | Å/min | 100 | 100 | 100 | 100 | 200 | 200 | 150 |

Table II indicates the strained heterojunctions grown by LP-MOCVD.

Table II. Strained heterostructure grown by LP-MOCVD

| Substrate | First epilayer | Second epilayer |
|---|---|---|
| InP | GaAs | |
| InP | InAs | |
| GaAs | InP | |
| GaAs | InAs | |
| InAs | InP | |
| InAs | GaAs | |
| InP | GaAs | InAs |
| InP | $Ga_{0.47}In_{0.53}As$ | $Ga_{0.49}In_{0.51}P$ |
| GaAs | $Ga_xIn_{1-x}As$ | |
| GaAs | InP | $Ga_xIn_{1-x}As_yP_{1-y}$ |

Featureless mirrorlike surfaces have been grown over a wide temperature range of 500 to 650°C. The X-Ray diffraction rocking curve about the (400) Kα reflection from InP epilayer on GaAs substrate and GaAs epilayer on InP substrate is shown in figure 1. The X-Ray diffraction rocking curve of heterostructure of InAs-GaAs on InP substrate is shown in figure 2. We have performed a study of LP-MOCVD growth of InP simultaneously on InP, GaAs and InAs substrates with orientations of (100) placed adjacent to one another within the reactor for a growth temperature of 550°C.

Figure 3 shows the photoluminescence spectra of these layers at 5 K, using a Helium gas-flow variable temperature cryostat. Luminescence was excited using a He-Ne laser, and was analyzed in a 60 cm grating-spectrometer and detected with a high sensitivity N-cooled Ge photodiode. A series of luminescence transitions, which can be attributed to recombination mechanisms such as free excitons, exciton bound to shallow impurities (such as Zn), donor-acceptor recombination were observed on these spectra. The exciton recombination energy of InP on InP, InP on GaAs and InP on InAs substrates are 1.419 eV, 1.423 eV and 1.421 eV respectively. Considering that lattice parameter of InP, GaAs and InAs are 5.869 Å, 5.653 Å, and 6.057 Å respectively. One expects the InP layer on GaAs substrate to be compressed, and InP layer on InAs substrate to be expanded. Usually these layers are pseudomorphic (i.e. the elastic straining of the deposited lattice produces a zero misfit with the substrate). The accomodation of the mismatch by elastic strains induces changes in the magnitude of the gap. Thus in the case of InP epilayer on InAs substrate, one expects lower energy for exciton recombination than InP epilayer on InP substrate. We have not any interpretation for these results yet.

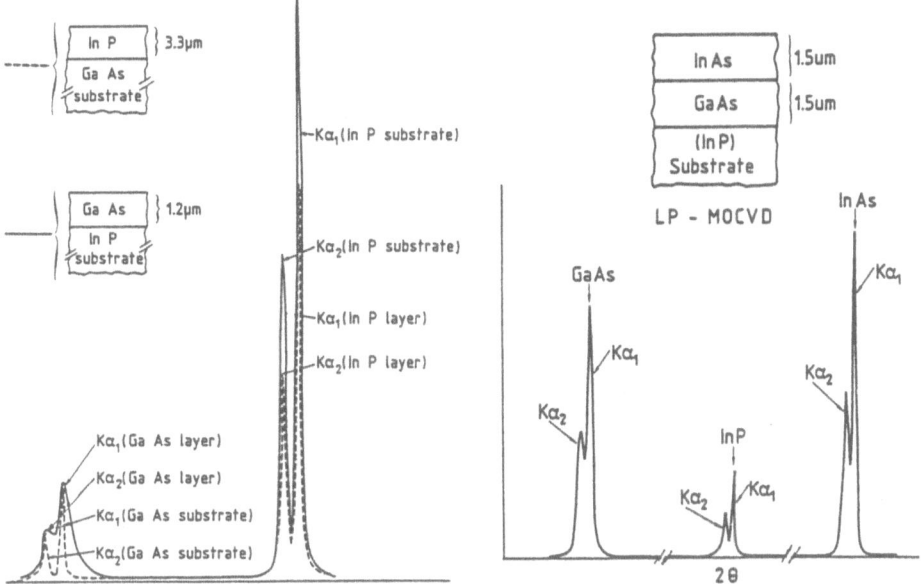

Fig.1. X-Ray diffraction rocking
curve of (400) CuKα reflection
from InP epilayer on GaAs
substrate and GaAs epilayer
on InP substrate.

Fig.2. X-Ray diffraction rocking
curve of (400) CuKα reflection
from InAs-GaAs-InP heterostructure grown by LP-MOCVD.

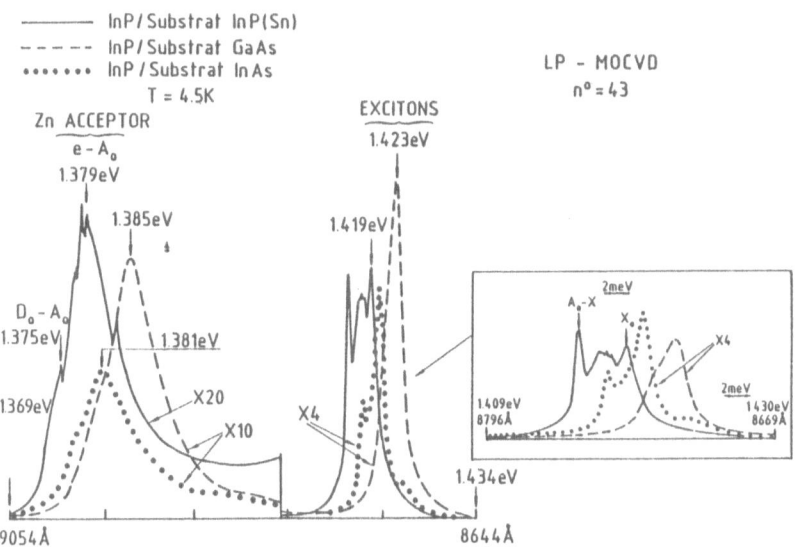

Fig.3. Photoluminescence spectrum of InP epilayer on InP, GaAs and InAs
substrates at 5 K by an He-Ne laser.

SIMS (Secondary Ion Mass Spectrometry) analysis are performed for
quantitative determination of impurities accumulated at the substrate-
epilayer interfaces. The analysis are carried out by a modified CAMECA
IMS 3F[32]. The surface of InP epitaxial layers are scanned with a focused
mass filtered oxygen ion beam ($I_p \sim 1.5$ μA at 10 keV). The scanned area

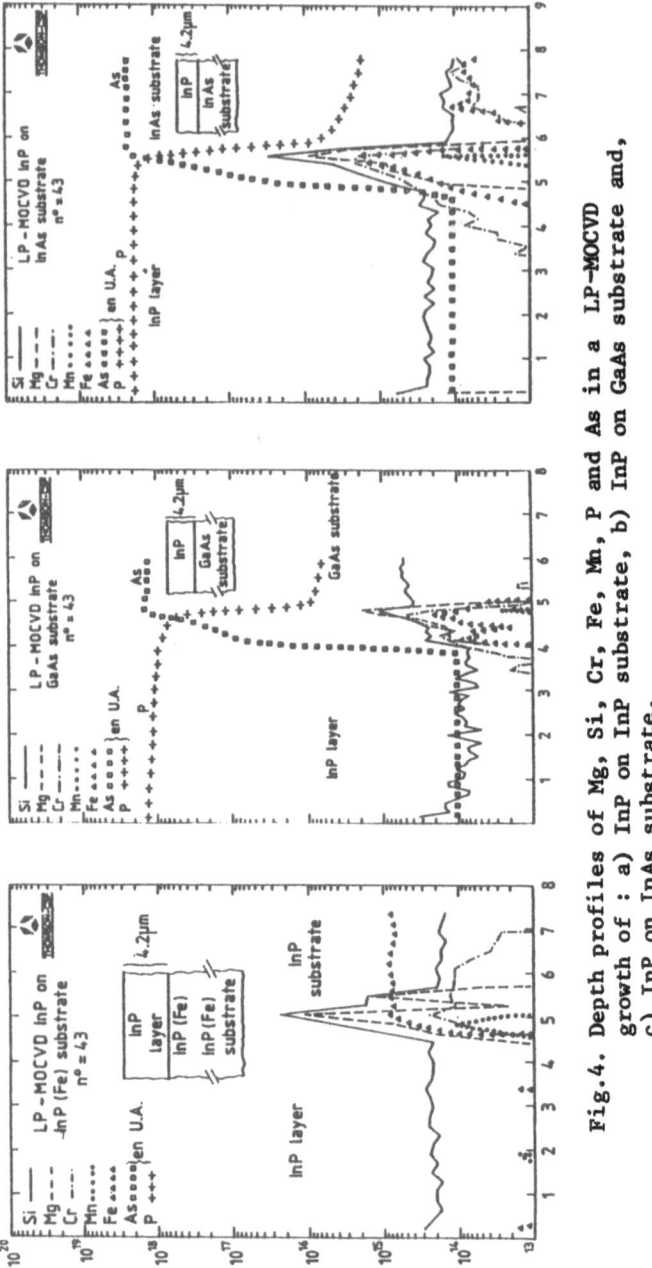

Fig.4. Depth profiles of Mg, Si, Cr, Fe, Mn, P and As in a LP-MOCVD growth of : a) InP on InP(Fe) substrate, b) InP on GaAs substrate and, c) InP on InAs substrate.

was 250 × 250 μm and the analysed region was 150 μm in diameter. Ion implanted samples were the standard for quantitative calibration of the instrument. The statistical results of various experiments show that the quantitative results of SIMS are given with an accuracy of ± 20 % above a concentration level of $1×10^6$ at.cm$^{-3}$. Below this level, results are less accurate, ± 50 %, at $1×10^{14}$ at.cm$^{-3}$. Talysurf measured depth precision is estimated at ± 10 %. The detection limit of the impurities which were measured are : Mg, Cr, Mn : $5×10^{12}$ at.cm$^{-3}$, Fe : $1×10^{13}$ at.cm$^{-3}$, and Si : $7×10^{13}$ at.cm$^{-3}$. Figure 4 (a,b,c) shows depth profiles of Mg, Si, Cr, Fe, Si, As and P in  InP layers grown on InP, InAs and GaAs substrates by LP-MOCVD. Each sample was analyzed in two different areas about 10 mm apart. Generally, analysis of two clean areas gives reproducible and representative results for the material. We have already shown that the major source of impurities at the interfaces is the adsorption of atoms on the substrate surface during chemical etching prior to epitaxy[32]. The pretreatment of the InP and InAs substrates are similar (see ref.8). So their SIMS profiles are identical. But the chemical etch of the GaAs substrate prior to epitaxy is $SOH_4+H_2O+H_2O_2$, and the concentration of impurities at the interface of InP-GaAs is lower than InP/InP or InP/InAs.

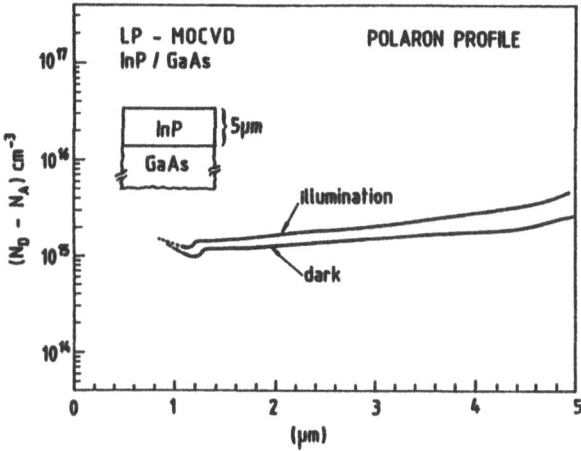

Fig.5. Electrochemical Polaron profile of
an InP epilayer on GaAs substrate.

These results show that the quality of InP layers far from the interfaces is independent of substrate origin. Figure 5 shows an automatic electrochemical profile through an InP layer grown on a GaAs substrate, with and without light (under illumination, for n-type semiconductors, to generate holes required for the reaction). The result shows that near the interface there are some perturbations, but far from the interfaces of epilayer-substrate the quality of the epilayer and the carrier concentration become similar to InP epilayer grown on InP substrate under the same conditions.

<u>Auger analysis</u>. The constituent concentration gradients at an InP/GaAs interface were determined by Auger analysis on a chemical bevel. The sample was chemically etched by using a methanol-bromine solution (15 % Br), in order to obtain a level having a mean amplification coefficient (M) of 2100 (measured with a Talysurf). This means that a change of one micron along the surface corresponds to a change of 4.75 Å in depth (z-direction). Figure 6 shows a schematic representation of the bevel. By scanning the incident electron beam four times along the bevel, the successive Auger profiles of the four elements P, As, In and Ga have been obtained. All the four profiles, shown on the figure 7 were obtained in 12 minutes but additional profiles following the inverse sequence (Ga, In, As, P) were obtained subsequently, in order to verify that there are no changes either in the intensity scale or in the position scale, all the four profiles follow the same line (same starting point) along the bevel. It can be observed that the In and P profiles are rather smooth but the Ga and As profiles show rather large fluctuations. Despite the use of a larger modulation bevel and a larger time for the acquisition of a scan, this is probably due to the poor Auger sensitivites of Ga and As relative to In and P. Due to the large magnification coefficient obtained by the chemical bevelling, the spatial broadening of the profiles related to the incident spot size and back scattering effects can be neglected because the incident beam diameter corresponds to an error in the depth of $\Delta Z \simeq d_o/M < 2$ Å. For the signal intensity for a given element, the change of the backscattering contribution when the incident electron beam is scanned along the interface can also be neglected due to the fact that the mean atomic number $\bar{Z}$ of the sample does not change when going from GaAs (Z(Ga)31+Z(As)33 = 64) to InP (Z(In)49+Z(P)15 = 64). Under the above simplifications, the relation between Auger intensity $I^A$ and concentration $C^A$ of an element A can be easily obtained by assuming that the interface region consists of n slices each of thickness "a" where the slices are numbered starting from the deepest one (at the end of the homogeneous GaAs concentration, here)[34], see figure 6.

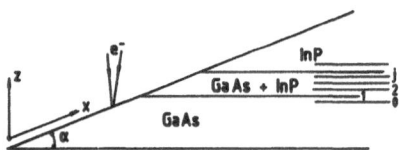

Fig.6. Schematic representation of the bevel of the InP/GaAs structure showing how a surface analysis along the bevel is converted into depth analysis.

The intensity due to the element A in the $J^{th}$ slice corresponds to the intensity flowing from outside the slice, $I^A_J$, minus the intensity coming from all the other deeper slices, $I^A_{J-1}$, which is attenuated by the factor K by travelling through the $J^{th}$ slice of thickness a. The interface composition (transition region) of InP-GaAs is $Ga_xIn_{1-x}As_yP_{1-y}$ where $0 \leq x,y \leq 1$.

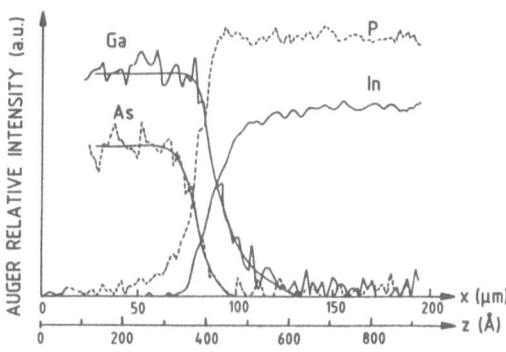

Fig.7. Corresponding Auger profiles relative to the four components : P,
In, As and Ga.

After Auger analysis, the transition region constituting the interface
can be subdivided into 3 parts (starting from the GaAs substrate) :
1) a region (thickness = 150 Å) where In is absent, its chemical composition is $GaAs_yP_{1-y}$ with $0.87 < y < 1$ which corresponds to the heating of
the GaAs substrate under $PH_3$ before growth. So it is possible to have the
adsorption of As and absorption of P. This can be remedied by heating the
GaAs substrate under $AsH_3$ pressure, before introducing $PH_3$ into the reactor.
2) The mid region (thickness = 120 Å) where all the four components are
present. Its chemical composition is $Ga_xIn_{1-x}As_yP_{1-y}$ with $0 \leq y \leq 0.87$,
$0.24 \leq x \leq 1$.
3) The region (thickness = 115-130 Å) where As is quite absent, its chemical composition is $PGa_xIn_{1-x}$ with $x < 0.24$.
Such analysis can be developed on any hetero-epitaxial structure if a good
composition of the chemical etchant is found[34].

Etch-pit-density (EPD). Figure 8 (a,b,c) shows photomicrographs of
these layers after forming a chemical bevel with very low angle and selective etching. The EPD of epitaxial layers and substrates are indicated in
table III.

Table III. EPD in epitaxial layers of InP on InP, InP on GaAs and InP on
InAs substrate

| Epilayer/substrate | EPD (epilayer) $cm^{-2}$ | EPD (substrate) $cm^{-2}$ |
|---|---|---|
| InP/InP | $8 \times 10^3$ | $8 \times 10^3$ |
| InP/GaAs | $6 \times 10^5$ | $2 \times 10^3$ |
| InP/InAs | $5 \times 10^5$ | $1 \times 10^2$ |

These results show that the EPD in InP/GaAs and InP/InAs interfaces is
independent of the EPD of the InAs or GaAs substrates.

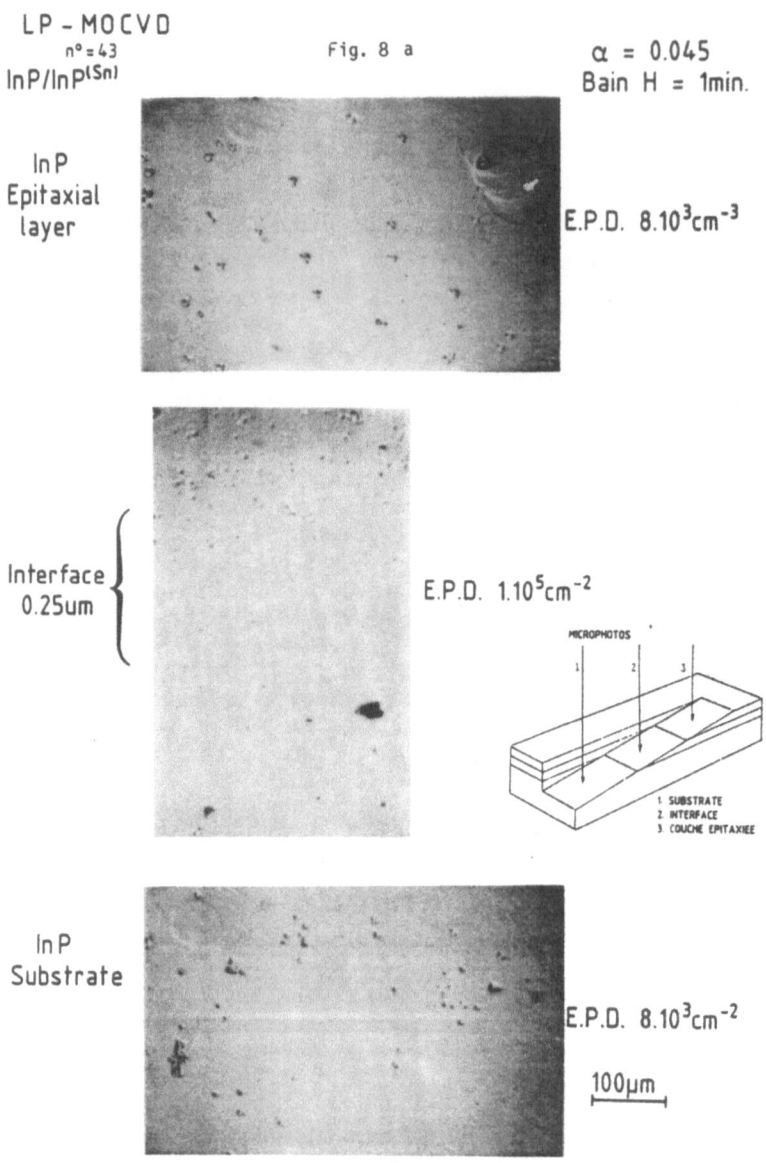

LP - MOCVD
n° = 43
InP/InP(Sn)

Fig. 8 a

$\alpha$ = 0.045
Bain H = 1min.

InP
Epitaxial
layer

E.P.D. $8.10^3 cm^{-3}$

Interface
0.25um

E.P.D. $1.10^5 cm^{-2}$

MICROPHOTOS

1. SUBSTRATE
2. INTERFACE
3. COUCHE EPITAXIEE

InP
Substrate

E.P.D. $8.10^3 cm^{-2}$

100µm

Fig. 8a

LP - MOCVD
InP/GaAs
n° ≈ 43

$\alpha$ : 0,070
Bain H : 2min.
Bain AB : 5min.

E.P.D.   a) layer : $\sim 6.10^5$ cm$^{-2}$
         b) substrate : $\sim 2.10^3$ cm$^{-2}$

1. SUBSTRATE
2. INTERFACE
3. EPITAXYAL LAYER

MICROPHOTOS HERE

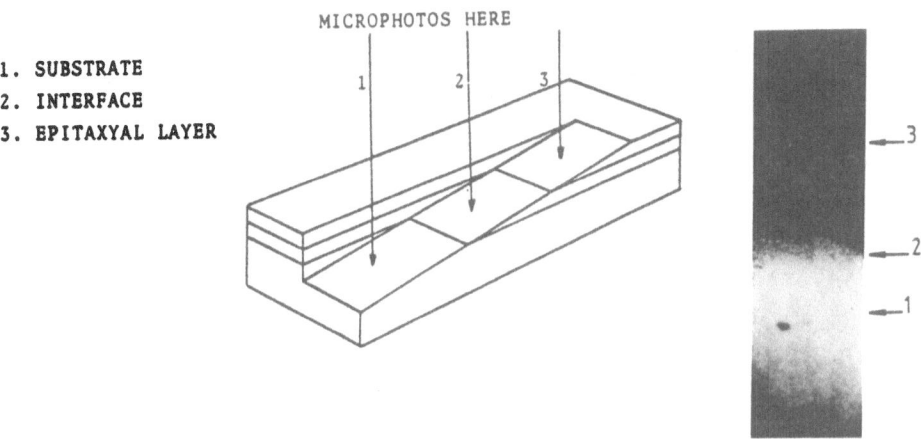

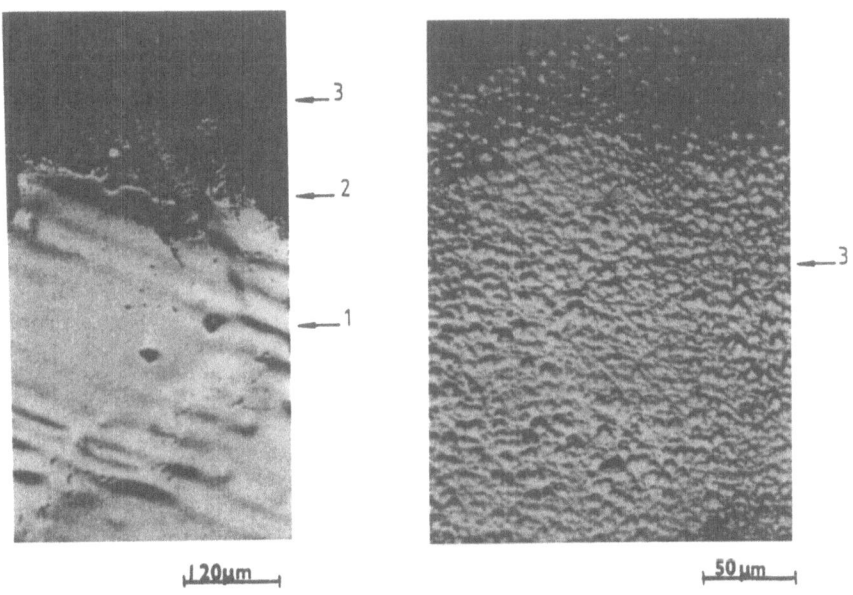

20μm

50μm

Fig. 8b

47

LP - MOCVD
InP/InAs
n° = 43

α : 0,0650
+ Bain H : 2min
Bain AB : 5min

<u>E.P.D.</u>   a) layer :~ 5 x $10^5$ cm$^{-2}$
      b) substrate :~ $10^2$ cm$^{-2}$

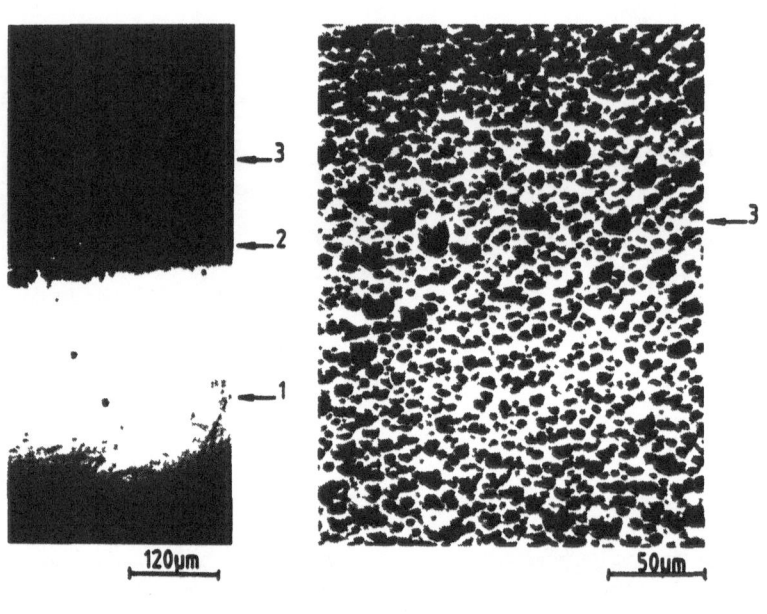

120µm

50µm

MICROPHOTOS HERE

1      2     3

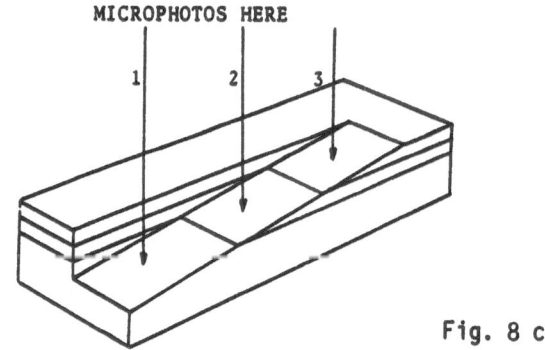

Fig. 8 c

1. SUBSTRATE
2. INTERFACE
3. EPITAXYAL LAYER

Fig. 8c

48

Long distance optical links use lasers emitting at 1.3 μm fabricated by material grown on InP substrate.

Unfortunately the technology of IC's (integrated circuits) for signal treatment is a lot more difficult on this material than on GaAs substrate. A solution would be to combine the advantages of these two materials which is against nature, owing to the large lattice mismatch.

However we have shown that it is possible to do it with the LP-MOCVD growth technique. $Ga_{0.25}In_{0.75}As_{0.5}P_{0.5}$-InP buried ridge structure (BRS) lasers emitting at 1.3 μm have been fabricated on GaAs substrates using the LP-MOCVD growth technique. The BRS laser structure was manufactured as follows ; first the following layers were successively grown by LP-MOCVD on a Si-doped (100) 2° off GaAs substrate :
- 2 μm InP confinement layer, sulphur doped with $N_D$-$N_A \simeq 10^{18}$ $cm^{-3}$
- 0.2 μm thick undoped GaInAs (composition 1.3 μm) active layer
- 0.2 μm thick Zn-doped ($N_A$-$N_D \simeq 2 \times 10^{17}$ $cm^{-3}$) InP layer in order to avoid
  the formation of defects near the active layer during the etching.
The morphology of these layers was excellent, and the photoluminescence intensity and PL half width were <u>the same</u> as for material grown on InP substrate. The details of growth conditions are given in ref.[26]. Next, a ridge of about 2 μm width was etched in the InP (P) and GaInAsP active layers through a photolithographic resist mask. With the aim of having a good control of the etching, we used a selective etchant composed of $H_2SO_4$, $H_2O_2$-$H_2O$ (1:8 = 40).

After removing the resist mask, the ridges were then covered with 1 μm of Zn-doped InP confinement layer and 0.5 μm Zn-doped GaInAs (with $N_A$-$N_D \simeq 10^{19}$ $cm^{-3}$) cap layer, grown by LP-MOCVD. In order to localize the injection current only in the buried-ridge active region, a deep proton implantation was performed through a 5 μm-wide photoresist mask after the metallization of the contacts. Further localization of the current in the buried ridge is achieved by the built-in potential difference between the P-N InP homojunction on each side of the active region and the N-P-InP-GaInAsP heterojunction of the active region.

Figure 9 shows schematically the resulting GaInAsP-InP-BRS laser. The devices were cleaved and sawn, producing chips of width 350 μm with cavity lengths of 300 μm. The laser chips were tested, unmounted under pulsed conditions at a repetition rate of $10^4$ Hz with a pulse length of 100 nsec.

Figure 10 shows the light-current characteristics of 7 LP-MOCVD laser diodes obtained from the same wafer. Pulse threshold current of 190 mA at room-temperature has been measured with an output power up to 10 mW.

GaInP/GaInAs/InP MESFETs

The $Ga_{0.47}In_{0.53}As$ lattice-matched to InP is a potentially important material for field effect transistors (FET) with high peak electron velocity[35] and high electron mobility for application in optoelectronic integration.

Metal-Schottky barrier heights on GaInAs are too low to be used as MESFET gates. FETs in lattice matched $Al_{0.48}In_{0.52}As/Ga_{0.47}In_{0.53}As/InP$, which exploit the increased Schottky barrier height provided by AlInAs[36,37], and lattice-mismatched GaAs gate/GaInAs structures have also been prepared[38].

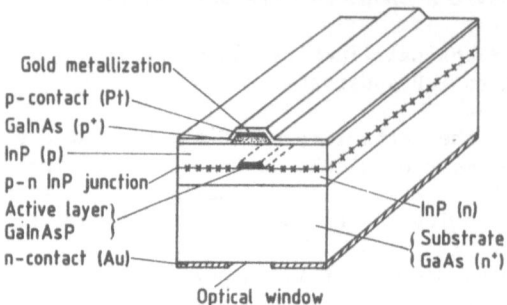

Fig.9. Schematic diagram of the cross-section of a GaInAsP-InP-BRS-laser
emitting at 1.3 μm grown by LP-MOCVD on a <u>GaAs</u> substrate.

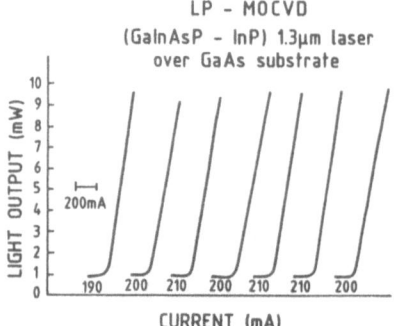

Fig.10. Light-current characteristics of laser diodes emitting at 1.3 μm
grown by LP-MOCVD on a <u>GaAs</u> substrate.

In this paper we report for the first time    the preparation of
an $Ga_{0.49}In_{0.51}P/Ga_{0.47}In_{0.53}As/InP$ MESFET fabricated from material grown
by LP-MOCVD. The energy gap of $Ga_{0.49}In_{0.51}P$ is 1.9 eV[39] and lattice
parameter is 5.65 Å.

The growth conditions are given in table I.

Figure 11 represents the FET device structure. Materials structures
consisting of n-type $Ga_{0.47}In_{0.53}As$ of 1500 Å thick, doped to $3 \times 10^{17}$ cm$^{-3}$
with sulphur and undoped $Ga_{0.49}In_{0.51}P$ of 800 Å thick with a carrier concentration ($N_D-N_A \simeq 10^{16}$ cm$^{-3}$) grown at 550°C onto (100) oriented Fe-
doped semi-insulating InP substrates.

Large geometry FETs with 2 μm gate lengths, 150 μm gate widths and
5 μm source drain spacing have been fabricated. The source and drain con-

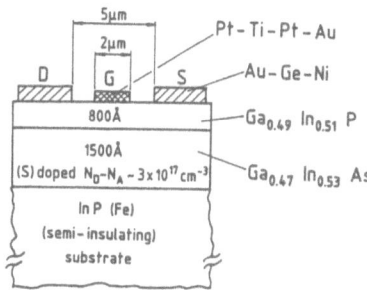

Fig.11. Schematic representation of the GaInP/GaInAs/InP MESFET structure.

tacts on the GaInP layer consist of evaporated Au–Ge–Ni. Pt/Ti/Pt/Au was used for gate contact. The gate pads were finally isolated from the active area by under–etching at the same time as the component isolation by mesa etching down to the semi–isolating InP substrate.

Figure 12 shows source–drain current–voltage characteristics of a GaInP–GaInAs–InP FET, gate bias step is 0.5 V, the transconductance of this device is $g_m \simeq 50$ ms/mm.

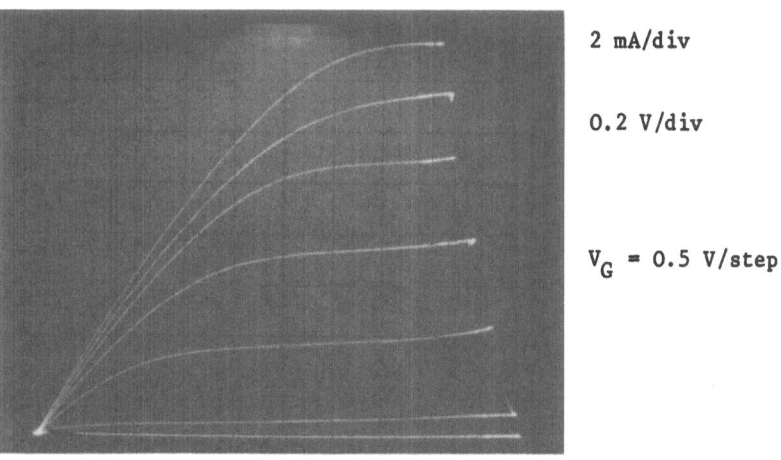

2 mA/div

0.2 V/div

$V_G$ = 0.5 V/step

Fig.12. D.C. drain current–voltage characteristics of a GaInP/GaInAs/InP
FET. Gate length 2 μm, gate width = 150 μm and source–drain spacing
is 5 μm. The transconductance of this device is $g_m$ = 50 ms/mm.

CONCLUSION

Growth and characterization of strained heterostructures of
InP ⇄ GaAs, InP ⇄ InAs and GaAs ⇄ InAs grown by LP–MOCVD have been descri-

bed. The results of photoluminescence, SIMS analysis, electron Auger measurements, etch-pit-density and Polaron profiles have been discussed.

In heterostructures, it is certainly desirable to select a pair of materials closely lattice-matched in order to minimize defect formation or stress. However, heterostructures lattice-mismatched to a limited extent can be grown with essentially no misfit dislocations, if the layers are sufficiently thin, because the mismatch is accommodated by a uniform lattice strain. Anyway, without the requirement of lattice matching, the number of available pairs for device applications and integrated circuits is greatly augmented. Buried ridge structure lasers emitting at 1.3 μm have been fabricated from the GaInAsP-InP double heterostructure grown on a GaAs substrate by a two step LP-MOCVD growth-technique.

Single heterostructure (with lattice mismatch) $Ga_{0.47}In_{0.51}P$/$Ga_{0.47}In_{0.53}As$/InP Schottky gate FETs were prepared by LP-MOCVD showing DC performance comparable to GaAs-GaInAs/InP structures grown by MBE.

ACKNOWLEDGMENTS

The authors would like to thank D. Leguen for technical assistance, Dr. A. Huber for SIMS analysis, J. Nagle for photoluminescence measurements, P. Etienne and Prof. Gazaux for Auger analysis, R. Blondeau for laser processing, G. Colomer and Prof. Decoster for FET processing.

We wish to express our cordial thanks to J.P. Duchemin, B. Winter for many valuable discussions.

We acknowledge financial support from the "Ministère de la Recherche et de la Technologie".

REFERENCES

1. M. Razeghi, P. Hirtz, R. Blondeau, B. de Crémoux and J. P. Duchemin, Electron. Lett., 17:597 (1981).
2. M. Razeghi, P. Hirtz, R. Blondeau, B. de Crémoux and J. P. Duchemin, Electron. Lett., 18:643 (1981).
3. M. Razeghi, R. Blondeau, P. Hirtz and J. P. Duchemin, Electron. Lett., 18:132 (1982).
4. Y. Guldner, J. P. Vieren, P. Voisin, M. Voos, M. Razeghi, Appl. Phys. Lett., 40:877 (1982).
5. M. Razeghi, M. A. Poisson, J. P. Larivain and J. P. Duchemin, M. Voos, Electron. Lett., 18:339 (1982).
6. M. Razeghi, M. A. Poisson, J. P. Larivain and J. P. Duchemin, Electron. Matter., 12:371 (1983).
7. M. Razeghi, P. Hirtz, R. Blondeau and J. P. Duchemin, Electron. Lett., 19:481 (1983).
8. M. Razeghi, S. Hersee, R. Blondeau, P. Hirtz and J. P. Duchemin, Electron. Lett., 19:336 (1983).
9. M. Razeghi, Rev. Thomson-CSF, 15:1 (1983).
10. M. Razeghi and J. P. Duchemin, J. Cryst. Growth, 69(1):76 (1983).
11. M. Razeghi and J. P. Duchemin, J. Vac. Sci. Technol., B1:262 (1983).
12. M. Razeghi, J. P. Hirtz, V. O. Ziemelis, C. Delalande and M. Voos, Appl. Phys. Lett., 43:585 (1983).
13. M. Razeghi, B. de Crémoux and J. P. Duchemin, J. Cryst. Growth, 68:389 (1984).
14. M. Razeghi, Rev. Thomson-CSF, 16:1 (1984).

15. M. Razeghi and J. P. Duchemin, "Solid-state sciences 53", ed : G. Bauer, F. Kuchar and H. Heinrich, Berlin, (1984).

16. M. Razeghi, "Light wave technology for communication", ed : W. T. Tsang and C. Beer, New York (1985).

17. M. Razeghi, R. Blondeau, B. de Crémoux and J. P. Duchemin, Appl. Phys. Lett., 45:784 (1984).

18. M. Razeghi, R. Blondeau, B. de Crémoux and J. P. Duchemin, Appl. Phys. Lett., 46:131 (1985).

19. M. Razeghi and J. P. Duchemin, J. Cryst. Growth, 70:145 (1984).

20. M. Razeghi, "Technology for chemicals and materials for electronics", ed : Howells, London, (1984).

21. M. Razeghi, R. Blondeau, J. C. Bouley, B. de Crémoux and J. P. Duchemin, Proceeding of the 9th IEEE international laser conference (1984).

22. M. Razeghi, J. P. Duchemin and J. C. Portal, Appl. Phys. Lett., 46:46 (1985).

23. M. Razeghi, R. Blondeau and J. P. Duchemin, Inst. Phys. Conf. Ser., 74:679 (1984).

24. M. Razeghi, J. Nagle and C. Weisbuch, Inst. Phys. Conf. Ser., 74:379 (1984).

25. M. Razeghi, R. Blondeau, J. C. Bouley and J. P. Duchemin, Inst. Phys. Conf. Ser., 74:451 (1984).

26. M. Razeghi, "Advances in solid state physics", 371 (1985).

27. M. Razeghi, P. L. Meunier and P. Maurel, J. Appl. Phys., 59:2261 (1986).

28. M. Razeghi, J. P. Duchemin and J. C. Portal, Appl. Phys. Lett., 48:712 (1986).

29. M. Razeghi, P. Maurel, F. Omnes and J. C. Portal, Appl. Phys. Lett., 48:1267 (1986).

30. M. Razeghi, J. Ramadani, H. Verriele, D. Decoster, M. Constant, Appl. Phys. Lett. (to be published) (1986).

31. M. Razeghi, P. Maurel, F. Omnes and J. C. Portal, J. Appl. Phys. (to be published) (1986).

32. A. M. Huber, M. Razeghi, G. Morillot, Inst. Phys. Conf. Ser. n°74, 223 (1984).

33. T. E. Gallon, Surf. Sci., 17, 486 (1969).

34. J. Gazaux, P. Etienne, M. Razeghi, to be published in J. App. Phys. (1986).

35. A. M. Littlejohn, J. R. Hauner, T. H. Glisson, Appl. Phys. Lett., 30, 242 (1977).

36. J. Barnard, H. Ohno, C. E. C. Wood and L. F. Eastman, IEEE Electron Device Letts EDL-1, n°9, 174 (1980).

37. M. D. Scott, A. H. Moore, I. Griffith, R. J. M. Griffith, R. S. Sussmann and C. Oxley, Inst. Phys. Conf. Ser. n°79, 475 (1984).

38. C. Y. Chen, A. Y. Cho, P. A. Garlinski, IEEE Electron Device Lett. EDL-6, n°1, 20 (1985).

39. C. Hilsum, Proc. 7th Intern. Conf. Semicond. Phys. Paris, p.1127, Dunod (1964).

# CRYSTAL QUALITIES AND OPTICAL PROPERTIES OF MBE GROWN

# GaSb/AlGaSb SUPERLATTICES AND MULTI-QUANTUM-WELLS

Seigo Tarucha

NTT Electrical Communications Laboratories
3-9-11, Midori-cho, Musashino-shi
Tokyo 180, Japan

## ABSTRACT

Crystal qualities and optical properties of MBE grown GaSb/AlGaSb superlattices and multi-quantum-wells are described.   Crystal qualities of
the superlattices strongly depend on the [V]/[III] beam ratio during the
MBE growth.   Deviation from the optimum value leads to drastic deterioration in the optical quality as well as in the crystallographical quality,
which have been evaluated by various kinds of analyses; X-ray diffraction,
Rutherford backscattering, cross-sectional TEM, Raman scattering, and
photoluminescence efficiency.   A GaSb/AlGaSb multi-quantum-well double-
heterostructure grown under the optimum [V]/[III] condition was prepared
for optical absorption measurement.   The measured spectrum exhibited a
well-defined double peak structure.   The lower and higher energy peaks
were assigned to be due to heavy hole excitons and light hole excitons,
respectively, based on a study of the polarization dependence of the
guided emission.   This assignment was confirmed by a biaxial strain
measurement using an asymmetric X-ray diffraction.

## INTRODUCTION

GaSb/AlGaSb material system is one of the candidates for preparing
semiconductor lasers in the long wavelength range of 1.3 - 1.7 μm, which
is the most important in optical communication.   As compared with the
other candidates such as GaInAsP/AlInAs and GaInAsP/InP, this system has
an advantage that strict control of composition in epitaxial layers is not
required.   The largest lattice mismatch in GaSb/AlGaSb system is 0.65 %
between GaSb and AlSb.   This value is  4 times larger than that between
GaAs and AlAs, but it is not too large for an epitaxial layer with device
quality to grow.   Most recently Ohmori et al.[1,2] have pointed out that
crystal quality of MBE grown GaSb/AlSb superlattices strongly depend on
the [V]/[III] ratio, where [V] and [III] are the impinging rates for the
group V and III elements, respectively.   They have succeeded in preparing
a current-injected GaSb/AlGaSb MQW laser diode continuously operating at
room temperature by optimizing the [V]/[III] ratio.

GaSb/AlSb superlattices were grown by MBE first by Naganuma et al.[3],
followed by Mendez et al.[4] and Griffith et al.[5]   They have demonstrated
that the MBE grown superlattices exhibit high optical quality and in

addition, several optical properties specific to GaSb quantum wells. Among those of the quantum size effect, excitons in quantum wells are attracting much attention in fundamental physics as well as in applied physics[6-8]. Recently, Voison et al.[7] observed exciton resonances in optical absorption spectra of GaSb/AlSb MQWs at low temperature. Based on the analysis of the absorption spectra, they concluded that the strain induced by the lattice mismatch between GaSb and AlSb causes the reversal of heavy and light hole exciton energy levels. Ploog et al.[8] have recently studied potoluminescence and photoluminescence excitation spectra of short period GaSb/AlGaSb MQWs, and observed the exciton resonance in the temperature range from 4 to 200 K.

This paper reviews recent investigations of the NTT superlattice research group on crystal qualities and optical properties of MBE grown GaSb/AlGaSb superlattices and MQWs on (001)-oriented GaSb substrates;
(1) First, influence of the [V]/[III] ratio on optical quality and crystallographical quality of MBE grown GaSb/AlSb superlattices is described. As reported previously, crystal quality of GaSb/AlSb superlattices strongly depends on the [V]/[III] ratio[2]. GaSb/AlSb superlattices grown under the various [V]/[III] ratios are analyzed by X-ray diffraction, Rutherford backscattering, Raman scattering, photoluminescence efficiency, and transmission electron microscopy (TEM).
(2) Second, optical absorption characteristics of a GaSb/AlGaSb MQW grown under the optimum [V]/[III] condition is described. The first room temperature observation of heavy and light hole excitons, and the influence of strain in the GaSb layer on the exciton absorption spectrum are mainly described.

## CRYSTAL QUALITIES OF GaSb/AlSb SUPERLATTICES

### Sample preparation

GaSb/AlSb superlattice epitaxial layers were grown on Te-doped (001)-oriented GaSb substrates by MBE. The substrate temperature during growth was 550 °C as determined by an infrared thermometer[1,9]. The growth rates of GaSb and AlSb were 20 and 6.7 nm/min, respectively, and were kept constant during all sequences. The [V]/[III] ratio was varied from 1.0 to 13.0 by changing the Sb vapor pressure. The GaSb well layer thickness $L_z$, the AlSb barrier layer thickness $L_B$, and the [V]/[III] ratio during the growth for the superlattice samples are summarized in Table 1. The total thickness of the superlattice epitaxial layers was 1 μm or more.

### X-ray diffraction (ω-scan)

Figure 1 shows the full width at half maximum (FWHM) of the 0-th order peak in an X-ray (400) reflection rocking curve as a function of

Table 1.  $L_z$, $L_B$ values, and [V]/[III] ratios of GaSb/ AlSb superlattices

| Sample No. | $L_z$ (nm) | $L_B$ (nm) | [V]/[III] ratio |
|---|---|---|---|
| 1 | 14.0 | 7.0 | 1.0 |
| 2 | 12.0 | 6.0 | 1.4 |
| 3 | 8.0 | 5.5 | 2.5 |
| 4 | 10.0 | 5.0 | 3.3 |
| 5 | 12.0 | 6.0 | 4.5 |
| 6 | 12.0 | 6.0 | 6.8 |
| 7 | 5.0 | 5.0 | 13.0 |

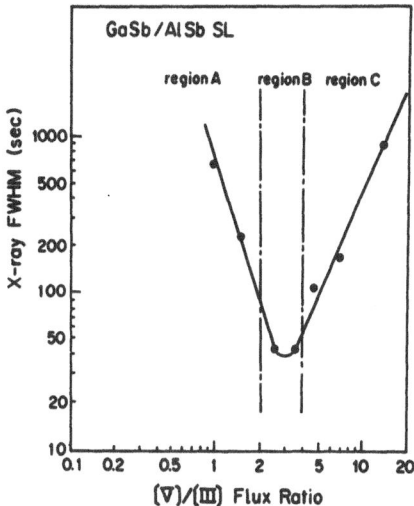

Fig. 1. Relative FWHM of ω scan X-ray (004) rocking curve as a function
    of impinging molecular beam ratio [V]/[III].

[V]/[III].  Here, the Ga arrival rate is used for convenience, as the impinging rate of the group III element [III].  The X-ray FWHM depends
strongly on [V]/[III].  There exists an optimum region of [V]/[III] = 2 -
4, where the X-ray FWHM becomes minimum around 40 sec.  Deviation from
this optimum region leads to a steep increase in X-ray FWHM, indicating
drastic deterioration of crystal quality.  Satellite diffraction peaks of
the superlattices were observed only in the optimum [V]/[III] region.
The optimum [V]/[III] region is called region B hereafter.  The region
whose [V]/[III] values are smaller and larger than that of region B are
called region A and C, respectively, hereafter.  The reason for the large
X-ray FWHM was further investigated by $\omega - 2\theta$ scan X-ray diffraction
measurement as described in the following.

X-ray diffraction ( ω - 2θ scan)

    Generally, there are two factors to broaden the FWHM of ω-scan X-ray
rocking curve.  One is a fluctuation, and the other is a spread of
reciprocal vector direction.  These two factors can be distinguished by
an X-ray diffraction measurement over a wide range of angles by ω-2θ coupling scan mode.  The X-ray sources were Cuk$\alpha_1$ and Cuk$\alpha_2$ lines.  Figure
2 shows the X-ray diffraction pattern for sample 7 in region C.  The diffraction signal  from the epitaxial layer is smaller than that from the
substrate, but {002}, {004}, and {006} diffractions were clearly observed.
However, {111}, {222}, {333}, and {115} diffractions were not observed.
This was also the case for the other samples in regions A and B.  This
result indicates that the epitaxial layers having the large X-ray FWHM
values are not polycrystals.  The FWHM values of the ω-2θ scan mode X-ray
diffraction measured for the  samples are 200 - 300 sec limited by the
measurement system, and they are smaller than those of ω-scan mode X-ray
diffraction for samples 1 and 7 with the large FWHM values.  This result
indicates that the fluctuation of 1/|G| = d (lattice constant) is small,
and that the orientation of reciprocal vector is spread over 300 - 900 sec
at least.

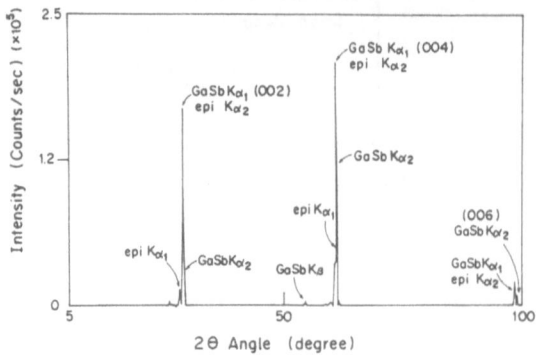

Fig. 2   X-ray diffraction pattern by ω-2θ coupling scan mode for sample 7.


## Rutherford backscattering

The lattice arrangement of the samples was investigated by Rutherford backscattering (RBS) measurement.   2 MeV He$^+$ was employed in the present measurement.   Figure 3 shows the random and aligned RBS spectra of the superlattice samples grown with different [V]/[III] ratios.   Figure 3(a) shows the spectrum for sample 1 in region A and Fig. 3(b) for sample 3 in region B, and Figs. 3(c) and (d) for samples 6 and 7 in region C, respectively.   The RBS $\chi_{min}$ value, which is defined by the minimum value of channeling to random scattering yield ratio, is plotted in Fig. 4. $\chi_{min}$ becomes minimum for the samples in region B and larger for the samples in region A and C.   The minimum $\chi_{min}$ value of 0.052 is nearly equal to that for GaAs of high quality, indicating the high quality of the GaSb/AlSb superlattice grown under the optimum [V]/[III] condition.   When the RBS spectra of Figs. 3(a) – (d) are compared with one another, the Ga (and Al) signal is larger in the aligned spectrum in Fig. 3(a), and the Sb signal is larger in the aligned spectra in Figs. 3(c) and (d).   These results   indicate that the lattice arrangement is not regular for the samples grown under non-optimum conditions.   Some of Ga (and/or Al) atoms

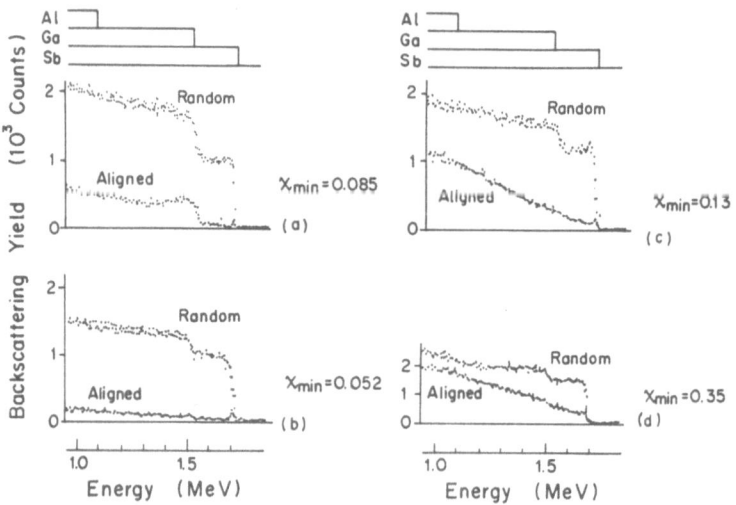

Fig. 3. Random and aligned RBS spectra for different values of  [V]/[III] ratios.   (a), (b), (c) and (d) are for samples  1, 3, 6, and 7, respectively.

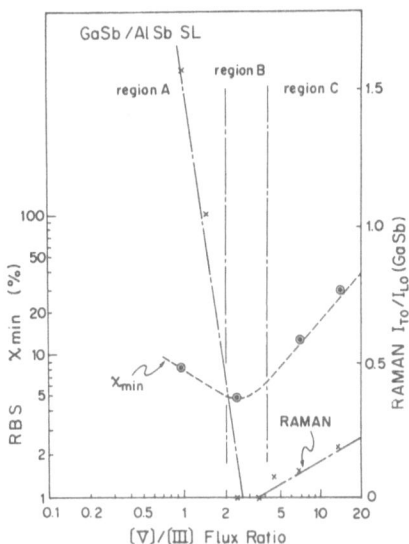

Fig. 4. RBS $X_{min}$ (solid line) and Raman GaSb-like TO phonon intensity
normalized by GaSb-LO phonon intensity (broken line) of MBE grown
GaSb/AlSb superlattices as a function of [V]/[III].

are in the dechanneling sites for the samples in region A as expected in
the epitaxial layers grown under a Sb-lack condition.   On the other hand,
some of Sb atoms are in the dechanneling sites for the samples in region
C, as expected in the epitaxial layers grown under a Sb-rich condition.
Probably, there are Ga (and/or Al) interstitial atoms in the  epitaxial
layer grown under the Sb-lack condition, and there are Sb  interstitial
atoms in the epitaxial layers grown under the Sb-rich condition.

### Raman scattering measurement

Raman scattering is a nondestructive method to give important infor-
mations about crystal orientations.   The measurements were performed at
300 K using an Ar laser ( $\lambda$ = 514.5 nm).   $Z(X,Y)\overline{Z}$ configuration was used
where the direction Z of the incident and the reflected beam is normal to
the epitaxial layer and the polarization X is parallel to the $\langle 011 \rangle$
direction.   TO phonon scattering is forbidden in this configuration.
Lukovsky et al.[10] performed the Raman scattering measurement on  bulk
$Al_xGa_{1-x}Sb$ with x from 0.0 to 1.0, and observed the Raman lines  due to LO
phonons at 235 cm$^{-1}$ and 339 cm$^{-1}$ for GaSb and  AlSb, respectively, and
those due to TO phonons at 225 cm$^{-1}$  and 318 cm$^{-1}$ for GaSb and AlSb,
respectively.

Figure 5 shows the Raman spectra for different GaSb/AlSb superlattice
samples grown with the different [V]/[III] ratios.   Figures 5(a) – (c)
are for sample 2 in region A (Sb-lack condition), sample 4 in region B
(Sb-optimum condition), and sample 7 in region C (Sb-rich condition),
respectively.   The Raman spectrum for sample 7 shows TO phonon peaks as
well as LO phonon peaks for GaSb and AlSb.   The existence of  the TO
phonon lines implies that the crystal orientation is  misaligned.
However, the GaSb- and AlSb-like LO phonon lines  for this sample appear
at the wave numbers exactly  corresponding to those in bulk GaSb and AlSb.
This result  suggests that the periodic structure of GaSb and AlSb layers
is preserved  without any compositional disordering between the GaSb and
AlSb layers.   TO phonon lines are seen also on the spectrum for sample 2,
but the phonon lines are slightly shifted to the lower wave numbers from
those in bulk crystal.   This shift is possibly attributed the composi-

tional disordering of GaSb and AlSb layers.    It is well known that AlSb-like LO and TO phonon lines shift to lower wave numbers in $Al_xGa_{1-x}Sb$ alloy with  decreasing x, and that GaSb-like LO and TO lines shift to lower wave numbers in $Al_xGa_{1-x}Sb$ alloy with increasing x[7].    327 $cm^{-1}$ (LO) and 316 $cm^{-1}$ (TO) AlSb-like phonon lines correspond to those in $Al_{0.5}Ga_{0.5}Sb$, and 233 $cm^{-1}$ (LO) and 223 $cm^{-1}$ (TO) GaSb-like  phonon lines correspond to those in  $Al_{0.05}Ga_{0.95}Sb$.    Therefore, the  Raman spectrum observed for sample 2 suggests a partial compositional disordering in the superlattice structure in which  Ga atoms from GaSb layers interdiffuse into AlSb layers more easily than in the reverse direction.

[V]/[III] ratio dependence of GaSb TO phonon line intensity normalized to GaSb LO-phonon intensity is shown in Fig. 4 by the broken line.  This figure shows that the abrupt increase in the TO phonon line intensity in the Sb-lack and Sb- rich regions is consistent with the results on the X-ray FWHM and RBS $X_{min}$ described previously.

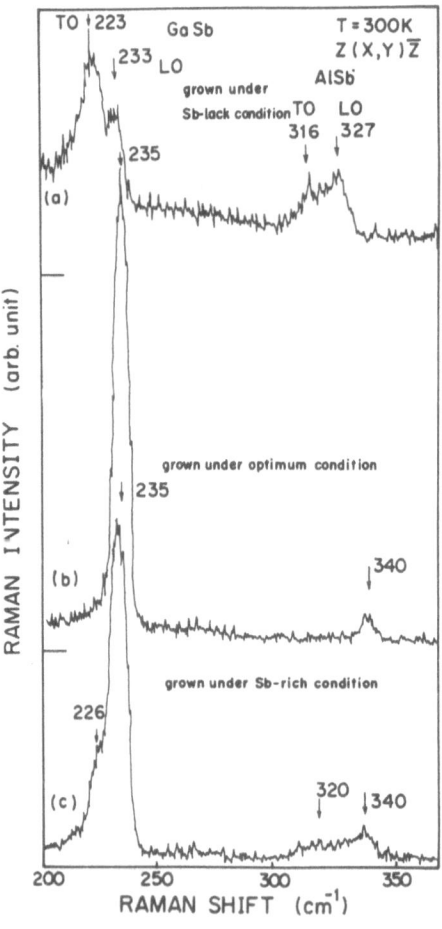

Fig. 5. Raman spectra for different GaSb/AlSb superlattices (a)sample 2, (b)sample 4, and (c)sample 7.

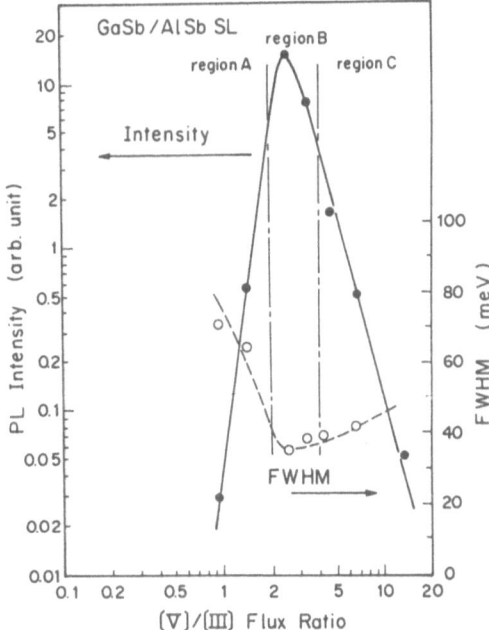

Fig. 6. Relative PL peak intensity (solid line) and FWHM of PL spectra (broken line) as a function of [V]/[III].

## Photoluminescence efficiency

Figure 6 shows PL peak intensity as well as FWHM of PL spectrum (PL-FWHM) as a function of [V]/[III]. PL was measured at room temperature using a Kr laser 647-nm line as an excitation light source. The excitation power density was 100 W/cm$^2$. The PL is associated with the transition between the n = 1 electron level and the n = 1 heavy hole level. Both the PL intensity and PL-FWHM strongly depend on the [V]/[III] ratio. PL intensity becomes maximum and simultaneously PL-FWHM becomes minimum for the samples in region B as expected from the other measurements described before. PL intensity decreases steeply with respect to either decrease or increase in [V]/[III]. On the other hand, PL-FWHM increases more steeply with decrease of [V]/[III] from the optimum value than with increase of [V]/[III]. Decrease of PL intensity and increase of PL-FWHM are attributed to deterioration of crystal quality as revealed by the X-ray diffraction, RBS, and Raman scattering measurements. However, a factor to increase the PL-FWHM seems to be different in regions A and C. This difference is more clearly interpreted by a microscopic analysis by TEM, and the large PL-FWHM for the samples in region A is found to be due to large well size fluctuation as described in the following.

## Cross-sectional transmission electron microscopy

The microscopic structures of the samples were investigated by cross-sectional TEM observation. Sample preparation for the TEM observation of the GaSb/AlSb superlattices is essentially the same as that of GaAs/AlGaAs superlattices reported previously[11]. In the procedure, Ar+ ion acceleration voltage was decreased from 5 kV (usually used for Ga-Al-As materials) to 3.5 kV because of the mechanical softness of Ga-Al-Sb materials. The TEM apparatus used here is a JEM-4000EX with 400 kV acceleration voltage. The incident beam is perpendicular to a (110) cleavage surface.

Fig. 7. Bright-field cross-sectional TEM micrograph of a GaSb/AlSb
        superlattice sample 4.

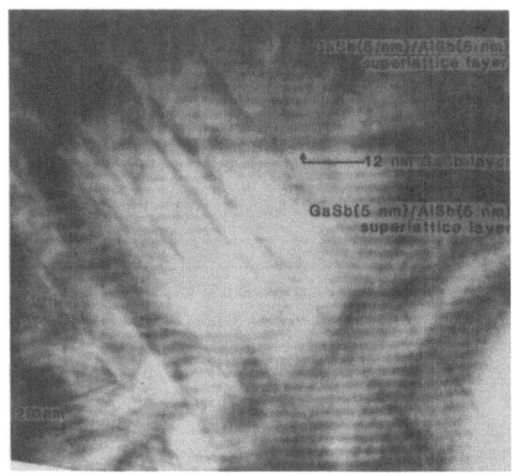 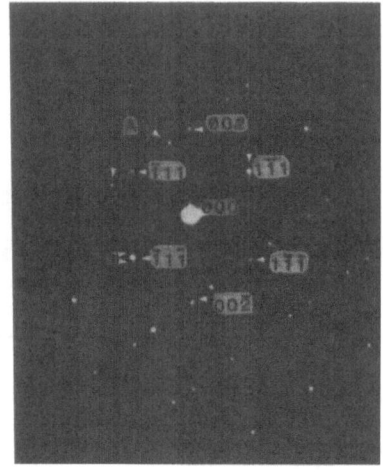

Fig. 8. Bright-field cross-sectional TEM    Fig. 9. Transmission electron
        micrographs of a GaSb/AlSb                 diffraction pattern for a
        superlattice sample 7.                     GaSb/AlSb superlattice sample 7.

A bright field TEM micrograph of sample 4 in region B (grown under
the optimum condition) is shown in Fig. 7.    Dark and bright stripes,
which can be distinguished with high contrast,   correspond to 8-nm GaSb
and 5-nm AlSb layers, respectively.    Good uniformity in both dark and
bright stripes was observed  without any feature.

Figure 8 shows a bright-field TEM micrograph of sample 7 in region C
(grown under Sb-rich condition).   The epitaxial layer structure consists
of a 5-nm GaSb and 5-nm AlSb superlattice including a 12-nm thick GaSb
layer in the midst, grown on a GaSb  buffer layer.  Many dislocations
(stacking faults and/or twins) were observed in the epitaxial layers.
The grain size formed by the dislocations  is evaluated as 20 - 100 nm in
length.    These dislocations are responsible for the deterioration of
crystal quality giving the large X-ray FWHM, small PL intensity, and etc.

Figure 9 shows a transmission electron diffraction pattern for this sample. Not only the spots of the fundamental diffractions {111}, {002} etc., but also those of additional diffractions indicated by (A) were observed. Spots (A) appeared at intermediate locations which are 1/3 and 2/3 of the interval between (002) and (111) spots etc., indicating the diffractions due to twins. Satellite diffractions accompanying the fundamental diffractions (111) and (111) etc., which are indicated by (s), were also observed. These satellite diffractions mean that a uniform period of superlattice structure is formed in the epitaxial layer.

Figure 10 shows a bright-field TEM micrograph of sample 1 in region A (grown under Sb-lack condition), which was designed as a 14-nm GaSb and 7-nm AlSb superlattice. A large bend of the layer structure with its lateral pitch of about 2 μm is seen. This layer bending results in a surface roughness giving a milky surface morphology, which was observed by

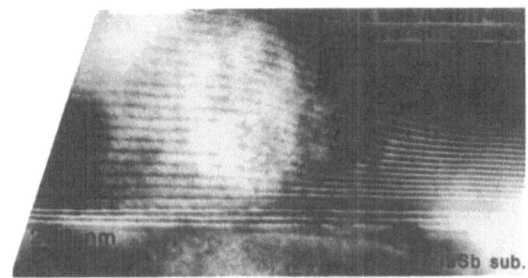

Fig. 10. Bright-field cross-sectional TEM micrographs of a GaSb/AlSb superlattice sample 1.

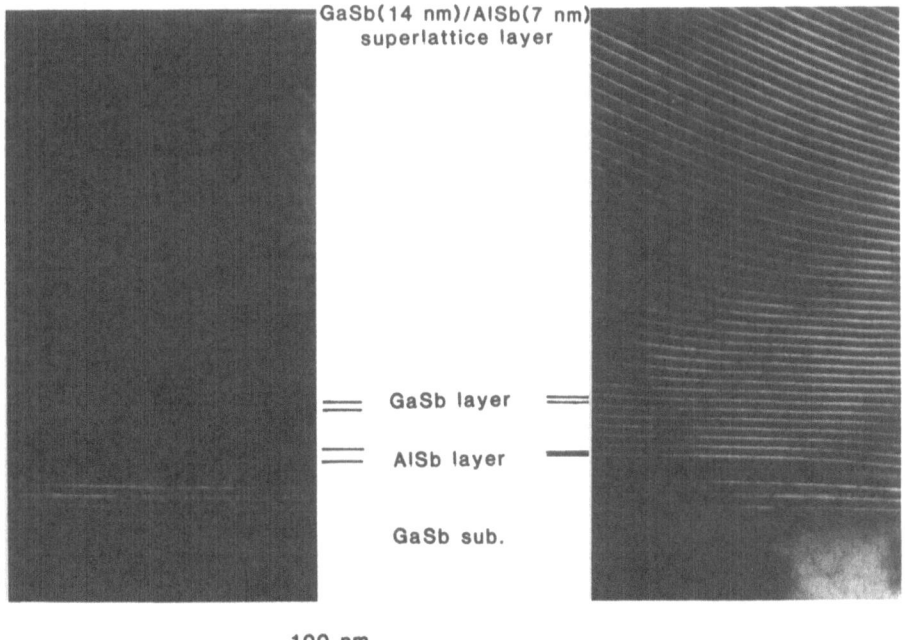

100 nm

Fig. 11. TEM micrographs magnifying Fig. 10.

a lower magnification TEM micrograph.   The large X-ray FWHM observed for the samples in region A is attributed to the bending of the layer structures.   Different sections of the same  sample under much higher magnification are shown in Fig. 11.   It is clearly seen that the stripe width differs between the two  sections, which is due to the bend of the layer structures.   The spread of the stripe width is equivalent to a large well size fluctuation, so that it explains well the large PL-FWHM for the samples in region A.

## Growth mechanism of MBE grown GaSb/AlSb superlattices

Simple models for the GaSb/AlSb superlattice growth in the three [V]/[III] regions are proposed, which are schematically shown in Fig. 12.

### A. Model I : growth under the Sb-lack condition (region A)

Surface migration of Ga during the MBE growth is large, and crystal growth like a vapor transport growth[12] takes place  in which Sb atoms are incorporated into group III melts.   Nucleation occurs as shown in Fig. 12(a).   Although the size of the cap-shaped nucleus is very large (about 2 μm), the growth is  similar to three-dimensional growth like Volmer-Weber mode[13],[14].

### B. Model II : growth under the optimum [V]/[III] condition (region B)

The growth is two-dimensional one like Frank-van der Merwe mode[14],[15], namely layer-by-layer growth, as shown in Fig. 12(b).

### C. Model III : growth under the Sb-rich condition (region C)

The growth is a layer-by-layer one like Frank-van der Merwe mode[14],[15], but surface migration is small because Sb atoms  suppress the migration.   Isolated growth islands occur site-by-site as shown in Fig. 12(c).   When the coalescence of islands  (the size of 20 – 100 nm in length) occurs[14], many twins and/or  stacking faults  are brought about. Probably, the growth  mechanism is similar to the growth at low substrate temperature.

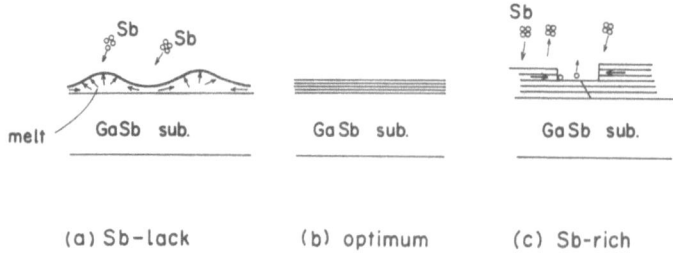

(a) Sb-lack          (b) optimum          (c) Sb-rich

Fig. 12.  Models of growth mechanism. (a)Sb-lack condition,(b)optimum condition, (c)Sb-rich condition.

## Sample preparation

Optical absorption of a GaSb/AlGaSb MQW grown with the optimum [V]/[III] ratio was studied based on photocurrent measurement. Photocurrent spectrum is almost an exact replica of the absorption spectrum[15]. Figure 13 shows a schematic diagram of the sample used here, which has a p-i-n diode structure identical to an MQW laser diode. The epitaxial layer was grown on a Te-doped (001)-oriented GaSb substrate. The active layer is unintentionally doped, and consists of 16 periods of 7-nm thick GaSb and 3-nm thick $Al_{0.25}Ga_{0.75}Sb$ MQW, and it is sandwiched by a 1500-nm thick $Al_{0.3}Ga_{0.7}Sb$ cladding layer doped with Te to $2 \times 10^{18}$ $cm^{-3}$ and a 1500-nm thick $Al_{0.3}Ga_{0.7}Sb$ cladding layer doped with Be to $2 \times 10^{18}$ $cm^{-3}$. The diode was etched into a high-mesa structure with a diameter of 690 μm using a mixture of $CH_3COOH : HF : HNO_3 = 40 : 1 : 8$ as shown in Fig. 13(b). A cap layer not covered by the electrode metal, whose diameter is 200 μm, was removed using the same etchant. Cr-Au and Au-Ge-Ni were used for the p- and n-electrode metals. The short circuit photocurrent of the diode was measured when the monochromatic light from a tungsten lamp was irradiated onto the diode surface as shown in Fig. 13(b).

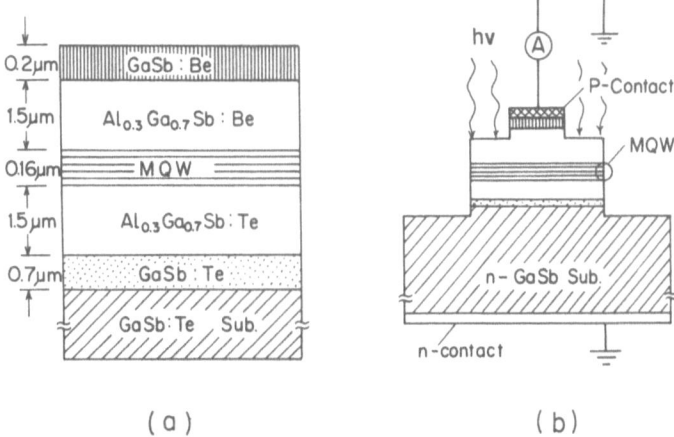

( a )                    ( b )

Fig. 13.   Schematic diagram of a diode for photocurrent measurements.
(a)the layer structure of MBE grown GaSb/AlGaSb MQW and
(b)cross-section of the diode.

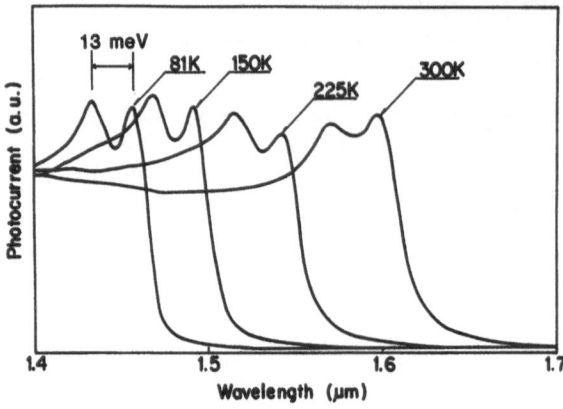

Fig. 14. Photocurrent spectra at various temperatures ranging from 81 to 300 K.

## Optical absorption spectrum

Figure 14 shows the photocurrent spectra at various temperatures ranging from 81 to 300 K. A well-resolved double-peak structure was observed at all temperatures. The double-peak structure is associated with the n = 1 heavy and light hole excitons. The same double peak structure was previously reported in GaSb/AlSb short period superlattices at temperatures below 200 K[8]. An important point in the present observation is that the double peak structure is clearly seen at temperatures as high as 300 K. The successful room temperature observation of exciton absorption peaks in GaSb quantum wells is considered to be due to the optimized [V]/[III] ratio during the MBE growth as described earlier.

## Assignment of exciton peaks

Quantum well structures consisting of zinc-blende crystal grown on a (001)-oriented substrate shows polarization dependence in the optical transition as reported by the previous authors[17]. Both electron to heavy hole transition and electron to light hole transition are allowed for the TE-polarized emission. On the other hand, only electron to light hole transition is allowed for the TM-polarized emission. Therefore, the TE-polarized emission is expected to appear in the lower energy range than the TM-polarized emission when the heavy hole level is below the light hole level with respect to hole energy, as is the case in GaAs/AlGaAs quantum wells. In contrast, both TE- and TM-polarized emissions are expected to appear in the same energy range when the light hole level is below the heavy hole level as in the case of GaSb/AlSb quantum wells with the hole band reversal[7]. The polarization dependence has been clearly observed in the guided spontaneous emission of GaAs/AlGaAs MQW laser diodes[17]. For assignment of the observed double peaks to heavy and light hole exciton absorption peaks, we fabricated a low-mesa type 60 x 200 μm stripe geometry MQW laser diode using the same epitaxial wafer, and measured the TE- and TM-polarization dependence of the guided spontaneous emission from the cleaved facet. The threshold current density of this laser was 9 kA/cm$^2$.

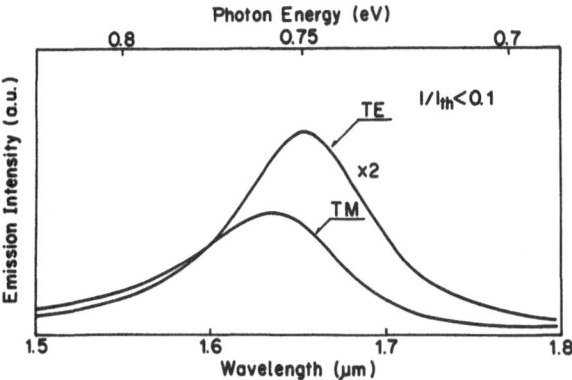

Fig. 15. TE- and TM-polarized component spectra of the guided spontaneous emission for a MQW laser diode at 293 K.

Figure 15 shows the guided emission spectra measured at 293 K when the injected current level is 0.1 times or less the threshold current density. The TM-polarized emission appears in the higher energy range than the TE-polarized emission. This indicates that the lower and higher energy peaks in the absorption spectra are due to heavy and light hole excitons, respectively. Consequently, the present GaSb/AlGaSb MQW does not exhibit the hole band reversal[7]. The lower energy shift of both of the emission peaks as compared with the absorption peaks is due to the selfabsorption effect[18].

Biaxial strain in GaSb layers

Voison et al.[7] pointed out that in a GaSb/AlSb MQW the strain induced by the lattice mismatch between GaSb and AlSb causes a shrinkage of the GaSb band gap accompanied by a splitting of $\Gamma_8$ valence band. With sufficient thickness of AlSb layers compared to GaSb layers, the splitting is large enough to make the energy difference between the n = 1 electron and heavy hole levels $E_{n=1e-hh}$ larger than that between the n = 1 electron and light hole levels $E_{n=1e-lh}$. They considered that the GaSb layers stretch in the x, y directions of the layer plane as a result of a biaxial tensile stress, while the AlSb layers are under compressive biaxial stress. According to this mode, fundamental light hole to conduction band gap and heavy hole to conduction band gap of the GaSb band parameters are given by

$$E_c - E_{1h} = E_g(GaSb) - 2a(S_{11}+2S_{12})x - b(S_{11}-S_{12}) \qquad (1)$$

$$E_c - E_{hh} = E_g(GaSb) - 2a(S_{11}+2S_{12})x + b(S_{11}-S_{12}), \qquad (2)$$

where x is the strength of strain, a and b are the deformation potentials, and $S_{11}$ and $S_{12}$ are the elastic compliance constants[7]. $E_g(GaSb)$ is the band gap of the bulk GaSb, which is 720 meV at 300 K. Equations (1) and (2) lead to the calculated values of ($E_g(GaSb)$ − 5.7 meV) and ($E_g(GaSb$ − 2.2 meV) for ($E_c - E_{1h}$) and ($E_c - E_{hh}$), respectively. Using these values, we calculated $E_{n=1e-hh}$ and $E_{n=1e-lh}$, which were 789.7 and 792.4 meV, respectively. Although these values do not give the reversal of the quantized heavy and light hole levels, the energy separation of 2.7 meV is as little as 20 % of the observed energy separation of 13 meV. This discrepancy indicates that much smaller biaxial stress is introduced in the present GaSb layers than the calculated value, although the calculated biaxial stress explains well the experimental results reported by Voison et al[7]. This contradiction is ascribed to the difference of the substrates on which the GaSb/AlGaSb MQWs were grown. Our GaSb/AlGaSb MQW was grown on a GaSb substrate while their GaSb/AlSb MQW was grown on a lattice-mismatched GaAs substrate. It has been reported that in GaAs/AlGaAs MQWs grown on GaAs substrates, the in-plane lattice constant of both GaAs and AlGaAs is aligned with that of the substrate, free from the strain[19]. Our result also suggests that GaSb wells in the present MQW grown on a GaSb substrate are free from the biaxial stress. This interpretation was confirmed by the in-plane strain measurement of a GaSb/AlSb MQW as described in the following.

The in-plane strain was determined based on a study of asymmetric X-ray double crystal diffraction[20]. The GaSb/AlSb MQW used in this study consists of 75 periods of 14.4-nm thick GaSb and 7.8 nm thick AlSb grown on a GaSb substrate. GaSb/AlSb MQWs grown on GaSb substrates are deformed by the lattice mismatch. The strain in the MQW epitaxial layer along the c-axis $\Delta a_n$ is simply determined from the (001) Bragg reflection angles using the equation

$$\Delta a_n/a_s = -\Delta\theta_{001}/\tan\theta_{001}, \tag{3}$$

where $\theta_{001}$ is the (001) Bragg reflection angle, $\Delta\theta_{001}$ is the difference of $\theta_{001}$ between the substrate and the epitaxial layer, and $a_s$ is the lattice constant of the substrate. On the other hand, the strain along the a-axis, $\Delta a_p$ is determined from the asymmetric (hkl) Bragg reflection angles using the equation

$$-d_{hkl}(\Delta\theta_{hkl}/\tan\theta_{hkl}) = 1/(h^2+k^2+l^2) \ (h^2+k^2)(\Delta a_p/a_s)+l^2(\Delta a_n/a_s), \tag{4}$$

where $d_{hkl}$ is the hkl lattice spacing, $\theta_{hkl}$ is the (hkl) Bragg reflection angle, and $\Delta\theta_{hkl}$ is the difference of $\theta_{hkl}$ between the substrate and the epitaxial layer. In the present experiment, (004) and (224) Bragg reflections were measured for determining the strains in the GaSb layers. The measured $\Delta\theta_{004}$ and $\Delta\theta_{224}$ were 404.1 and 398.2 sec, respectively. When these values are substituted into eqs. (3) and (4), $\Delta a_n$ and $\Delta a_p$ become $3.4 \times 10^{-3} \pm 3.0 \times 10^{-4}$ and $4.8 \times 10^{-4} \pm 3.0 \times 10^{-4}$, respectively. On the other hand, $\Delta a_p$ calculated according to the Voison model is $2.2 \times 10^{-3}$. This value is much larger than the above experimental value. This result gives us good reason to conclude that the heavy and light hole levels of GaSb do not reverse in GaSb/AlGaSb and GaSb/AlSb MQWs grown on GaSb substrates. The possible observation of the hole band reversal in GaSb/AlSb MQWs grown on GaSb substrates might be achieved with much thicker AlSb layers.


## CONCLUSIONS

Crystal qualities and optical properties of MBE grown GaSb/AlGaSb superlattices and MQWs were studied.

(1) Optical quality as well as crystallographical quality of GaSb/AlSb superlattices strongly depend on the [V]/[III] ratio during MBE growth, which were investigated by various kinds of measurements. The obtained results are as follows.

## A. Growth under Sb-lack condition

The epitaxial layer was an imperfect single crystal having a spread orientation of crystal axis. Rutherford backscattering measurement exhibited more Ga (and/or Al) atoms in the dechanneling sites. The layer structure of superlattices was confirmed, but compositional disordering might occur, as suggested from Raman spectroscopy. The layer structure was bent, and the epitaxial layer showed a milky surface morphology. The local inhomogeneity of the growth rates, which is the reason for the large PL-FWHM, was observed.

## B. Growth under optimum condition

The epitaxial layer was uniform and no dislocation was observed. (004) X-ray FWHM showed relatively small value and no disordering of crystal orientation was observed. The lattice arrangement is regular and RBS $\chi_{min}$ showed relatively small value. Therefore, the epitaxial layer showed optically as well as crystallographically good quality.

## C. Growth under Sb rich condition

The epitaxial layer was an imperfect single crystal having a spread orientation of crystal axis. Rutherford backscattering measurement exhibited more Sb atoms in the dechanneling sites. The layer structure of superlattices was confirmed by TEM, but many dislocations are included in the epitaxial layers.

(2) A GaSb/AlGaSb MQW grown under the optimum growth condition exhibited a clear double peak structure in the optical absorption spectrum at room temperature. The lower and higher energy peaks were assigned to heavy and light hole exciton absorptions, respectively, based on a study of the polarization dependence of guided emission of the MQW laser diode. This assignment was further confirmed by the in-plane strain measurement using asymmetric X-ray diffraction. The measured strain was negligibly small in the GaSb layers. This result suggests that hole band reversal rarely occurs in GaSb MQWs grown on GaSb substrates.

## ACKNOWLEDGEMENT

This review would be impossible without a great deal of contributions by Drs. Yoshifumi Suzuki, Yutaka Ohmori, Takeo Miyazawa, and Hiroshi Okamoto in the NTT superlattice research group which I belong to. I would like to express my great thanks to them.

## REFERENCES

1. Y. Ohmori, Y. Suzuki and H. Okamoto, Jpn. J. Apply. Phys. 24, L657 (1985).
2. Y. Suzuki, Y. Ohmori, and H. Okamoto, J. Appl. Phys. 59, 3760 (1986).
3. M. Naganuma, Y. Suzuki and H. Okamoto, Inst. Phys. Conf. Ser. No. 63, 125 (1982).
4. E. E. Mendez, C-A. Chang, H. Takaoka, L. L. Chang, and L. Esaki, J. Vac. Sci. Technol. B1, 152 (1983).
5. G. Griffith, K. Mohammed, S. Subbana, H. Kroemer, and J. L. Merz, Appl. Phys. Lett. 43, 1509 (1983).
6. T. Miyazawa, S. Tarucha, Y. Ohmori, Y. Suzuka, and H. Okamoto, Jpn. J. Appl. Phys. 25, L200 (1986).
7. P. Voison, C. Delalande, M. Voos, L. L. Chang, A. Segmuller, C-A. Chang, and L. Esaki, Phys. Rev. 30, 2276 (1984).
8. K. Ploog, Y. Ohmori, and H. Okamoto, Appl. Phys. Lett. 47, 384 (1985).
9. Y. Ohmori, S. Tarucha, Y. Horikoshi, and H. Okamoto, Jpn. J. Appl. Phys. 23, L94 (1984).

10. G. G. Lukovsky, K. Y. Cheng and G. L. Pearson, Phys. Rev. 12, 4135 (1975).
11. Y. Suzuki and H. Okamoto, J. Appl. Phys. 58, 3456 (1985).
12. R. A. Sigsbee, J. Appl. Phys. 42, 3904 (1971).
13. M. Volmer and A. Weber: J. Phys. Chem. 119, 277 (1925).
14. As a review, E. Bauer and H. Poppa: Thin Solid Films 12, 167 (1972).
15. F. C. Frank and J. H. van der Merwe: Proc. Roy. Soc. A198, 205 (1949).
16. S. Tarucha, H. Iwamura, T. Saku, and H. Okamoto, Jpn. J. Appl. Phys. 24, L442 (1985).
17. H. Iwamura, T. Saku, H. Kobayashi, Y. Horikoshi, J. Appl. Phys. 54, 2692 (1983).
18. S. Tarucha, Y. Horikoshi, and H. Okamoto, Jpn. J. Appl. Phys. 8, L482 (1983).
19. A. Segmuller, P. Krishna, and L. Esaki, J. Appl. Cryst. 10, 1 (1979).
20. W. J. Nartels and W. Nijiman, J. Cryst. Growth 44, 518 (1978).

# THE MBE GROWTH OF InSb-BASED HETEROJUNCTIONS AND LDS

C.R. Whitehouse

Royal Signals and Radar Establishment
St. Andrews Road, Malvern, Worcs., UK

Low-dimensional structures based on InSb are attracting increasing interest not only in terms of fundamental physics, but also in view of their potentially important device applications [1,2]. For example, HEMT-type structures based on the CdTe/InSb material combination are theoretically predicted [1] to exhibit an order of magnitude higher electron mobilities than is observed in corresponding GaAs/GaAlAs devices whilst CdTe/InSb quantum-well structures are also attractive for use as infra-red sources and detectors. Studies of LDS based on the InAs/InSb and InAlSb/InSb material systems have also been reported [3,4].

Despite these attractions, however, there are surprisingly few reports of the MBE growth of InSb-based LDS, or indeed even of the growth of InSb itself. The present paper will therefore briefly review the existing publications and will then describe data which has been obtained in the author's laboratory relating to the MBE growth of both InSb and also CdTe/InSb multilayer structures. Detailed studies of the electrical and optical properties of the individual component materials have been performed and the CdTe/InSb interface assessed using cross-sectional TEM (XTEM) and SIMS. Whilst structures exhibiting sharp interfaces have been successfully grown, angle-resolved photoelectron spectroscopy measurements [5] have revealed the existence of a complex interface reaction which could well affect the performance of some devices based on this material combination.

As a conclusion to the presentation, the future prospects of the CdTe/InSb system as well as other InSb-based LDS will be appraised.

REFERENCES

1. R.G. van Welzenis and B.K. Ridley, Solid State Electronics 27:113 (1984).
2. G.M. Williams, C.R. Whitehouse, N.G. Chew, G.W. Blackmore and A.C. Cullis, J.Vac.Sci.Technol. B3:704 (1985).
3. G.S. Lee, Y. Lo, Y.F. Lin, S.M. Bedair and W.D. Laidig, Appl.Phys.Lett. 47:1219 (1985).
4. F. Cerdeira, A. Pinczuk, T.H. Chiu and W.T. Tsang, Phys.Rev. B32:1390 (1985).
5. K.J. Mackey, P.M.G. Allen, W.G. Herrenden-Harker, R.H. Williams, C.R. Whitehouse and G.M. Williams, accepted for publication in Appl.Phys.Letts. (August 1986).

# OPTICAL PROPERTIES OF HgTe-CdTe SUPERLATTICES

T. C. McGill and G. Y. Wu

T. J. Waston, Sr. Laboratory of Applied Physics
California Institute of Technology
Pasadena, California 91125

## I. Introduction

One of the primary applications of narrow band gap semiconductors is IR sources and detectors. In recent years, it has been predicted that superlattices offer a number of potential advantages over alloys of HgTe-CdTe for application in IR detectors and sources.[1,2] In the alloy the band gap is controlled by the relative composition of Hg to Cd, while in the superlattice, the band gap is controlled by the thickness of the layers making up the superlattice.[2] Studies of the growth of alloys and superlattices by molecular beam epitaxy indicate that it will be much easier to control the band gap of the superlattice than the band gap of the alloy.[3] In alloys, the effective mass is strongly coupled with the band gap. The effective mass is proportional to the gap. Hence, small band gap implies small effective mass. This has lead to serious difficulties with leakage currents in various p-n junction-device structures.[2] Finally, superlattices may be more stable than the alloys. Alloys of HgTe and CdTe are thought to be structurally unstable while the superlattices could represent a lower free energy state of the system, and, hence, be more stable.

In this manuscript, we will provide a brief review of some of the more important developments in the optical properties of HgTe-CdTe superlattices. In Section II, we review some of the important properties of HgTe and CdTe. In Section III, we

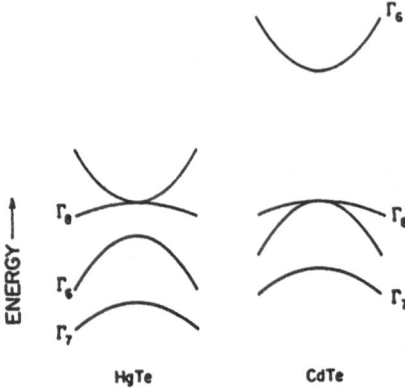

**FIG. 1.**    The electronic band structures for HgTe and CdTe near the valence band and conduction band edges. For CdTe, the valence band edge is at the state labelled $\Gamma_8$ and the conduction band edge is at the state labelled $\Gamma_6$. For HgTe, the conduction and valence band edge occur at the 4 fold degenerate state labelled $\Gamma_8$ resulting in a zero bandgap semiconductor.

present the results of some recent calculations of the optical properties. In Section IV, we present some recent experimental measurements of the optical properties of these superlattices, and, finally, we will summarize and indicate directions for future work in section V.

## II. Relevant Properties of HgTe-CdTe

The electronic properties of HgTe and CdTe are well known.[5] In Fig. 1, we have presented the results of the band structure near $k = 0$. CdTe is a typical zinc blende semiconductor with a valence band edge occurring at $\Gamma_8$ and the conduction band edge occurring at $\Gamma_6$. The band gap of the CdTe is on the order of 1.6 eV. HgTe has an inverted band structure with $\Gamma_8$ corresponding to both the valence band edge and conduction band edge and $\Gamma_6$ being far below the $\Gamma_8$ level. Hence, HgTe is a zero band gap semiconductor.

One of the basic concepts in heterojunction physics is that of a band offset. As illustrated in Fig. 2, the band offsets are the difference in position between the valence band edge or the conduction band edge as one moves across an interface between two materials. There is a great deal of discussion about the appropriate value for HgTe-CdTe. Simple arguments[6] and some experimental studies[7] have lead to the conclusion that the valence band offset is small, $\Delta E_v \approx 0eV$. On the other hand, recent experimental results[8] and new speculations[9] about the band offset have lead to the conclusion that the valence band offset is substantially larger than zero, $\Delta E_v = 0.3 - 0.5eV$ with the valence band of HgTe above that for CdTe. In these calculations, we will assume that the band offset is small, but future work will be required to provide more accurate results once the correct values of the band offsets have been determined.

## III. Theory of the Optical Properties of Superlattices

In the superlattices, the electrons and holes near the band edges are primarily in the HgTe. Since in HgTe the electrons and holes both have $\Gamma_8$ symmetry and the $\Gamma_8 \longrightarrow \Gamma_8$ transition is very, very weak in zinc blende semiconductors, one might ask whether the superlattice will absorb light strongly. Hence, we have carried out theoretical calculations of the optical properties of these systems.[10] Calculations are based on an eight-band $k \cdot p$ model. More details on the calculations are contained

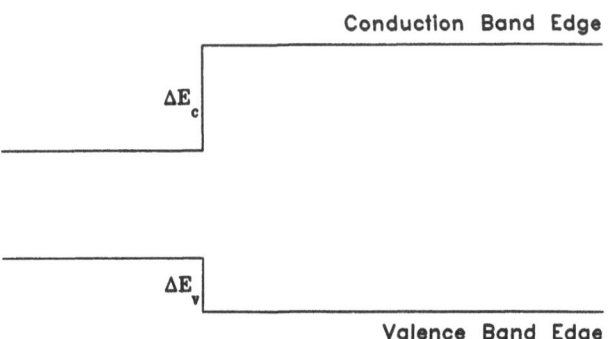

**FIG. 2.** Schematic diagram illustrating band offsets in the valence band $\Delta E_v$ and conduction band $\Delta E_c$ at a heterojunction between two semiconductors.

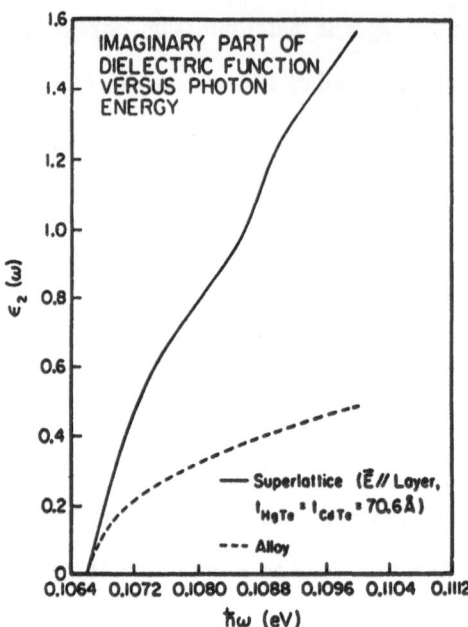

**FIG. 3.** The imaginary part of the complex dielectric constant for a superlattice and for an alloy with the same bandgap as a function of the frequency. For the superlattice the electric field is chosen to be polarized parallel to the layers (after Ref. 10).

in Ref. 10. In Fig. 3, we present the calculated values of the imaginary part of the dielectric constant $\epsilon_2(\omega)$ for an electric field polarized parallel to the layers. Also, included in the figure are the values of $\epsilon_2$ for an alloy whose composition has been selected to have approximately the same band gap as a superlattice. As can be seen from this figure, the superlattice has values of $\epsilon_2$ that are comparable to those for the alloy. These comparable values of $\epsilon_2$ would lead to comparable values of the optical absorption for the superlattices and alloys for the polarizations of the electric field described above.

One of the important issues is how do the optical properties vary with band offset. In particular, one question is how does the band gap vary? In Fig. 4, we present the results of a simple calculation of the band gap as a function of valence band offset.[11] From this figure, one can see that the band gap decreases as the valence band offset deviates from zero. While small variations in the valence band offset around zero will result in small changes in the band gap, values of the valence band offset as large as 0.3-0.5 eV would result in decreases of the band gap by a factor of two or more.

One might also ask about the influence of the band offset on the optical absorption of a superlattice. A simple model of the $\epsilon_2(\omega)$ that is valid near the band edge gives

$$\epsilon_2(\omega) = A \frac{\sqrt{\hbar\omega - E_{gap}}}{(\hbar\omega)^2},$$

where $E_{gap}$ is the band gap of the superlattice and A is a constant that indicates the strength of the absorption. In Fig. 5, we have plotted this constant A as a function of the valence band offset. From this figure, one can see that we expect little change in the absorption with variation in the band offset.

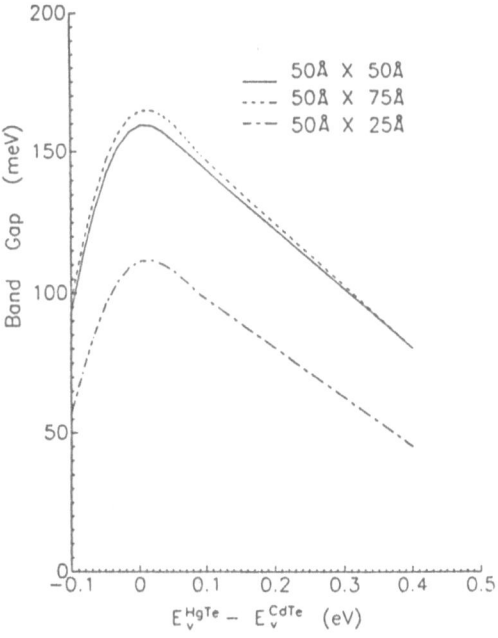

FIG. 4.    The variation of the band gap with valence band offset for a HgTe-CdTe superlattice (after Ref. 11).

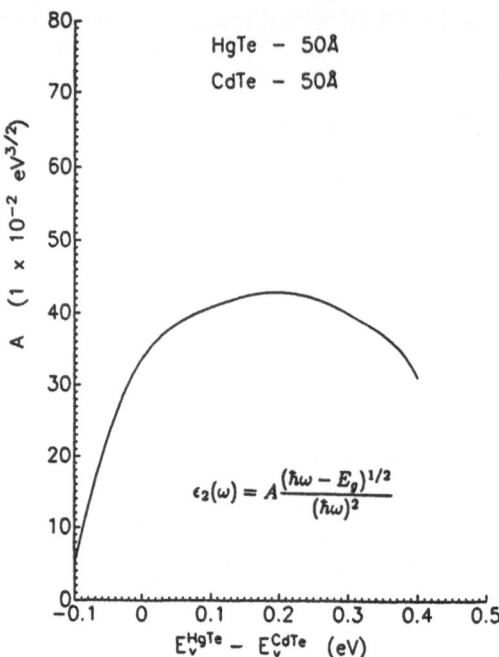

**FIG. 5.** The variation of the constant A with valence band offset for a HgTe-CdTe superlattice. The constant A is defined by the expression $\epsilon_2(\omega) = \frac{A(\hbar\omega - E_G)^2}{\hbar\omega}$ (after Ref. 11).

## IV. Experimental Results

As in the study of quantum wells and superlattices made from other materials, photoluminescence experiments[12] have been very important in verifying the existence of superlattices with a band gap that is in approximate agreement with theory. In Fig. 6, we present the photoluminescence spectra from an alloy and two superlattices. Sample 1 (grown by J. P. Faurie[13]) consisted of 250 repeats of a layered structure made up of 38-40 Å of HgTe followed by 18-20 Å of CdTe. Sample 2 (grown by P.Chow[14]) consisted of 75 repeats of 50Å of HgTe followed by 50Å of CdTe. The alloy had a composition that is approximately the same as the average Hg to Cd ratio in the superlattice labelled Sample 1. The important points to be made from this figure are: First, the photoluminescence signal of the superlattice occurs at lower energy than that from the corresponding alloy. Second, photoluminescence signals are observed from more than one superlattice.

From the temperature dependence of the peak positions in the photoluminescence spectrum, we can attempt to gauge the temperature dependence of the band
gap. In Fig. 7, we have plotted these temperature dependence peak positions along
with the temperature dependence of the band gap for an alloy with a composition that is approximately that of the superlattice. For comparison, we have also
included the variation of the band gap predicted by a simple calculation that incorporates the temperature dependence of the bulk band gaps of HgTe and CdTe into
a simple theory of the band gaps of the superlattices.[15,16] The data in this figure
show that the peak position of the photoluminescence signal from the superlattice
occurs at energies substantially below that of the alloy–the direction one expects
for superlattices. The data also show that the peak positions are in relatively good
agreement with the temperature dependence of the band gap predicted for a superlattice with layer thicknesses around the values that were planned for during
the growth. The calculations assume that the valence band offset is zero. Since
these results were obtained, photoluminescence signals have been observed from a
number of superlattices.[16,17]

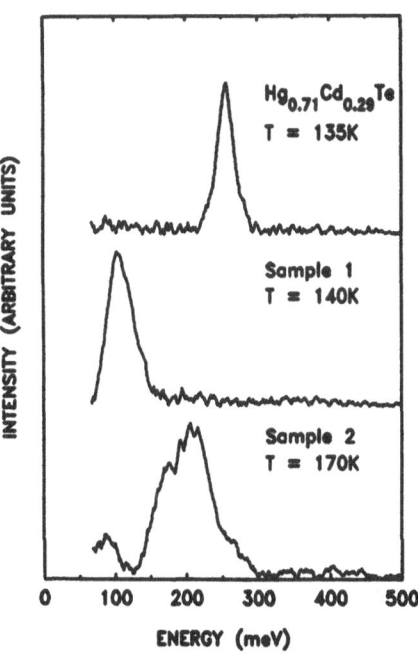

FIG. 6.   The photoluminescence spectra from a HgTe-CdTe superlattice (lower
curve) and that from a superlattice (upper curve) (after Ref.12).

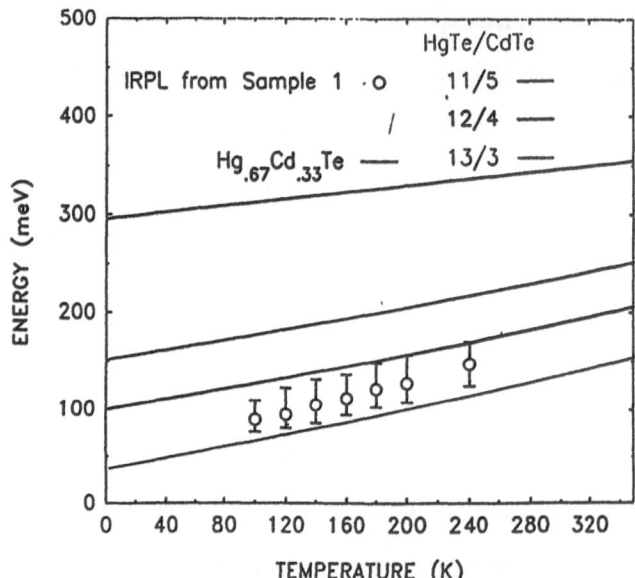

**FIG. 7.** The temperature dependence of the peak of the photoluminescence spectrum from sample one. The temperature dependence of the band gap of an alloy with approximately the same ratio of Hg to Cd as is found in the superlattice is included for comparison. Lines are also given showing the theoretical values of the band gap as a function of the temperature with the number of HgTe and CdTe layers in the repeated structure indicated by HgTe/CdTe in the figure.

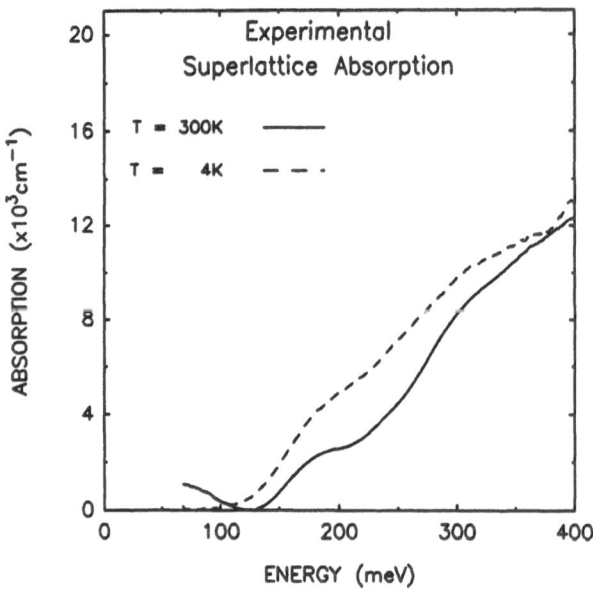

**FIG. 8.** The absorption curve for a HgTe-CdTe superlattice at two different temperatures. The absorption is computed by taking the transmission at some wavelength where we expect little absorption $T_0$ and then setting $\alpha(\omega) = \frac{\ln \frac{T_0}{T(\omega)}}{2t}$ where $t$ is taken to be the thickness (after Ref. 20)

Infrared absorption measurements have been carried out on superlattices.[19,20] The measurements are somewhat difficult to interpret, since they contain Fabry-Perot oscillations due to the interference between waves reflected from the superlattice and the underlying substrate and buffer layers. In Fig. 8, we present the results of measurement of the infrared absorption of Sample 1 described in the preceding paragraph. The data show the effect of the Fabry-Perot oscillations. While it is difficult to determine the precise location of the band gap from such data, the approximate position of the band gap is consistent with the band gap obtained from the photoluminescence spectrum. The infrared absorption shifts in the correct direction with decreasing temperature. The magnitude of the absorption is in agreement within a factor of two of that obtained from a theoretical calculation like that described above.

Finally, in Fig. 9, we present the photoconductivity spectrum[20] for the superlattice labelled Sample 1. This photoconductivity spectrum shows that the superlattice responds electrically to infrared radiation. The onset of the photoconductive signal occurs at a frequency in approximate agreement with the onset of absorption and the peak in the photoluminescence spectrum from this sample.

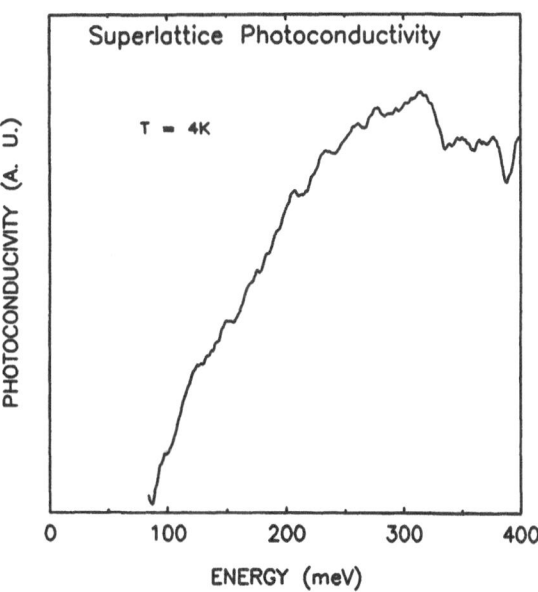

FIG. 9. The photoconductivity spectra for a HgTe-CdTe superlattice taken at 4K (after Ref. 20).

## V. Conclusions

We have reviewed the current level of understanding of the properties of HgTe-CdTe superlattices. Our current level of understanding is rather limited since we are just beginning to carry out systematic experimental investigations. Some of the important areas that still need to be addressed include: the values of the band offsets, the interdiffusion of HgTe and CdTe during superlattice preparation,[21] the transport of electrons and holes, and both normal to the layers and in the plane of the layers,[22] the configuration in which the superlattices could be employed in infrared devices. The current situation is that the brief studies of HgTe-CdTe superlattices suggests that these man-made materials could be very interesting for application in the infrared.

The author has profited from numerous technical discussions with his colleagues, particularly those with J. N. Schulman (HRL), D. L. Smith (LANL), and J. P. Faurie (UI). The work has been supported by the Army Research Office under Contract No. DAAG29-83-K-104.

## REFERENCES

1. J. N. Schulman and T. C. McGill, The CdTe/HgTe Superlattice: Proposal for a New Infrared Material, **Appl. Phys. Lett.** 34:663(1985).

2. D. L. Smith, T. C. McGill and J. N. Schulman, Advantages of the HgTe-CdTe Superlattice as an Infrared Detector Material, **Appl. Phys. Lett.** 43:180(1983).

3. J. Reno and J. P. Faurie, Experimental Relation Between Cut-Off Wavelength and HgTe Layer Thickness for HgTe-CdTe Superlattices , **Appl. Phys. Lett.** 49:409(1986).

4. Tse Tung, Ching-Hua Su, Pok-Kai Liao, and R. F. Brebrick, Measurement and Analysis of the Phase Diagram and Thermodynamic Properties in the Hg-Cd-Te System , **J. Vac. Sci. Technol.** 21: 117 (1982).

5. R. Dornhaus and G. Nimtz, in: "Narrow Band-Gap Semiconductors" Springer-Verlag, Berlin (1983).

6. W. A. Harrison, Elementary Theory of Heterojunctions, **J. Vac. Sci. Technol.** 14: 1016(1977).

7. Y. Guldner, G. Bastard, J. P. Vieren, M. Voos, J. P. Faurie, and A. Mil-

lon, Magneto-Optical Investigations of a Novel Superlattice:HgTe-CdTe, **Phys. Rev. Lett.** 51: 907(1983).

8. Steven P. Kowalczyk, J. T. Cheung, E. A. Kraut, and R. W. Grant, CdTe-HgTe (111) Heterojunction Valence-Band Discontinuity: A Common-Anion rule Contradiction,, **Phys. Rev. Lett.** 56:1605(1986).

9. J. Tersoff, Band Lineups at II-VI Heterojunctions: Failure of the Common-Anion Rule, **Phys. Rev. Lett.** 56:2755(1986).

10. G. Y. Wu, C. Mailhiot, and T. C. McGill, Optical Properties of HgTe-CdTe Superlattices, **Appl. Phys. Lett.** 46:72(1985).

11. G. Y. Wu and T. C. McGill, Band Offsets and the Optical Properties of HgTe-CdTe Superlattices, **J. Appl. Phys.** 58: 3914(1985).

12. S. R. Hetzler, J. P. Baukus, A. T. Hunter, J. P. Faurie, P. P. Chow, and T. C. McGill, Infrared Photoluminescence Spectra from HgTe-CdTe Superlattices, **Appl. Phys. Lett.** 47:260(1985).

13. J. P. Faurie, A. Million, and J. Piaguet, **Appl. Phys. Lett** 41:713(1982).

14. P. P. Chow, D. K. Greenlaw, and D. Johnson , **J. Vac. Sci. Technol.** A1:562(1983).

15. Y. Guldner, G. Bastard, and M. Voos, Calculated Temperature Dependence of the Band Gap of HgTe-CdTe Superlattices, , **J. Appl. Phys.** 57:1403(1985).

16. T. C. McGill, G. Y. Wu, and S. R. Hetzler, Superlattices:Progress and Prospects, **J. Vac. Sci. Technol.** A4:2091(1986).

17. K. A. Harris, S. Hwang, D. K. Blanks, J. W. Cook, Jr., J. F. Schetzina, N. Otsuka, J. P. Baukus, and A. T. Hunter, Characterization Study of a HgTe-CdTe Superlattice by Means of Transmission Electron Microscopy and Infrared Photoluminescence, **Appl. Phys. Lett.** 48:396 (1986).

18. J.P. Baukus, A. T. Hunter, J. N. Schulman, and J. P. Faurie, Photoluminescence of HgTe-CdTe Superlattices: Comparison of Theory and Experiment, **J. Vac. Sci. Technol.** (to be published).

19. C. E. Jones, J. P. Faurie, S. Perkowitz, J. N. Schulman, and T. N. Casselman, Infrared Properties and Band Gaps of HgTe/CdTe Superlattices, **Appl. Phys. Lett.** 47:140(1985).

20. J. P. Baukus, A. T. Hunter, O. J. Marsh, C. E. Jones, G. Y. Wu, S. R. Hetzler, T. C. McGill, and J. P. Faurie, Infrared Absorption Measurement and Analysis of HgTe-CdTe Superlattices, **J. Vac. Sci. Technol.** A4: 2110(1986).

21. D. K. Arch, J. L. Staudenmann, and J. P. Faurie, Layer Intermixing in HgTe-CdTe Superlattices, **Appl. Phys. Lett.** 48:1588(1986).

22. J. P. Faurie, S. Sivananthan, and J. Reno, Present Status of Molecular Beam Expitaxial Growth and Properties of HgTe-CdTe Superlattices, **J. Vac. Sci. Technol.** A4:2096(1986).

# OPTICAL PROPERTIES OF INAS-GASB AND GASB-ALSB SUPERLATTICES

Paul. Voisin
Groupe de Physique des Solides de l'Ecole Normale Superieure
24 rue Lhomond F75005 Paris

We review here some of the remarkable optical properties of two GaSb-based systems, the "type II" InAs-GaSb superlattices in which the electron and hole wave functions are spatially separated, and the strained-layer GaSb-AlSb superlattices where the ground valence state is the first light-hole state.

Parallel to the progressive sophistication of optical studies devoted to the conventional "type I" GaAs-Al$_x$Ga$_{1-x}$As heterostructures, two systems based on GaSb have been investigated, namely the InAs-GaSb "type II" superlattices (SL) and the GaSb-AlSb "strained layer" quantum well structures. Both systems displayed in an exemplary way new and stimulating properties such as the spatial separation of the electron and hole wavefunctions in InAs-GaSb, or the strain-induced reversal of the ground light and heavy valence subbands in GaSb-AlSb. We review here some of the fundamental optical properties of these two systems, with emphasis on the characteristic differences with respect to the standard type I configuration. In InAs-GaSb semiconductor superlattices, the main consequences of the carrier separation are the relative weakness of the absorption coefficient, the existence of specific parity selection rules and the absence of excitonic features. These properties are put into evidence from the study of optical absorption and band to band recombination. Another consequence of the non-zero charge distribution is the existence of a new, macroscopic quantum photovoltaic effect. On the opposite, the magnitude of the absorption coefficient in GaSb-AlSb multi-quantum well structures proves the type I nature of the band line-up in this system. However the bandgap of the heterostructure may be smaller than that of bulk GaSb, which evidences the importance of misfit strains in this system. Furthermore, the detailed analysis of the transmission spectrum shows that the ground valence state in the quantum well is the first light hole state and it also indicates strongly that the valence band offset is rather small. In the same sample, interband magnetooptical absorption gives an experimental insight into the in-plane dispersion relations of the valence subbands, exhibiting a very heavy in-plane effective mass for the ground light hole subband, and a rather light one for the first heavy hole subband. In contrast again with other materials, the luminescence in this system seems to be generally dominated by acceptor-related recombination. The paper is organized in two almost independant sections devoted respectively to each of these systems.

I- InAs-GaSb

The comparison of the electronic affinities of InAs and GaSb led Esaki et al (1977)[1] to the prediction that the Γ bandedge line-up in these materials should exhibit a new feature, which they called the type II configuration, i.e. the quantum well for the conduction electrons is in the InAs layers while that for the valence states is in the GaSb layers. In fact, the interpretation of the optical absorption edges in a series of InAs-GaSb superlattices[2,3] indicated that the conduction band of InAs overlaps the valence band of GaSb by (150± 50) meV. It is remarkable that this estimate is not strongly model-dependant and was found consistent with all subsequent investigations. Figure 1 displays the actual bandedge configuration, and a few absorption spectra are shown in Fig.2.

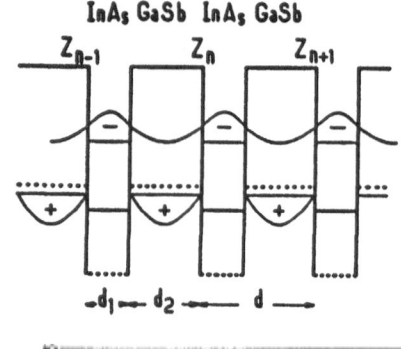

Fig.1: Conduction (solid line) and Valence (dotted line) bandedge profiles in an InAs-InSb super-lattice. The shape of the ground state wavefunctions is indicated schematically.

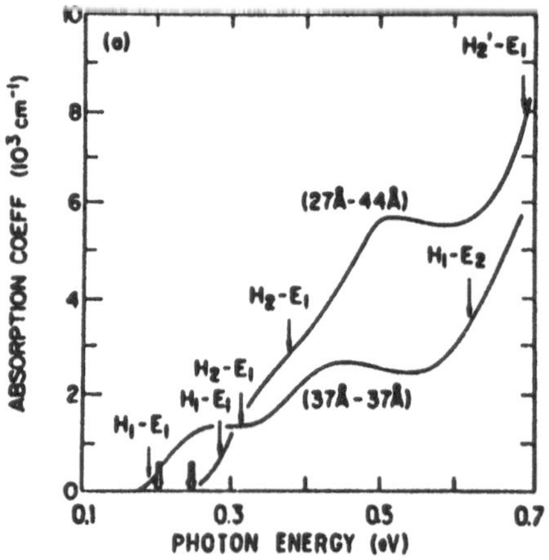

Fig.2: Low temperature absorption spectra of two InAs-GaSb SL's (from reference 3). The absorption coefficient is typically five times smaller than that of an equivalent type I superlattice.

It was clear from the begining[1,4] that a proper band structure calculation should take into account non-parabolicity, or more precisely a multiband description of the host's band structure. This requirement is conveniently fulfilled within the envelope function formalism[5,6]. The results are simple only when the in-plane wave vector $k_\perp = (k_x, k_y)$ is equal to zero. The band structure at $k_\perp = 0$ as a function of the period d is shown in Fig.3 for the case of equal layer thicknesses. Two characteristics are noteworthy:(i) The width of the first conduction (and light hole) subband $\Delta E_1$ ($\Delta LH_1$), which arises from the tunneling through the GaSb (InAs) layers, does not vanish at large d, but instead, it keeps values of the order of 20 meV. This is a remarkable consequence of the $\Gamma_6$ and $\Gamma_8$ band mixing. (ii) The first conduction subband $E_1$ and heavy hole subband $HH_1$ cross for a period of 160Å this crossing gave support to the idea of a semiconductor to semimetal transition which was found consistent with a variety of experimental results[7,8,9]. However, as was shown by Altarelli[6], the k.p coupling of all these subbands leads to anticrossing behavior at finite $k_\perp$, and the notion of semimetallic regime is probably incorrect in most cases. Despite the attempt to fit[11] the magneto-optical data[9,10] with the appropriate calculation, the situation remains unclear from the experimental point of view and we shall, in the following, restrict ourselves to the d<150Å regime where the semiconductor nature of the superlattice is clearly established.

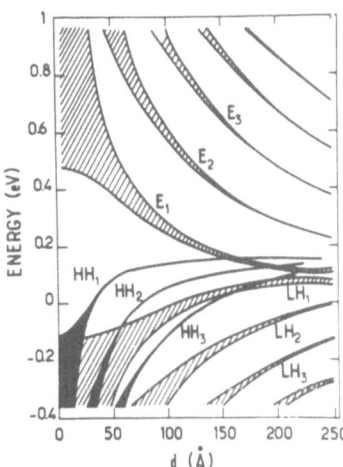

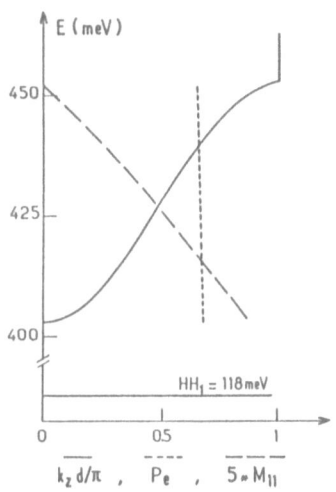

Fig.3: Band structure of InAs-GaSb SL's at $k_\perp$ = 0 as a function of the period in the case of equal layer thickness

Fig.4: Theoretical conduction band profile, electron confinement, and overlap with the ground heavy hole state for a 30-50 Å InAs-GaSb SL.

Though the SL symmetries are basically the same for the type I and type II systems, the optical properties, and more specifically the optical selection rules, are quite different[12,13]. The optical selection rules at $k_\perp$ = 0 for a type I SL are essentially that of the isolated quantum well, i.e. only the subbands having the same index are strongly optically coupled; the optical matrix element does not depend appreciably on the wavevector along the SL axis $k_z$. In sharp contrast, for a type II SL, the transitions with $\Delta n$ even are parity allowed at $k_z$ =0 and they become parity forbidden at $k_z$ =$\pi$/d, while the transitions with $\Delta n$ odd are parity forbidden at $k_z$ =0 and they become parity allowed at $k_z$ =$\pi$/d. Transitions with $\Delta n=0$ or $\Delta n \neq 0$ have a-priori equivalent strengths. The squared modulus of the optical matrix element obeys an approximate $k_z$-dependance given by:

$$M_{nm}(k_z) = 0.5\, M_{nm}(0).(1+(-1)^{n-m}\cos k_z d ) \qquad\qquad (1)$$

A consequence of the spatial separation of the carriers is the relative weakness of $M_{nm}(0)$. All this is illustrated on Fig.4 which shows the profile$E(k_z)$ of the first conduction subband, the probability $P_e$ for the electron to be in the InAs layers and the squared modulus $M_{11}$ of the overlap with the first heavy hole wavefunction in the case of the 30Å-50Å InAs-GaSb SL. These features are also apparent in , and they do explain, the shape of the spectra shown in Fig.2. In particular, the small overlap also weakens the Coulomb matrix elements, which explains the absence of excitonic features. On the other hand, it should be kept in mind that the band mixing at finite $k_\perp$ may at least weaken the parity selection rule.

In spite of the reduced optical matrix element, semiconductor InAs-GaSb superlattices luminesce[14]. The low temperature luminescence spectra of a 30Å-50Å SL excited by a $Kr^+$ laser at various incident powers is shown in Fig.5. It consists of a single line accompanied with a low energy tail which tends to saturate when increasing the excitation level. The energy-position, close to the calculated bandgap, and both the temperature- and excitation-dependence of the spectrum support the interpretation of the main line in terms of band to band recombination, the low energy tail being attributed to recombination processes involving shallow defects, which should be essentially thickness fluctuations of one or two monolayers in the InAs layers and acceptors in the GaSb layers; the other defects, donors in the InAs layers and GaSb layer thickness fluctuations, are much shallower. These defects, even if there existed only a few of them, form a continuum of bound states which accounts for the observed low energy tail. It is seen in Fig.5 that the high energy side of the line is relatively smooth, which indicates that the carriers have an effective temperature higher than the bath temperature.

A quantitative information is obtained through a lineshape analysis[15]. The fit shown in Fig.6 thus yields a carrier effective temperature $T_c$ = 43K. The luminescence signal is still observed at room temperature, as shown in Fig.7. At this temperature, the electron distribution thermally occupies a large part of the Brillouin zone (in the z direction !), and the lineshape reflects the $k_z$-dependance of the optical matrix element which blurs out the Van Hove discontinuity of the density of states at $E_g+\Delta E_1$. This effect is shown on Fig.8 .

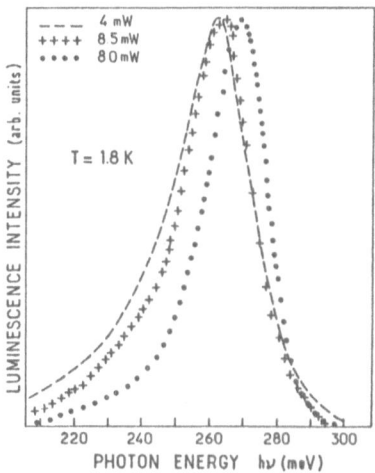

Fig.5: Low temperature luminescence spectra of a 30-50 Å InAs-GaSb SL at various excitation powers.

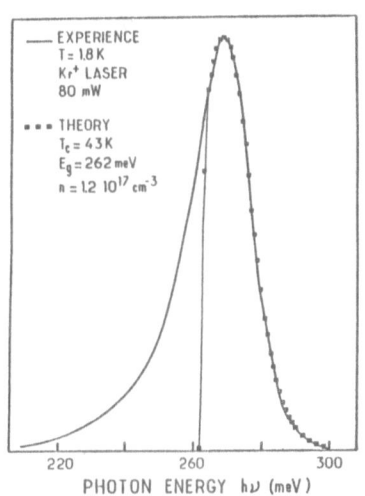

Fig.6: Experimental (solid line) and theoretical (dotted line) low temperature luminescence lineshapes, yielding an effective temperature $T_c$= 43 K.

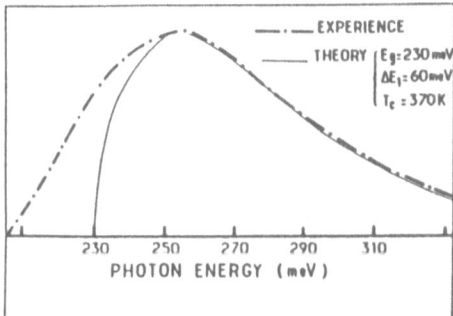

Fig.7: Experimental (dashed-and-dotted line) and theoretical (solid line) room temperature luminescence lineshapes.

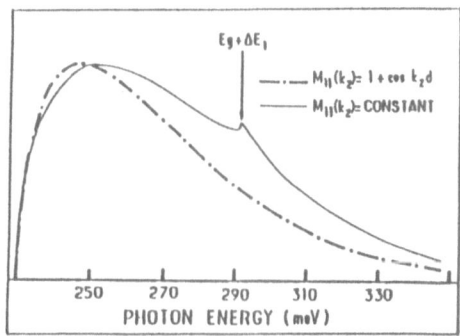

Fig.8: Theoretical 300K luminescence lineshapes for a type I (solid line) and type II (dashed-and-dotted line) superlattice.

Another consequence of the spatial separation of the electron and hole ground state wave functions is the existence of a macroscopic quantum photovoltaic effect[15,16]. The physical picture is the following: photocreated electrons and holes thermalize on a very short time scale towards their respective ground states, which are spatially separated. This results in a non-zero charge distribution. The SL unit cell, as schematized in Fig.1, behaves as a quantum capacitor with the positive charge at the center of the GaSb layer and the negative charge at the center of the InAs layer. The voltages across these quantum capacitors obviously add, as the superlattice itself is a <u>series</u> of unit cells. If N is the number of period in the SL, and $n_s$ the areal density of injected carriers, elementary algebra gives the voltage drop across the SL as[15,16]:

$$\Delta V = N \, n_s e \, d \, (P_e + P_h - 1)/ \, 2\varepsilon\varepsilon_0 \qquad\qquad (2)$$

where $P_{e\,(h)}$ is the probability for the electron (hole) to be in the InAs (GaSb) layer, d the SL period and $\varepsilon$ the relative dielectric constant. This yields $\Delta V = 0.2$ V for a 100 period, 30Å-50Å SL with a density of injected carrier $n = n_s/d = 10^{17} cm^{-3}$. However, this macroscopic <u>quantum</u> effect cannot be observed under DC illumination, for the <u>classical</u> reason that the residual conductance of the sample parallel to the growth axis tends to cancel it through a global charge transfer at the terminating planes of the structure. Note that this effect presents many analogies with the piezoelectricity of ionic crystals. It was recently put into evidence in a series of InAs-GaSb SL's[17].

## II- GaSb-AlSb

InAs-GaSb SL's may be a promising material for the 4 μm wavelength range; conversely, the GaSb-AlSb quantum well structures could be an interesting alternative for the 1.5 μm and 1.3 μm ranges.

Figure 9 shows the transmission spectra at low temperature obtained in two GaSb-AlSb multi-quantum-well (MQW) structures S1 and S2[18]. The thicknesses of the GaSb ($d_1$) and AlSb ($d_2$) layers are respectively 181Å and 452 Å for S1 and 84 Å and 419 Å for S2; both samples have ten periods and were grown on semi-insulating GaAs substrates. The spectra clearly exhibit the steplike behavior characteristic of the two-dimensional density of states and marked excitonic peaks at the onset of the first absorption steps. The type I nature of the system is definitely established from the consideration of the order of magnitude of the absorption coefficient which is not far from the characteristic value[12,19] of 0.5% per transition and per quantum well. However, two surprising features are observed: The onset of the absorption occurs at an energy about 50 mev smaller than expected for a simple GaSb quantum well and the intensities of the two first excitons are reversed compared to the usual GaAs-AlGaAs case, as easily observed for sample S1 which presents an absorption edge below the bandgap of bulk GaSb. In fact, these features evidence the importance of misfit strains[20,21] in this system: as a striking consequence of the strain, the bandgap of the QW material is reduced and the first light hole state becomes the ground valence state of the system.

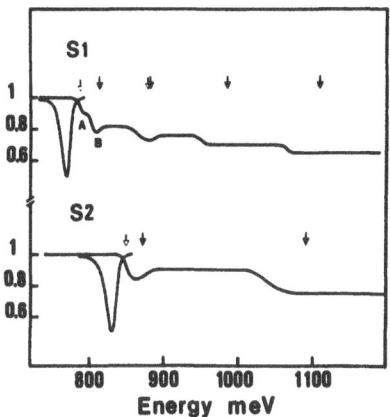

Fig.9: Low temperature transmission (solid lines) and luminescence (dashed lines) spectra in two GaSb-AlSb MQW's, S1 (181 Å - 452 Å) and S2 (84 Å - 419 Å).

In our SL's, the epilayer relaxes from the strongly lattice-mismatched GaAs substrate and is in self-mechanical equilibrium[22], as was evidenced from the study of X-ray double diffraction spectra[18,23]. The GaSb (AlSb) layers experience a biaxial tensile (compressive) stress in such a way that their in-plane lattice parameter becomes the same. As the elastic constants are essentially the same, the SL in-plane lattice parameter $a_\perp$ is simply given by:

$$a_\perp(d_1 + d_2) = a_1 d_1 + a_2 d_2 \qquad (3)$$

Where $a_{1(2)}$ is the lattice parameter of bulk GaSb (AlSb). In our case, the lattice mismatch is essentially accommodated by straining the thin GaSb layers by $\varepsilon_\perp = (a_\perp - a_1)/a_1$. $\varepsilon_\perp$ of the order of $(a_2 - a_1)/a_1 = 0.65\%$. This stress configuration, which may be analyzed as the sum of a hydrostatic dilatation and a uniaxial compression in the z direction, induces changes in the band structure of the host material, namely the band gaps between the conduction and heavy, light and split-off hole bands become:

$$E_c - E_{HH} = E_g + 2aS\varepsilon_\perp - bS'\varepsilon_\perp$$
$$E_c - E_{LH} = E_g + 2aS\varepsilon_\perp + bS'\varepsilon_\perp - 2b^2S'^2\varepsilon_\perp^2/\Delta \qquad (4)$$
$$E_c - E_{SO} = E_g + \Delta + 2b^2S'^2\varepsilon_\perp^2/\Delta$$

In these equations, $E_g$ and $\Delta$ are the bulk bandgap and spin-orbit coupling respectively; $S=(S_{11}+2S_{12})/(S_{11}+S_{12}) \simeq 0.5$ and $S'=(S_{11}-S_{12})/(S_{11}+S_{12}) \simeq 2$ are combinations of the elastic compliance constants, and $a \simeq -8\,eV$ and $b \simeq -2\,eV$ are the deformation potentials. In sample S1, the GaSb layers are thus characterized by a fundamental <u>light hole to conduction</u> bandgap $E_c - E_{LH} = 756\,meV$, and a larger heavy hole to conduction bandgap $E_c - E_{HH} = 791\,meV$, as illustrated in Fig.10. In other words, in this strain configuration, there is a competition between the effect of the quantum confinement and the effect of strain, which results in a reversal of the energy positions of the heavy- and light-hole excitons.

The full and open arrows in Fig.9 show the transitions involving the heavy- and light-hole subbands respectively, calculated within the three band envelope function model, using standard values of GaSb band parameters[24]. From the common anion rule, considered here as a chemical argument, the valence band offsets $\Delta E_{HH}$ and $\Delta E_{LH}$ are certainly small compared to the the conduction band offset $\Delta E_c$. It follows that the energies of the calculated transitions depend only very weakly on the value of $\Delta E_c$, while the <u>number</u> of expected transitions depend crucially on $\Delta E_{HH}$, $\Delta E_{LH}$[18]: The number of confined conduction subbands is always larger than the number of valence subbands accommodated in the heavy- and light-hole quantum wells, which will therefore fix the number of observable transitions. This leads to a maxima-minima argument (at least 4 HH→E transitions and at most 2 LH→E transitions for sample S1) which give an astonishing precision (a few meV) in the determination of the band offsets: $\Delta E_{HH} = 40$ meV, $\Delta E_{LH} = 70$ meV and $\Delta E_c \simeq 1350$ meV. Note that the same analysis leads to the same result for sample S2, which has completely different parameters, and that a sample S3[25] having a GaSb layer thickness of 207 Å shows definitely more transitions than S1, which gives an a-contrario support to our analysis. On the other hand, from the study of the resonant Raman scattering at the E1 gap of GaSb in samples grown on GaSb substrates, Tejedor et al[26] concluded that the valence band offset in these samples should be larger than 300 meV; the same experiment performed in our samples[27] did not lead to the same conclusion, which raises an important question: can the band offsets depend <u>strongly</u> on the strain? Finally, let us remember that the recent discussion of the band offsets in GaAs-AlGaAs recommends prudence with this matter.

It can also be observed in Fig.9 that the fit of the higher lying transitions is unsatisfactory, say for confinement energies larger than 200 meV. This effect can be understood by examining Fig.11, which shows the band structure of GaSb: The potential barrier between the $\Gamma$ and L valleys is less than 1 eV high, and in fact, significant deviation from a Kane model should be expected for energies in the conduction band larger than $\simeq 200$ meV; actually, the energy increases slower than predicted by the Kane model, and thus, the actual confinement energies are certainly smaller than predicted by our calculations.

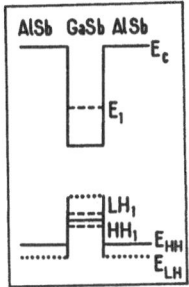

Fig.10: Band-edge profiles in a strained-layer GaSb-AlSb QW structure.

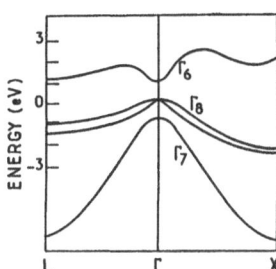

Fig.11: Schematic band structure of unstrained bulk GaSb.

As can be observed in Fig.11, in bulk GaSb, the conduction band minimum at the L point lies 84 meV only above that at the $\Gamma$ point. When the motion along a (100) axis is quantized, the energy minimum in the L valleys rises, in a first approximation, as $\hbar^2 k_z^2/2m_L d_1^2$; the "confinement mass" $m_L$ is equal to $(2m_t + m_l)/3 \simeq 0.51 \, m_0$, where $m_{t(l)}$ is the transverse (longitudinal) mass in the L valley[28,29]. Because of the symmetry mismatch of the atomic part of the Bloch wavefunctions, the quantized states built from the L minima mix only weakly with those built from the $\Gamma$ minimum. These L-originating states will not be seen in an optical absorption experiment, at least because, as they lie in the (110) directions of the SL Brillouin zone, the corresponding bandgaps remain indirect. However, because of the large value of $m_L$, these L-originating states become the fundamental conduction states for small GaSb layer thicknesses. The critical thickness $d_c$ at which the SL becomes an indirect bandgap material depends on the strain state: indeed, under the biaxial tensile strain, the $\Gamma$-L energy separation increases as $\Delta(E_c^L - E_c^\Gamma) = 2E_1 S\varepsilon_\perp$, where $E_1 \simeq 5$ eV[24]. In the thick AlSb barriers limit ($d_2 \gg d_1$), $E_c^L - E_c^\Gamma$ thus becomes equal to 118 meV.

Neglecting the non-parabolicity at the L minima, we thus estimate $d_c \simeq 60$ Å. Due to the above mentionned deviation from the Kane model, the actual value of $d_c$ should be slightly smaller. According to these remarks, the $\Gamma$–L crossover is likely to be the reason for the sharp decrease of the direct gap luminescence reported by Griffiths et al[28]. Recent time-resolved spectroscopy measurements brought further confirmation of this interpretation[30].

As already mentionned, the SL band structure at $k_\perp \neq 0$ results from the coupling of all the valence subbands, which depends on their spacing and ordering at $k_\perp = 0$. In this respect, the strain-induced light and heavy hole subbands reversal in our samples is particularly interesting. Figure 12 shows magnetooptical transmission spectra in sample S1, recorded either at a constant photon energy or at a constant magnetic field[31]. These spectra show many transmission minima which correspond to the transitions $LH_1^N \rightarrow E_1^N$ or $HH_1^N \rightarrow E_1^N$ between the Landau levels associated with the first light-hole, heavy-hole and conduction subbands. These transmission minima, or absorption maxima are reported in the usual transition energy versus magnetic field plot shown in Fig.13. This plot exhibits two distinct fan diagrams, eye-marked by the solid and dashed lines, which extrapolate towards $E_h = 799$ meV and $E_H = 829$ meV respectively. In addition, there are two dashed-and-dotted lines having a non linear behavior, which have been drawn through the exciton data points. They extrapolate to 795 meV and 820 meV for the light- and heavy-hole excitons respectively.

A quantitative analysis of these data requires a model calculation of the Landau level energies, and, in fact, of the oscillator strengths associated with the different transitions[32], which represents a considerable amount of theoretical work. In a first attempt to interpret the data, we have used a considerably simpler semi-empirical approach; we have: (i) discarded any spin effect, because the overall polarization dependance of the data is weak –even though there is a strong polarization dependance of the excitonic absorption– ; (ii) evaluated the energies $E_1^N$ from the semi-classical quantization rule $k_\perp^2 \rightarrow (2N+1)eB/\hbar$, using the simplified in-layer dispersion relations obtained by Bastard[33], which would be exact if the heavy hole mass were infinite. These $k_\perp$-dispersion relations are likely to be quite accurate for the conduction subbands in a relatively large gap material. Note that the accuracy in the evaluation of the $E_1^N$ energies is a crucial point of the interpretation, as they are the dominant contribution to the observed transitions. (iii) estimated finally the energies $LH_1^N$ and $HH_1^N$ in the same semi-classical approximation, using empirical parabolic $k_\perp$-dispersion relations with the in-plane effective masses $m_{LH}^\perp$ and $m_{HH}^\perp$ as fitting parameters.

This procedure leads to the fan diagrams shown in Fig.13, with a very heavy mass $m_{LH}^{\perp} = 0.8\ m_0$ (solid lines) for the ground light hole subband $LH_1$ and a rather light mass $m_{HH}^{\perp} = 0.11\ m_0$ (dashed lines) for the first heavy hole subband, respectively. We feel that the overall agreement witnesses that the involved hole subbands are not strongly non parabolic in the energy range of interest, which in turn partly justifies our method. However, it should be kept in mind that the experimental data reveal the oscillator strength rather than the transitions themselves. In case of strong band mixing, large differences may exist between these two quantities[32].

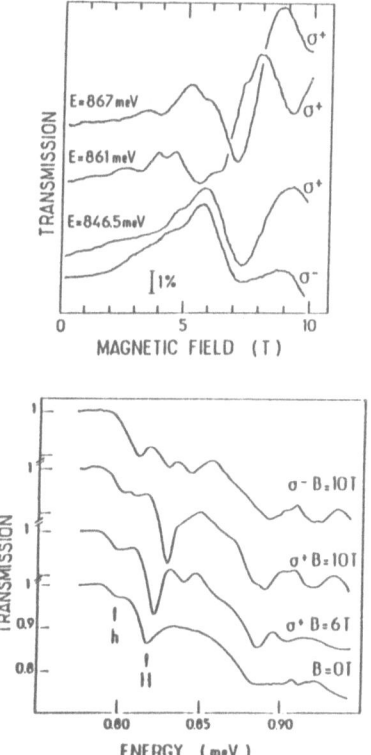

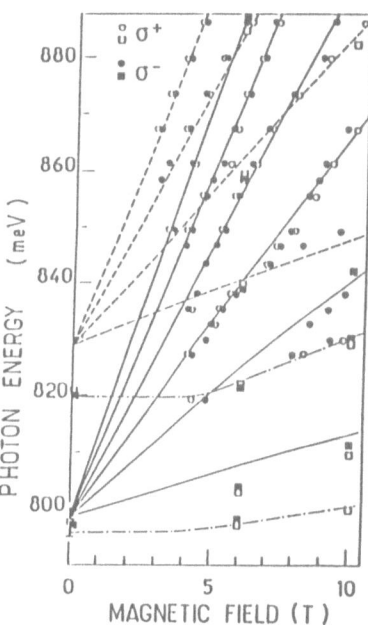

Fig.12: Magneto-optical transmission spectra in sample S1, recorded at a constant photon energy (upper part) or at a constant magnetic field (lower part)

Fig.13: Plot of the transition energies (transmission minima) versus magnetic field; black (open) circles and squares correspond to data obtained in the $\sigma^-$ ($\sigma^+$) polarization. The dashed-and-dotted lines correspond to the exciton data points

Figure 9 also shows the luminescence spectra (dashed lines) observed at low temperature in our GaSb-AlSb SL's. They lie 30 to 50 meV below the exciton peak seen in the absorption spectra and are typically 20 meV broad. This clearly indicates that this luminescence is related to shallow defects, which are most probably residual acceptors. In fact, it is very likely that the observed luminescence corresponds to electron-to-acceptor recombination, as often observed in moderate quality bulk GaSb. On the other hand, Ploog et al[34] reported recently a luminescence and excitation spectroscopy study of GaSb_AlGaSb SL's in which the confinement barrier consist of a short-period or pseudo-alloy GaSb-AlSb SL; the samples were grown on GaSb substrates. They attribute the 11 mev Stokes shift between luminescence and excitation to the trapping of excitons on interface fluctuations[35], and conclude that the luminescence of their samples is essentially excitonic. However, the observed Stokes shift has precisely the magnitude of the binding energy of the exciton on neutral acceptor in bulk GaSb, so that the contribution of these bound excitons in their data seems difficult to exclude.

III Conclusion

We have described, without pretending to completeness, the optical properties of two GaSb-based superlattices. Both systems displayed rather original features which were quite exciting from the point of view of fundamental material science. Up to now, however, their potentialities as device materials are not definitely established. Recently, successful fabrication of MQW GaSb-AlGaSb laser diodes operating continuously at room temperature was reported[36], which is very promising. To our knowledge,no attempt was made to observe laser emission from InAs-GaSb SL's, but the application of the quantum photovoltaic effect to the realization of fast pulse detection is of current interest.

Acknowledgements I am indebted to my colleagues G. Bastard and M. Voos and to Drs L.L. Chang and L. Esaki for their constant interest in the present investigations. I have also benefited from many fruitful discussions with Drs. J.A. Brum,C. Delalande, J.K. Maan and J.Y. Marzin.

References

1  G.A. Sai-Halasz, R. Tsu and L.Esaki, Appl. Phys. Lett. 30, 651 (1977)
2  G.A. Sai-Halasz, L.L. Chang, J.-M. Welter, C.-A. Chang and L. Esaki, Solid State Comm. 27, 935 (1978)
3  L.L. Chang, G.A. Sai-Halasz, L. Esaki and R.L. Aggarwal, J. Vac. Sci. Techn. 19, 589 (1981)

4   G.A. Sai-Halasz, L. Esaki and W.A. Harrison, Phys. Rev. B 18, 2812 (1978)

5   G. Bastard, Phys. Rev. B 24, 5693 (1981) ; G. Bastard, this volume.

6   M. Altarelli, Phys. Rev. B 28, 842 (1983) ; M. Altarelli, this volume.

7   L.L. Chang, E.E. Mendez, N.J. Kawai and L. Esaki, Surf. Sci. 113, 306 (1982)

8   L.L. Chang, N.J. Kawai, E.E. Mendez, C.-A. Chang and L. Esaki, Appl. Phys. Lett. 33, 30 (1981)

9   Y. Guldner, J.P. Vieren, P. Voisin, M. Voos, L.L. Chang and L. Esaki, Phys. Rev. Lett. 45, 877 (1980)

10  J.C. Maan, Y. Guldner, J.P. Vieren, P. Voisin, M. Voos, L.L. Chang and L. Esaki, Solid State Comm. 39, 683 (1981)

11  A. Fasolino and M. Altarelli, Surf. Sci. 142, 322 (1984)

12  P. Voisin, G. Bastard and M. Voos, Phys. Rev. B 29, 935 (1984)

13  P. Voisin, Surf. Sci.142, 460 (1984)

14  P. Voisin, G. Bastard, C.E.T. Gonçalves da Silva, M. Voos, L.L. Chang and L. Esaki, Solid State Comm. 39, 982 (1981) ; see also P.Voisin, These de Doctorat, Paris, 1983 (unpublished)

15  P. Voisin, J.A. Brum, M. Voos, L.L. Chang and L. Esaki, Proc. Int. Conf. Modulated Semiconductor Structures (MSS II) (Kyoto, Sept. 1985), to appear in Surf. Sci. (1986)

16  J.A. Brum, P. Voisin and G. Bastard, Phys. Rev. B 33, 1063 (1986)

17  G. Abdel-Fattah, P. Voisin, M. Voos, L.L. Chang and L. Esaki, unpublished

18  P. Voisin, C. Delalande, M. Voos, L.L. Chang, A. Segmuller, C.-A. Chang and L. Esaki, Phys. Rev. B 30, 2276 (1984)

19  P. Voisin, Winterschool "Semiconductor Heterojunctions and Superlattices", Les Houches (1985); to be published by Springer-Verlag

20  G.C. Osbourn, this volume; J.Y. Marzin, this volume

21  P. Voisin, Surf. Sci. 168, 546 (1986)

22  For samples grown on GaSb substrates, the critical layer thickness for the onset of plastic relaxation is not very small compared to the epilayer thickness, and the amount of plastic relaxation may be far from 100%. See: M. Sauvage, C.Delalande, P. Voisin, P. Etienne and P Delescluses, Proc. MSS II (Kyoto, Sept. 85); to appear in Surf. Sci. (1986)

23  M.C. Joncourt, Private Communication

24  Landolt-Börnstein,Numerical data and functional relationships in Science and technology, edited by O. Madelung, Group III,Vol. 17 (Springer-Verlag, 1982)

25  J. Bleuse, P. Voisin, M. Voos, L.L. Chang and L. Esaki, unpublished

26  C. Tejedor, J.M. Calleja, F. Meseguer, E.E. Mendez, C.-A. Chang and L. Esaki, Phys. Rev. B 32, 5303 (1985)

27  J.M. Calleja, F. Meseguer, C. Tejedor, E.E. Mendez, C.-A. Chang and L. Esaki, Surf. Sci. 168, 558 (1986)

28 G. Griffiths, K. Mohammed, S. Subbana, H. Kroemer and J. Merz, Appl. Phys. Lett. 43, 1059 (1983)

29 H. Kroemer, Private Communication

30 A. Forchel, U. Cebulla, G. Trankle, H. Kroemer, S. Subbana and G. Griffiths, Proc. MSS II (Kyoto, Sept. 1985); to appear in Surf. Sci. (1986)

31 P. Voisin, J.C. Maan, M. Voos, L.L. Chang and L. Esaki, Surf. Sci. 170, 651 (1986)

32 J.A. Brum and P. Voisin, unpublished

33 G. Bastard, Phys. Rev.B 25, 7584 (1982)

34 K. Ploog, Y. Ohmori, H. Okamoto, W. Stolz and J. Wagner, Appl. Phys. Lett. 47, 384 (1985)

35 G. Bastard, C. Delalande, M.H. Meynadier, P.M. Frijlink and M. Voos, Phys. Rev. B 29, 7042 (1984)

36 Y. Ohmori, Y. Suzuki and H. Okamoto, Japanese J. Appl. Phys. 24, L657 (1985)

# STRAINED LAYER SUPERLATTICES OF GaInAs-GaAs

J-Y Marzin

Centre National d'Etudes des Telecommunications
Laboratoire de Bagneux,* 196 av. Henri Ravera
92220 Bagneux, France

## INTRODUCTION

Since the pionneering work of Matthews and Blakeslee [1] on GaAs-GaAsP strained semiconductor superlattices, the improvement of the Molecular Beam Epitaxy (M.B.E.) and Metalorganic Vapor Phase Epitaxy (M.O.V.P.E.) has allowed the growth of numerous strained systems. This development is due to the potential interest of these structures for device applications. Their use broadens the choice of epitaxial materials on a given substrate by removing the drastic condition of lattice matching. In turn, the thicknesses of the sublayers must be kept small enough so that the mismatch can be elastically accomodated inside the structure. By adapting the design parameters of the superlattices, their band gap and mean lattice parameter can be independently varied in wide ranges [2]. Their ability to prevent the propagation of dislocations make it possible to grow good quality thick strained superlattices on a graded buffer layer matching this mean parameter. For the same reason, they constitute interesting buffer layers between two largely mismatched materials [3]. Moreover, the modifications of the properties of the grown semiconductors by the built-in strains they experience make them potentially useful.

Among these systems, structures built with InGaAs alloys grown either on GaAs or InP have been one of the first to be investigated [4,5] and the aim of this paper is to discuss some of its properties and potential applications. We first deal with moderately strained such superlattices in which the deformations are of the order of 1% like $In_{.15}Ga_{.85}As$-GaAs on GaAs before giving some data obtained on more severely strained systems like InAs-GaAs on InP or on GaAs substrates.

## I MODERATELY STRAINED SUPERLATTICES

In this part we discuss mainly the most studied strained layer superlattices which are $In_x Ga_{1-x} As$-GaAs with In composition x ranging from 10 to 30%. They are epitaxied on

001 oriented substrates, either by M.B.E. or M.O.V.P.E.. The
strains inside the sublayers are kept lower than 2%.In this
case, the structural properties are rather simple. The
feasibility of a large number of devices have been
demonstrated in this system, including FET's [6], lasers
[7,8] and photodiodes [9,10].

I-1 STRAIN ANALYSIS

As it is well established experimentally [4,11-13], a
superlattice grown from two mismatched materials can be
nearly free from misfit dislocations if the sublayers
thicknesses are small enough. At the mechanical equilibrium,
the two semiconductors have the same in-plane parameter,
which is the mean lattice parameter of the structure (for
simplification, we admit that they have the same elastic
constants).They are then quadratically strained, one in
biaxial tension (GaAs) and the other in compression
(InGaAs).This "critical thickness" has been evaluated by
Matthews and Blakeslee [1] and other authors [14] and the
result obtained in Ref. 1 is given in Fig. 1.These

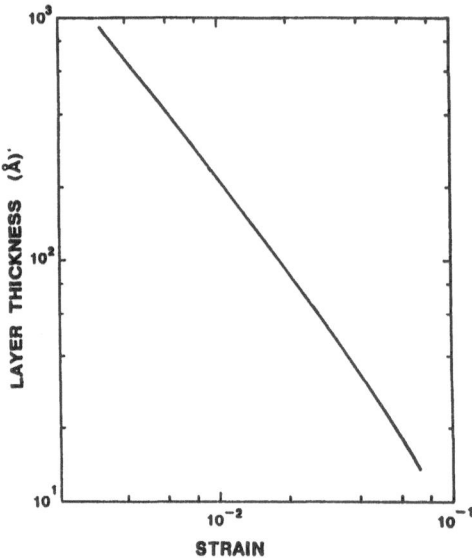

<u>Fig. 1</u> Critical layer thickness as a fonction of strain, calculated from Ref. 1. For thicknesses
smaller than this value, the strain can be elastically accomodated.

calculations are based on the comparison between the elastic
energy inside the sample with and without dislocations, or on
the bending of pre-existing dislocations by the strain
fields. They do not take into account the energetic barrier
to overcome in the creation of such defects, whose number in
the substrate is highly insufficient to relax important
mismatches. Similar approaches [15] lead for a single
strained layer to an underestimation of the critical
thickness. For superlattices, the agreement with the
available experimental data is more satisfying. For example,
Fritz et al. [16] from X-ray, transport and photoluminescence
analysis have reported results very close to the curve of
Fig. 1.

For the epitaxy of such a superlattice on a substrate, two techniques are used. The first one consist in growing it on top of a graded composition InGaAs buffer layer which parameter is slowly varied from that of the substrate to the equilibrium parameter of the superlattice (Fig. 2 a) ). Thick superlattices can then be grown on the buffer layer: the first few periods block the propagation of the dislocations present in the buffer layer and good quality epitaxial material can be obtained. The buffer layer may also consist in a single alloy layer with the mean In composition of the InGaAs-GaAs superlattice, and thick enough to take its own lattice parameter.

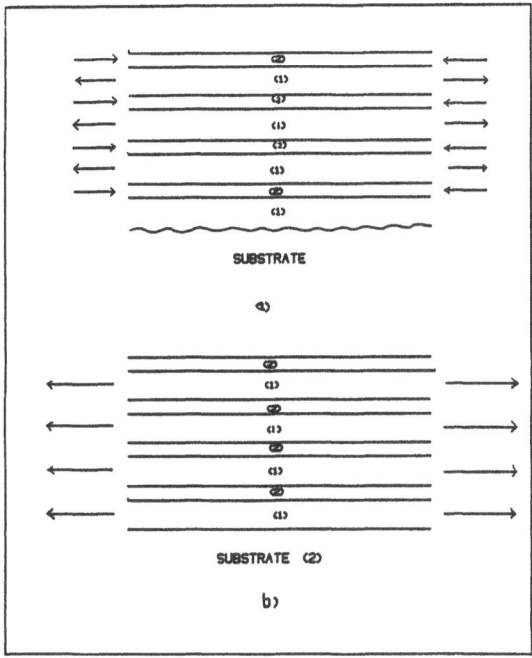

Fig. 2 Schematic diagram for two configurations in a superlattice built with materials 1 and 2 ,of different parameters.

a) The superlattice has taken its equilibrium parameter, and there is no resultant force on the surfaces of the superlattice.

b) The superlattice is strained on the material 2 substrate.Only the material 1 layers are strained. This planar configuration is not the equilibrium, because of the forces on the superlattice and the sustrate ( not represented) have a non vanishing torque.

The second technique allows to obtain structures as free as possible from defects: the superlattice is grown on a GaAs buffer layer and the whole structure is then elastically strained if the superlattice is thin enough. It takes an in-plane parameter which is now equal to

$$a_{//} = \bar{a} = \frac{d_s\, a_s + d_i\, a_i}{d_s + d_i}$$

where ds and di are the total thicknesses of GaAs (substrate and epilayers) and InGaAs , respectively, and as and ai their unstrained lattice parameters. It is very close to GaAs parameter. The sample is bent by the non vanishing torque of the stresses [17], but this bending, which is observed in X-ray topographies [18], is small (curvature radius of several meters) and can be neglected as far as the electronic properties are concerned. Consequently, under these assumptions, only the ternary layers are strained, whereas the GaAs ones are not (Fig. 2 b) ). This second method is generally preferred when a few periods are sufficient and were used successfully to grow quantum well lasers [7] emitting around 1 µm and low current thresholds. When the overall superlattice thickness is larger than its critical thickness, when epitaxied on the GaAs buffer layer, misfit dislocations which are generated are mostly confined at the first interfaces with the buffer layer as it was confirmed by Joncour et al. [19] in X-ray topographies on beveled structures as shown in Fig. 3, extracted from their paper.

1mm

Fig. 3 220 Mo K$_\alpha$ topograph of a sample consisting in 10 periods of alternating In Ga As layers (100 Å thick), and GaAs layers (200 Å thick), covered with a 1000 Å GaAs cap layer, M.B.E. grown on GaAs. The structure was beveled on its lower part from A to B. A very small strain relaxation occurs in this sample, and each dislocation line of the topograph can be followed from the unetched part of the sample to the first interface between the superlattice and the substrate, where they are thus located.

In both cases, the deformation in the two sublayers is simply given by:

$$\varepsilon_{xx}^i = \varepsilon_{yy}^i = \frac{a_{//} - a_i}{a_i} \quad , \quad \varepsilon_{zz}^i = -2\frac{C_{12}}{C_{11}} \varepsilon_{xx}^i$$

$$\varepsilon_{xy}^i = \varepsilon_{yz}^i = \varepsilon_{xz}^i = 0$$

the Cij's being the elastic constants.

X-ray diffraction analysis constitutes a very powerful tool for the study of these strained superlattices [13,20]: it gives a precise determination of the period, but also of the individual layers thicknesses, of the ternary layers composition and indications on the overall relaxation of the superlattice. The lattice parameter in the growth axis direction z is different in InGaAs and GaAs layers and is then modulated with the superlattice period. This was evidenced also by ion channeling studies by Picraux et al. [11,12] .In X-ray double diffraction on these planes, the patterns obtained for the two sublayers inside one period will be centered on the corresponding Bragg positions, as it is schematized in Fig.4, from Ref. 13, (curves C2 and C3), and the diffraction profile of one period results from the interferences between them (curve C1). When several periods are piled up to build the superlattice, diffraction coming from successive periods will be constructive only at Bragg angles corresponding to the new periodicity. The resulting

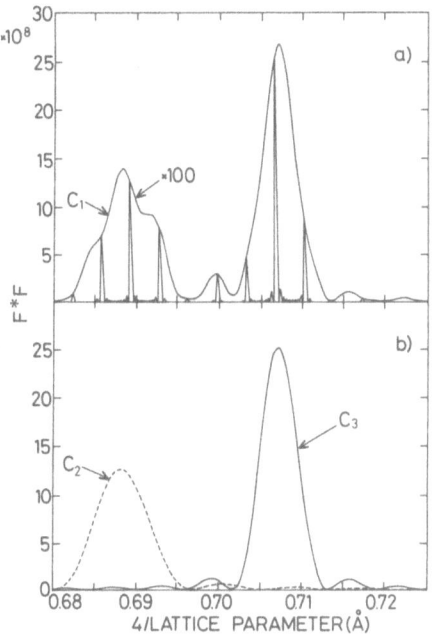

Fig. 4 a) Calculated profile for a N = 10 period superlattice. Each period consist in 30 $In_{.3}Ga_{.7}As$ monolayers and 60 GaAs monolayers. The envelope curve C1 is the diffracted intensity for 1 period, with a N = 100 scaling factor.

b) Profile obtained for a single $(In_{.3}Ga_{.7}As)$ layer (C2) and a single (GaAs) layer (C3). Comparing a) and b) shows that, in this case, each sets of peaks refers almost exactly to one of the materials, as if they were not spatially related.

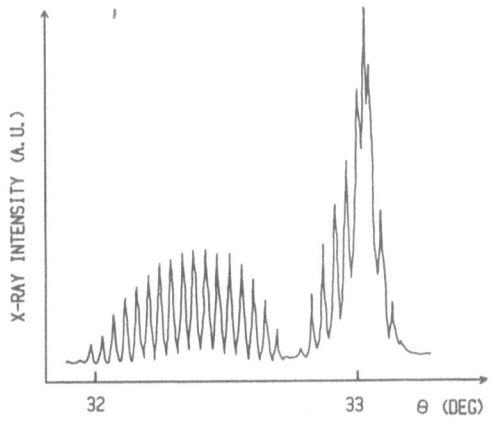

Fig. 5 Experimental 004 double diffraction profile obtained on a large period sample. The superlattice has 10 periods of $In_{.16}Ga_{.84}As$ (63 Å thick) and GaAs (1300 Å thick). The envelope functions corresponding to the diffraction of the 2 sublayers of a single period (see Fig. 4) can be easily extracted from such a profile.

103

satellites, whose envelope is given by curve Cl are shown in Fig.4. We want to point out the fact that their intensity are due more to the differences in the lattice parameters than to those in the structure factors of the starting materials: the satellites are much more intense than in "unstrained" structures, and are observed already in samples containing 4 periods. The fitting of their positions and intensities allows to obtain the structural parameters of superlattice. In case of large enough indium compositions and width of the sublayers, the profiles consist of two well separated groups of satellites (Fig.5), because the diffraction coming from the two materials merely interferes.

This technique is now widely used to characterize these structures

## I-2 BAND STRUCTURE AND OPTICAL PROPERTIES

The effects of biaxial strains on the semiconductors band structure are equivalent to those of the sum of a uniaxial stress on the Z direction and of a hydrostatic pressure (which should be of the order of 10 kbar for 1% strains). Both are well known in the case of bulk materials [21]. Fig.6 shows the modifications of the near zone center valence and conduction bands. The main ones are the

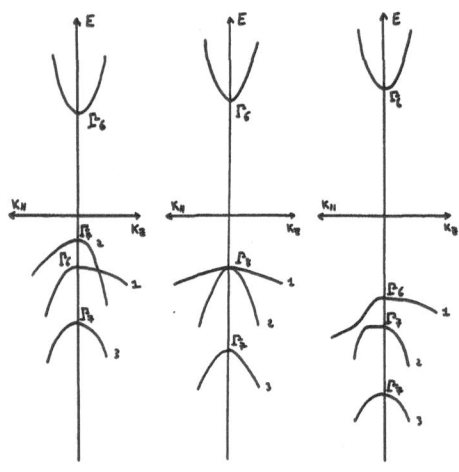

TENSION     UNSTRAINED     COMPRESSION

Fig. 6 Schematic band structure of a bulk III-V semiconductor under biaxial strain, around the point. The biaxial strain is in the (x,y) plane.$k_{//}$ refers to an in an (x,y) plane direction.

Bands labelled 1 2 and 3, correspond to heavy hole, light hole and split-off valence bands, for the unstrained material

following:
i) a change in the band gap due to the hydrostatic part of the strain, accompanied by small changes in the effective masses.
ii) a change in the symmetry from cubic to quadratic of Z axis, which entails the splitting of the $|3/2,\pm 3/2>_z$ heavy holes states from the $|3/2,\pm 1/2>_z$ light holes ones. The strains also couple the $|1/2,\pm 1/2>_z$ split-off band to these light holes states, whereas in the $k_z$ direction, the $|3/2,\pm 3/2>_z$ states stay uncoupled from them. The sign of the

splitting is such that for a material in biaxial compression (InGaAs in our case) the |3/2,+3/2>$_z$ band extremum lies higher in energy than the other valence band extrema.

For small strains, we have:

$$E_g(\varepsilon) - E_g(\emptyset) = a(\varepsilon_{xx} + \varepsilon_{yy} + \varepsilon_{zz}) - |b(\varepsilon_{zz} - \varepsilon_{xx})| \quad , \text{ for the band gap}$$

and

$$E(1) - E(2) = -2b(\varepsilon_{zz} - \varepsilon_{xx})$$

for the valence band zone center splitting (see Fig. 6) where a and b are the deformation potentials.

In GaAs, a = -8.6 eV and b = - 2 eV [22].

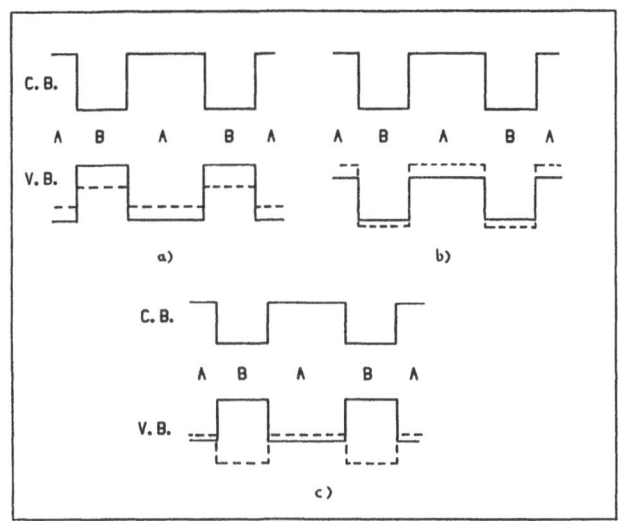

Fig. 7 Three possible band extrema configurations for a strained superlattice built with material A in tension and B in compression:

  a) the superlattice is of type 1 for "heavy" and "light" holes

  b) it is of type II for "heavy" and "light" holes.

  c) the system is of type I for electrons and "heavy" holes and of type II for electrons and "light" holes.

In the following, in "heavy" or "light", the quotation marks indicate the bands of the superlattice mainly built with the corresponding bulk valence bands. Due to the decoupling of the |3/2,±3/2>$_z$ states for $k_x = k_y = 0$, the effective masses in the kz direction stay nearly unchanged by the strains. In the superlattice, the eigenstates are obtained easily (again for $k_x = k_y = 0$), with a three band envelope function model [23,24] for light particles, and the independant treatment of the |3/2,±3/2>$_z$ like states. When non parabolicity effects are negligible, they are obtained by considering the three superpotentials schematized in Fig.7. According to the relative positions of the two sets of extrema in the constituent materials, and to the sign of the strain, a large variety of situations are observed. In InGaAs-GaAs structures, the valence band configuration corresponds to Fig.7 c). The valence band superlattice states emerging from "light" particles are confined more in the GaAs layers. This is also observed in the InGaAs-InGaAs superlattices, as it will be discussed in the next part. The selection rules for superlattice zone center optical properties are the same as in GaAs-GaAlAs superlattices (because the quadratic symmetry is maintained), but the transitions between electron states (confined in InGaAs) and states confined in GaAs are weaker than in this system, due to the small overlap of the envelope functions. This case corresponds to a type II superlattice (as InAs-GaSb), for which selection rules have been discussed in detail by Voisin et al. [25].

For $k_x$ or $k_y \neq 0$, the situation is somehow simpler than in GaAs-GaAlAs, where the superlattice subbands are close enough, so that the k.p coupling between the heavy and light hole states mixes them efficiently. In InGaAs-GaAs, the strain induced splitting of the valence band enlarges the energy distance of the first "heavy" and "light" hole states. It reduces the efficiency of their k.p coupling, so that the in plane effective masses for these "heavy" hole states is just given by the diagonal part of the Luttinger Hamiltonian [26].

This in plane effective mass is small. Values of 0.13 $m_o$ have been measured by Shubnikov-deHaas [27] and magneto-luminescence [28] studies, for the first "heavy" hole state. This value justifies the term of mass reversal in this situation. For states lying at lower energies, the problem needs a more complicated approach, similar to the calculations of Altarelli et al. [29], including the strain effects. This is the case in particular for valence band "light" particles states which are always close in energy to "heavy" hole excited subbands. Osbourn gives a discussion of these effective masses in strained superlattices in Ref.30. The incidence of this small value of the in-plane effective mass for the highest energy valence band on the hole in-plane mobility in p type modulation doped structures makes this system very promising for microelectronic applications [31].

I-3 OPTICAL PROPERTIES

Many groups investigated the optical properties of InGaAs-GaAs superlattices [32-44], which we will discuss now in some detail. As in GaAs-GaAlAs structures, the absorption spectra obtained on such samples [40] reveal strong excitonic features, which are persistant up to room temperature. Fig. 8, from Ref. 40, shows such spectra obtained on a series of samples, consisting in ten periods of alternating $In_{.15}Ga_{.85}As$ layers of thicknesses Lt and GaAs layers (200 Å thick). They were grown by M.B.E. on GaAs, without graded buffer layer. They reveal the high optical quality of these samples. In the configuration discussed above , the first excitonic transition corresponds to the creation of HH1-E1 excitons.

The effects of the strains are clearly seen here, because no light hole transition is observed in the immediate vicinity of this first peak, as it should be in GaAs-GaAlAs structures with the same design parameters, despite the similar effective masses for the electrons and holes in $In_{.15}Ga_{.85}As$ and GaAs. The second clear strain effect is to increase the shift of the superlattice band gap with respect to the bulk ternary one, this shift being too important to be attributed to confinement energies alone. Such modifications, (too large or too small band gaps, and changes in the distance between the first light and heavy hole excitons), are very often the signature of the existence (eventually non intentional) of strains inside the superlattices.

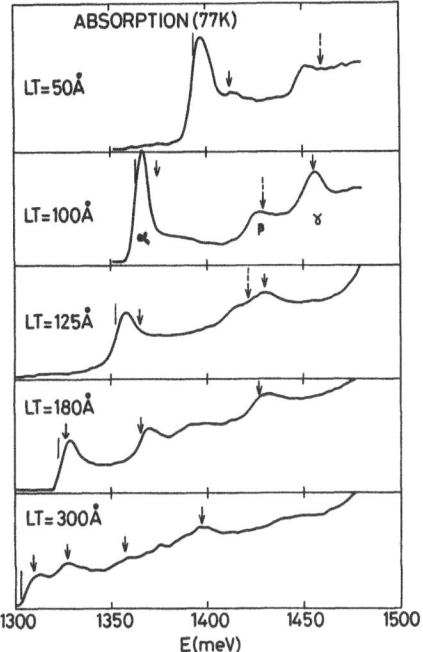

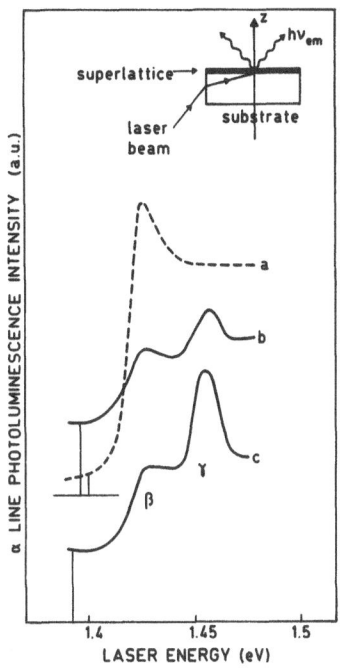

Fig. 8 77 K absorption spectra of $In_{.15}Ga_{.85}As$-GaAs superlattices, where the In Ga As layer thickness Lt is varied from sample to sample. The vertical bars indicate the energies of the maximum of the photoluminescence spectra obtained at the same temperatures. The arrows mark the calculated transitions associated to the "heavy" and "light" holes states (full and dashed lines, respectively). For Lt values of 50, 100 and 125 Å, the calculations assume the superlattice has taken the GaAs lattice parameter, in the layers plane, whereas it is supposed to be the superlattice alone equilibrium parameter for the 2 remaining samples.

Fig. 9 On-edge excitation spectra obtained in the Lt = 100 Å sample (Fig. 8), for light polarizations parallel (a) and perpendicular (b) to the Z axis. The experimental set-up is schematized in the inset. (c) Excitation spectrum recorded in the standard configuration, where the laser beam is focused on the sample surface. Comparison between a) and b) idicates the heavy hole character of transition $\gamma$, which is forbidden for a light polarization along the Z axis.

The nature of the higher energy transitions was established by on edge excitation of the photoluminescence, taking advantage of the transparency of the GaAs substrate in the energy range of interest. The result of this experiment is given in Fig.9, also from Ref.40, for the two highest energy transitions observed in absorption for the sample with Lt = 100 Å. From their selection rules, $\gamma$ and $\beta$ were assigned to HH2-E2 and LH1-E1 transitions, respectively. The fitting of the energies of these transitions allowed to determine the relative positions of the band extrema of GaAs and InGaAs. It corresponds to the situation of Fig. 7 c), where the system electron, "light" valence band particles is of type II. The transition is however observed due to the small confinement effect on these states. The low temperature photoluminescence spectra consist of one dominant sharp line, whose position is indicated in Fig.8 for the different samples. It corresponds to the first exciton HH1,E1 and has a full width at half maximum of about 7 to 10 meV for the 3 first samples. The arrows indicate the calculated transition energies for these samples , assuming an in plane lattice parameter equal to the GaAs one, and not taking into account the exciton binding energies which should be around 8 meV , for the well thicknesses of these samples. For samples with Lt = 200 Å and 300 Å, the absorption exciton peaks are less resolved and the agreement with the calculation assuming a perfectly strained superlattice on the substrate is poor. Moreover, for these samples, X-ray double diffraction profiles and topographies indicate that partial relaxation of the strains occurs. This partial relaxation is accompanied by an decrease of the superlattice band gap. For Lt = 300 Å, despite confinement energies it is even smaller than the band gap that would have bulk $In_{.15}Ga_{.95}As$ feeling the deformations corresponding to a perfectly strained superlattice. For these two samples, at least the critical thickness for the superlattice as a whole is reached. For Lt = 300 Å, the photoluminescence intensity decreased significantly, indicating the presence of non radiative defects inside the superlattice: relaxation may occur in this case, not only at the first interface, but between the sublayers themselves. If such is the case, then the in-plane parameter varies in the superlattice, resulting in the broaden of the absorption lines, and in a shift between the first absorption transition and the luminescence line, which is experimentally observed. The arrows in these two spectra indicate the calculated energy transitions assuming that the superlattice has taken its equilibrium parameter, which assumption gives a better agreement with the experimental data.

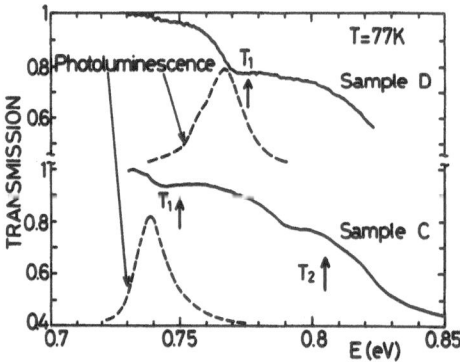

Fig. 10 Transmission spectra obtained at 77 K on 2 $In_{.4}Ga_{.6}As$- $In_{.6}Ga_{.35}As$ superlattices grown on InP, with equal sublayers thicness d.d = 90 Å for sample D and d = 175 Å for sample C. The transitions T1 and T2, are assigned to HH1-E1 and HH2-E2 respectively, and they merge from a bacground spectrum due to the $In_{.53}Ga_{.47}As$ buffer layers on which they are grown. The 77K photoluminescence spectra are also shown for comparison.

As already mentioned the critical thicknesses have been determined experimentally in this system from the degradation of the photoluminescence spectra obtained on thick In $_x$ Ga$_{1-x}$As-GaAs superlattices [37], and of in-plane transport or structural properties [16], so that the limits of this system are known precisely, though the exact relaxation dynamics are still to be examined.

In$_x$Ga$_{1-x}$As-In$_y$Ga$_{1-y}$As strained layer superlattices [45], M.B.E. grown on InP are promising structures for optical devices operating at wavelengths longer than 1.6 µm. The presence of dislocations here can be avoided by choosing the two In compositions and the layers thicknesses so that the equilibrium parameter of the superlattice matches that of the InP substrate. Absorption data obtained on such structures [46] are displayed in Fig.10, together with the associated luminescence spectra. The absorption transitions related to the superlattice merge from the spectra of the In$_{.53}$Ga$_{.47}$As buffer layers on which they were deposited in these two samples. The arrows indicate the energies of the calculated "heavy" hole related transitions, with no adjustable parameter. In these samples, both materials are strained, and the valence band configuration is again of a mixed type. We think that the "light" hole related transitions are not observed here because of their stronger confinement in the larger gap ternary alloy layers. The limiting case of this system consists in InAs-GaAs superlattices we will examine in the next part.

## II HIGHLY STRAINED STRUCTURES

### II-1 InAs-GaAs on InP

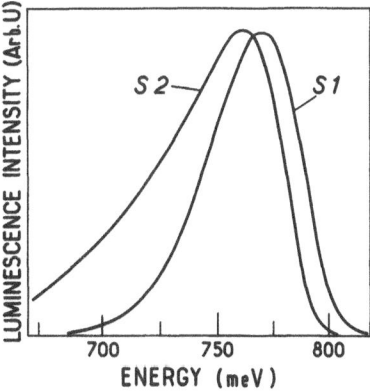

Fig. 11 2K photoluminescence spectra obtained on two InAs-GaAs superlattices grown on InP, with equal sublayers thickness d.d = 10 Å and 20 Å for sample S1 and S2 respectively, which contain 15 and 10 periods.

The first idea in realizing InAs-GaAs short period superlattices was to substitute them to the lattice matched ternary material on InP, $In_{.53}Ga_{.47}As$, hoping that the interface scattering will be smaller than the alloy one. They were grown successfully by several groups [47-49]. The low temperature photoluminescence spectra of two such M.B.E. grown samples, are shown in Fig.11, from Ref.50. These samples were of good crystalline quality and the photoluminescence shows essentially no spatial variation, revealing their good homogeneity. The energy of the transitions is very close to the corresponding alloy band gap and varies slowly with the sublayers thicknesses (which are the same for InAs and GaAs ones) equal to 10 and 20 Å for S1 and S2 samples respectively, indicating that the pseudo-alloy regime is reached in these samples. An envelope function model calculation led us to assign them to electron to light holes transitions, although for these very small thicknesses and very high strains, this approach can be questionned. Many problems are still open on this system: the residual valence band splitting is still to be observed in this quadratic pseudo-alloy and the transport properties to be examined.

II-2 InAs-GaAs on GaAs

When deposited on GaAs, the InAs-GaAs structures involve highly strained InAs layers [51-57]. In this case, the strain field appears to influence the growth process itself: it reaches rapidly a 3-dimensional regime [58], where InAs bulk diffraction character is observed in the RHEED pattern. Such surface degradation is much less severe for the less strained structures described in part I, although one can imagine that the quick damping of the RHEED oscillations (Fig. 12), during the growth of the ternary layers, may traduce a surface roughness constituting the premices of such 3D phenomenon. However, we could consider, for the moderatly strained systems, the superlattice layers as being planar. The strains could then be relaxed, from this starting situation, by the formation of structural defects such as dislocations, whose presence lowered the total elastic energy. For higher strains, in our opinion, the tendency of developing 3D growth shows the important role played by the local density of elastic energy.
During the growth of one InAs layer on top of a GaAs substrate, there is a competition between the tendency for In atoms to satisfy several bonds (leading usually to 2D growth), and the deformations that are imposed on these bonds by the underlying GaAs substrate. Even in the case of the 2D growth of one InAs monolayer, if one imagines it as the lateral growing of 2D islands (one monolayer high), the presence of these islands will result in the existence of strain fields: the GaAs between them will be in compression, thus offering a local "parameter", which is more distant from the InAs one in these regions. Such phenomena, here evoked with hand-waving arguments should be reconsidered including surface reconstruction, exchange reactions and In segregation, but will push anyway the system towards a 3D growth. For a given lateral size of such InAs islands, the strains will be relaxed more and more easily (elastically) when their thickness increases. During the subsequent growth of GaAs, In-rich clusters will be formed from these InAs islands or clusters.
The statistics of these defects concerning the layer thickness and composition depend on the dynamics and

thermodynamics of the growth, so that substrate temperature, growth rate, use of growth interuptions are essential parameters. Such compositional inhomogeneities in InAs-GaAs

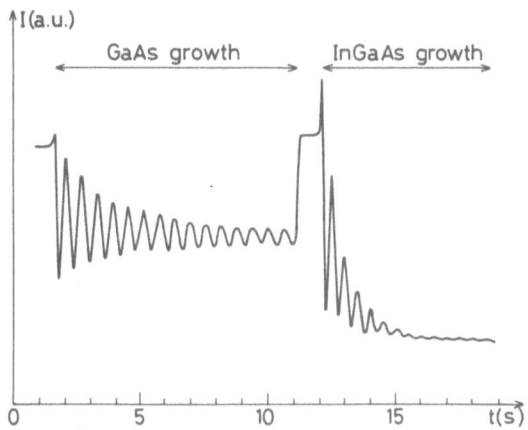

Fig. 12 Recording of the specular beam intensity of the RHEED, during the M.B.E. growth of GaAs and In$_{.15}$Ga$_{.85}$As. The growth was interupted for a few seconds, between the 2 layers.

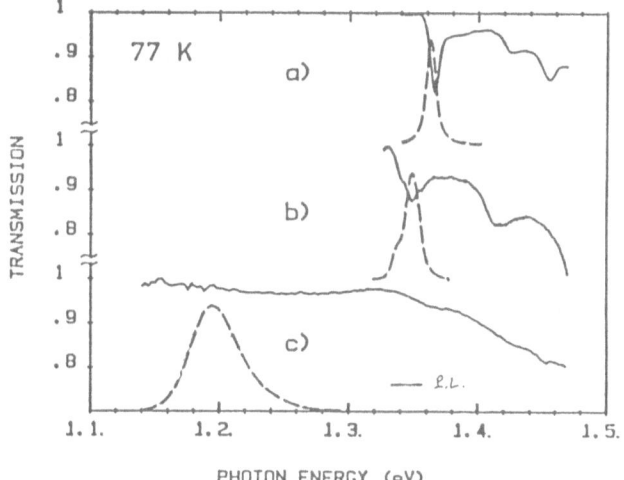

Fig. 13 77 K transmission and photoluminescence spectra obtained on M.Q.W. samples a) b) and c) (see text). Sample a) and b) spectra are similar, whereas the spectra of sample c) are deeply modified by the existence of defects.

structures grown on GaAs, have been observed by T.E.M. [57]. It revealed the associated non-uniform strain fields, which were detected also in High Resolution Microscopy [59], and the absence of dislocations, showing that samples containing these In-rich clusters can still be elastically strained. The last and noticeable structural property linked to the inhomogeneities we want to mention is the memory of the position of the clusters, through the long range associated strain fields, during the growth of several InAs layers separated by thick GaAs layers. The piling up of In-rich regions in layers separated by 200 Å thick GaAs layers was reported [55].

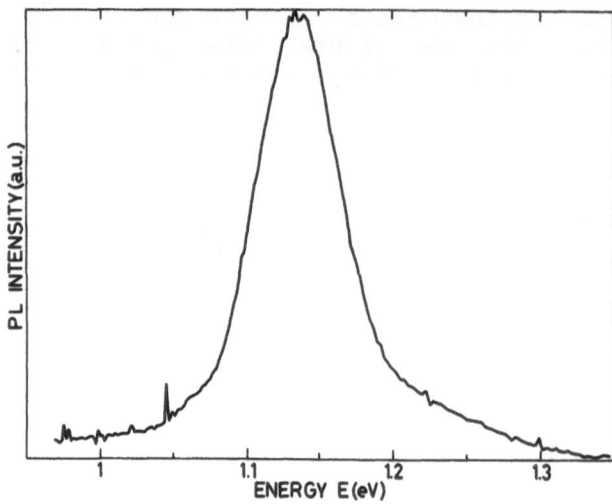

<u>Fig. 14</u> Photoluminescence spectrum obtained at 77 K in a InAs-GaAs superlattice with very thin InAs layers ( 2 monolayers), and 200 A GaAs layers, grown on GaAs.

Up to our knowledge, recombination processes associated with such defects dominate the photoluminescence properties of the InAs-GaAs structures grown on GaAs, at high substrate temperatures, without growth interuptions. Fig. 13, from Ref.60, shows the absorption and luminescence spectra of such samples M.B.E. grown at 540 C. They consisted of 10 quantum wells, 100 A wide, separated by 200 A GaAs layers. The wells are built either with In Ga As alloy (sample a)) or short period (InAs)m (GaAs)n superlattices, with the same average In content. m = .5 , n = 2.5 for sample b and m =1 , n = 5 for sample c. Sample b spectra are very similar to the alloy case. On the contrary, we note for sample c a large shift towards lower energy of the photoluminescence line and a degradation of its absorption spectrum. The In-rich clusters which are thought to be responsible for these modifications were observed in T.E.M.. The photoluminescence intensity was the same as in samples a and b, confirming the low density of non-radiative defects.

Structures containing 1 or 2 InAs monolayers separated by thick GaAs layers, studied in order to understand the very beginning of the growth in this system, show, at the present moment, similar optical properties, as it is illustrated in Fig.14, from Ref. 55. The emission lines remain rather broad, and the GaAs associated spectrum is shifted with respect to the bulk typical one. These preliminary results are tentatively assigned to the coexistence of several processes:
i) persistence of the existence of InAs clusters
ii) inhomogenious strain fields
iii) In segregation and exchange reactions


CONCLUSION

InGaAs-GaAs strained layer superlattices are very attractive systems. The moderately strained structures have to be considered as new materials with flexible characteristics, available for the design of opto and micro-electronic devices. Some care should be taken not to submit them to too high excitation rates or temperatures, to avoid degradation. They enlarge the family of possible epitaxial layers on GaAs or InP, and in particular they constitute the lower band gap materials on these substrates.

On the other hand, the successful growth of highly strained structures make it possibie to study the relationship between growth dynamics and the existence of large strain fields. It should also allow to compare the optical and transport properties of the In Ga As alloys with those of short period superlattices. For that purpose, the quality of such structures has to be further improved.

## ACKNOWLEDGEMENTS

Most of the work presented here in some detail have been performed in the C.N.E.T laboratory. The samples have been grown by L. Goldstein, M.N. Charasse, M. Quillec and J.L. Benchimol; the X-ray diffraction was studied by J. Burgeat, M. Quillec, J. Primot, G. Le Roux, and M.C. Joncour; and the T.E.M. observations were done by F. Glas, P. Henoc and C. d'Anterroches. The author also want to thank P. Voisin, G. Bastard, M. Voos, D. Paquet, F. Houzay, J.M. Moison and M. Quillec for their fruitful comments and discussion.

## REFERENCES

*Laboratoire de Bagneux is a laboratory associated to the Centre National de la Recherche Scientifique (LA 250).

1.  J.W. Matthews and A.E. Blakeslee, J.Cryst.Growth 32:265 (1976), and references therein.
2.  G.C. Osbourn, J.Appl.Phys. 53:1586 (1982).
3.  R. Fisher, D. Neuman, H. Zabel, H. Morkoc, C. Choi and N. Otsuka, Appl.Phys.Lett. 48:1223 (1986).
4.  L. Goldstein, M. Quillec, E.V.K. Rao, P. Henoc, J.M. Masson and J.Y. Marzin, J.Phys. (Paris) 12, C5:201 (1982).
5.  I.J. Fritz, L.R. Dawson, G.C. Osbourn, P.L. Gourley and R.M. Biefeld, Int.Phys.Conf.Ser. 65:241 (1982).
6.  T.E. Zipperian, L.R. Dawson, G.C. Osbourn and I.J. Fritz, Proc. of IEEE Int.Electron. Devices Meet. 696 (1983).
7.  W.D. Laidig, P.J. Caldwell, Y.F. Lin and C.K. Peng, Appl.Phys.Lett. 43:560 (1983).
8.  W.D. Laidig, P.J. Caldwell and Y.F. Lin, J.Appl.Phys. 57:33 (1985).
9.  D.R. Myers, T.E. Zipperian, R.M. Biefeld and J.J. Wiczer, Proc. of IEEE Int. Elect. Devices Meet. 700 (1983).
10. L.R. Dawson, G.C. Osbourn, T.E. Zipperian, J.J. Wiczer, C.E. Barnes, I.J. Fritz and R.M. Biefeld, J.Vac.Sci.Technol. B2:179 (1984).
11. S.T. Picraux, L.R. Dawson, G.C. Osbourn, R.M. Biefeld and W.K. Chu, Appl.Phys.Lett. 43:1020 (1983).
12. S.T. Picraux, L.R. Dawson, G.C. Osbourn and W.K. Chu, Appl.Phys.Lett. 43:930 (1983).
13. M. Quillec, L. Goldstein, G. Le Roux, J. Burgeat and J. Primot, J.Appl.Phys. 55:2904 (1984).
14. R. People and J.C. Bean, Appl.Phys.Lett. 47:322 (1985).
15. J.H. Van der Merwe, J.Appl.Phys. 34:123 (1962).
16. I.J. Fritz, S.T. Picraux, L.R. Dawson, T.J. Drummond, W.D. Laidig and N.G. Anderson, Appl.Phys.Lett. 46:967 (1985).
17. F.K. Reinhart and E.A. Logan, J.Appl.Phys. 44:3171 (1973).

18. J.F. Petroff, M. Sauvage, P. Riglet and H. Hashizume, Phil.Magaz. A42:319 (1980).

19. M.C. Joncour, R. Mellet, M.N. Charasse and J. Burgeat, J.Cryst.Growth 75:295 (1986).

20. Y.P. Khapachev, A.A. Dyshekov and D.S. Kiselev, Phys.Stat.Soli. B126:37 (1984).

21. G.L. Bir and G.E. Pikus, "Symmetry and strain induced effects in semiconductors", (Wiley, New York, 1974).

22. C.M. Chandrasekhar and F.H. Pollack, Phys.Rev. B15:2127 (1977).

23. G. Bastard, Phys.Rev. B24:5693 (1981).

24. S.R. White and L.J. Sham, Phys.Rev.Lett. 47:879 (1983).

25. P. Voisin, G. Bastard and M. Voos, Phys.Rev. B29:935 (1984).

26. J.M. Luttinger, Phys.Rev. 102:1030 (1956).

27. J.E. Schirber, I.J. Fritz and L.R. Dawson, Appl.Phys.Lett. 46:187 (1985).

28. E.D. Jones, H. Ackermann, J.E. Schirber, T.J. Drummond, L.R. Dawson and I.J. Fritz, Solid State Comm. 55:525 (1985).

29. M. Altarelli, U. Ekenberg and A. Fasolino, Phys.Rev. B32:5138 (1985), and references therein.

30. G.C. Osbourn, Superl. and Microst. 1:223 (1985).

31. I.J. Fritz, T.J. Drummond, G.C. Osbourn, J.E. Schirber and E.D. Jones, Appl.Phys.Lett. 48:1678 (1986).

32. J.Y. Marzin and E.V.K. Rao, Appl.Phys.Lett. 43:560 (1983).

33. P.L. Gourley and R.M. Biefeld, Appl.Phys.Lett. 45:749 (1984).

34. M.D. Camras, J.M. Brown, N. Holonyak, M.A. Nixon, R.W. Kaliski, M.J. Ludowise, W.T. Dietze and R.C. Lewis, J.Appl.Phys. 54:6183 (1983).

35. M. Nakayama, K. Kubota, H. Kato and N. Sano, Solid State Comm. 51:343 (1984).

36. H. Kato, M. Nakayama, S. Chika and N. Sano, Solid State Comm. 52:559 (1984).

37. N.G. Anderson, W.D. Laidig and Y.F. Lin, J. Electron. Mater. 14:187 (1984).

38. K. Kubota, T. Mizuta, M. Nakayama, H. Kato, N. Sano, Solid State Comm. 52:333 (1984).

39. W.D. Laidig, D.K. Blanks and T.F. Scherzina, J.Appl.Phys. 56:1791 (1984).

40. J.Y. Marzin, M.N. Charasse and B. Sermage, Phys.Rev. B31:8298 (1985).

41. N.G. Anderson, W.D. Laidig, G. Lee, Y. Lo and M. Ozturk in "Layered structures and Epitaxy", J.M. Gibson and L.R. Dawson, editors, Material Research Society, Pittsburg (1985).

42. U. Das and P.K. Bhattacharya, J.Appl.Phys. 58:341 (1985).

43. U. Das, P.K. Bhattacharya and S. Dhar, Appl.Phys.Lett. 48:1507 (1986).

44. I.J. Fritz, B.L. Boyle, T.J. Drummond, R.M. Biefeld and G.C. Osbourn, Appl.Phys.Lett. 48:1606 (1986).

45. G.C. Osbourn, Phys.Rev. B27:5126 (1983).

46. M. Quillec, J.Y. Marzin, J. Primot, G. Le Roux, J.L. Benchimol and J. Burgeat, J.Appl.Phys. 59:2447 (1986).

47. M.C. Tamargo, R. Hull, H. Greene, J.R. Hayes and A.Y. Cho, Appl.Phys. Lett. 46:569 (1985).

48. Y. Matsui, H. Hayashi, M. Takahashi, K. Kikushi and K. Yoshida, J.Cryst.Growth 71:280 (1985).

49. Y. Matsui, H. Hayashi, K. Kikushi and K. Yoshida, 2nd Int.Conf. on Modulated Semicon. Structures, Kyoto 1985 (to be published in Surface Science 1986).

50. P. Voisin, M. Voos, J.Y. Marzin, M.C. Tamargo, R.E. Nahory and A.Y. Cho, Appl.Phys.Lett. 48:1476 (1986).

51. W.J. Schaffer, M.D. Lind, S.P. Kowalczyk and R.W. Grant, J.Vac.Sci. Technol. B1:688 (1983).

52. B.F. Lewis, F.J. Grunthaner, A. Madhukar, R. Fernandez and J. Maserjian, J.Vac.Sci.Technol. B2:419 (1984).

53. R.A.A. Kubiak, E.H.C. Parker and S. Newstead, Appl.Phys. A35:61 (1984).

54. F.J. Grunthaner, M.Y. Yen, R. Fernandez, T.C. Lee, A. Madhukar and B.F. Lewis, Appl.Phys.Lett. 46:983 (1985).
55. L. Goldstein, F. Glas, J.Y. Marzin, M.N. Charasse and G. Le Roux, Appl.Phys.Lett. 47:1099 (1985).
56. H. Terauchi, K. Kamigaki, H. Sakashita, N. Sano, H. Kato and M. Nakayama, 2nd Int. Conf. on Modulated Semicon. Structures, Kyoto 1985 (to be published in Surface Science 1986).
57. M.Y. Yen, A. Madhukar, B.F. Lewis, R. Fernandez and J.F. Grunthaner, 2nd Int. Conf. on Modulated Semicon. Structures, Kyoto 1985 (to be published in Surface Science 1986).
58. F. Houzay, C. Guille, J.M. Moison, P. Henoc and F. Barthe, 4th Int. Conf. on Molecular Beam Epitaxy, York 1986 (to be published in J.Cryst.Growth).
59. C. d'Anterroches, J.Y. Marzin, G. Le Roux and L. Goldstein, 4th Int. Conf. on Molecular Beam Epitaxy, York 1986 (to be published in J.Cryst.Growth).
60. J.Y. Marzin, L. Goldstein, F. Glas and M. Quillec, 2nd Int. Conf. on Modulated Semicon. Structures, Kyoto 1985 (to be published in Surface Science 1986).

# PROPERTIES OF PbTe/Pb$_{1-x}$Sn$_x$Te SUPERLATTICES

G. Bauer[*] and M. Kriechbaum[§]

[*]Institut für Physik, Montanuniversität Leoben
A-8700 Leoben,Austria
[§]Institut für Theoretische Physik, Universität Graz
A-8010 Graz, Austria

## INTRODUCTION

Heterostructures of the IV-VI compounds PbTe/Pb$_{1-x}$Sn$_x$Te have been used for a number of years to produce efficient tunable laser diodes in the mid-infrared region of the electro-magnetic spectrum. Either liquid phase epitaxy (LPE), hot-wall epitaxy(HWE) or molecular beam epitaxy (MBE) have been used to grow double hetero junction lasers (Preier, 1979). Recently, also PbTe/PbSnTe multiquantum well (MQW) lasers for pulsed operation at 6 μm and temperatures up to about 200 K were fabricated. Together with other single or multiquantum well systems like e.g. PbEuSeTe/PbTe (Partin, 1984) the laser characteristics obtained, make these devices attractive for long wavelength fiber optic sensor/communication systems (Partin, 1985a,Ishida et al., 1986).

These narrow gap systems are also of interest for some properties which are quite different from III-V or II-VI compound MQW's or superlattices (SL's). The lead chalcogenides crystallize in the NaCl structure, their direct gap is at the L-point of the Brillouin zone and the surfaces of constant energy for electrons and holes have ellipsoidal shape with the <111> directions as the main axes (Fig.1). Due to a tendency towards a structural phase transition (O$_h$→C$_{3v}$) their static dielectric constants are huge (of the order $\varepsilon_s$ ≅ 1000 at T=4.2K) and depend strongly on temperature. As far as the PbTe/Pb$_{1-x}$Sn$_x$Te system is concerned, the rather high carrier concentrations (n,p ≅ 10$^{16}$...10$^{17}$cm$^{-3}$) originate from deviations from stoichiometry, the group IV element vacancies being responsible for holes and group VI element vacancies for electrons (Nimtz and Schlicht, 1983). Due to the huge static dielectric constants long range Coulomb scattering does not limit the mobility at low temperatures (Fig.2) but rather the scattering on short range potentials of lattice defects. Due to the absence of Coulomb scattering and due to the fact that usually in the smaller energy gap compound PbSnTe the deviations from stoichiometry are more severe than in PbTe, the usual concepts of modulation doping employed in III-V compounds cannot be applied.

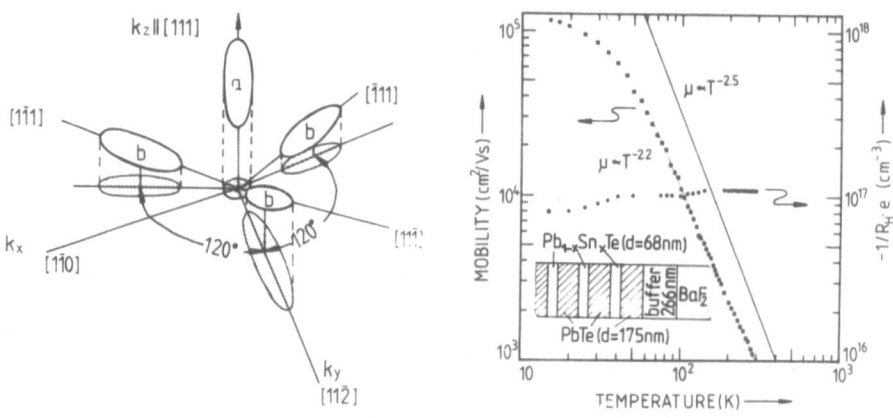

Fig.1: Schematic diagram illustrating three and two dimensional Fermi-surfaces in PbTe.

Fig.2: Temperature dependence
of the mobility of a PbTe/
$Pb_{1-x}Sn_xTe$ superlattice.

One of the difficulties encountered with PbTe/$Pb_{1-x}Sn_xTe$
superlattices (SL's) MQW's or SQW's is the fact that e.g. for
x=0.18 the total energy gap difference is about 100 meV. Therefore the conduction and valence band discontinuities are of the
order of several tens of meV's and therefore quite small in absolute values. In addition, the narrow gap causes small masses
which in turn are responsible for small subband spacings of
carrier states within the wells (of the order few meV), Fig.3.
Thus, in such a system always several subbands will be populated.

A problem encountered in the growth of PbTe/$Pb_{1-x}Sn_xTe$
MQW's even by molecular beam epitaxy in UHV conditions is the
role of oxygen at the interface. Whereas rather low temperatures are required to remove oxygen from PbTe surface (=200 $^{O}$C)
for $Pb_{1-x}Sn_xTe$ (x=0.14) surfaces even at 410 $^{O}$C oxygen is
still present (Partin, 1981). On the other hand for such concentrations and temperatures the $Pb_{1-x}Sn_xTe$ pressure is already about $10^{-7}$ mbar (Northrop, 1971) and about 50 Å of PbTe
evaporate at 380 $^{O}$C within 10 minutes. Since results obtained
by Grandke and Cardona (1980) using photoelectron spectroscopy
indicate a drastic change of the surface band structure by
oxygen uptake, the oxygen background pressure present during
growth might be of some importance on the values of band discontinuities in the PbTe/$Pb_{1-x}Sn_xTe$ system. Indeed, two groups
have proposed different types of band line-up for this systems:
Kriechbaum et al. (1984) suggested a type I system (straddling
system)whereas Murase et al. (1985) suggested a type I' or
staggered system. According to the first suggestion the electrons and holes are confined within the PbSnTe wells and the
PbTe layers form the barriers. In a type I' model the electrons
are within the PbTe wells, the holes within the PbSnTe wells
(minimum indirect energy gap in real space). We would like to
point out however, that the small band discontinuities can be
altered by rather small energy values of the order of 50 meV
from a type I to a type I' model.

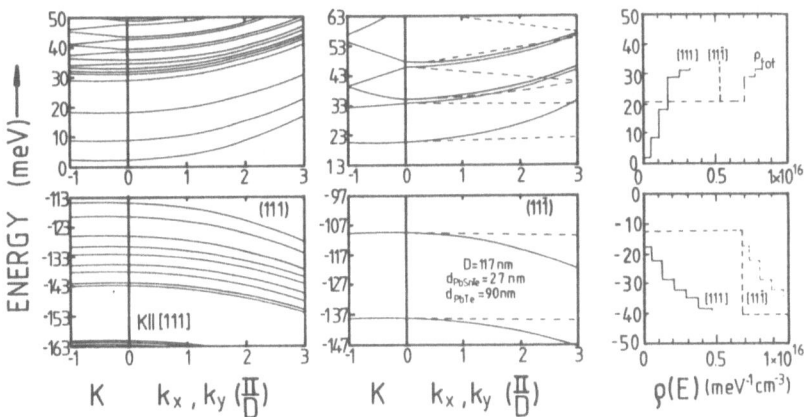

Fig.3: Energy vs $K \| |111]$ and $k_x$, $k_y$ dispersion of PbTe/ $Pb_{1-x}Sn_xTe$ (x=0.135, $d_{PbTe}$=90nm, $d^y_{PbSnTe}$=27nm) for the $|111|$ and <11$\bar{1}$> valleys and corresponding densities of states.

In this paper we first describe the structural charac-terization of the PbTe/$Pb_{1-x}Sn_xTe$ heterostructures. Then follows a section on magnetooptical intraband and interband experiments and the determination of electronic g-factors by a CARS-type experiment. The data are analysed within the framework of an envelope function approximation (EFA) inclu-ding quantizing magnetic fields and effects of strain.

## Structural Properties of the Samples

The PbTe/$Pb_{1-x}Sn_xTe$ samples are grown on (111) cleaved $BaF_2$ surface as described by Clemens et al. (1983): Trans-mission electron microscopic investigations have shown (Pon-gratz et al. 1985) that for buffer thicknesses exceeding 100 nm, the original island growth of the IV-VI films on $BaF_2$ is terminated. The islands are all oriented with their [111] axis parallel to [111] $BaF_2$ and with the inplane directions [1$\bar{1}$O] PbTe || [$\bar{1}\bar{1}$2] $BaF_2$. After the islands grow together a smooth two dimensional growth of the semiconductor film is possible.

The period of all of the MQW's and SL's was checked by high angle X-ray interference measurements (Fantner 1985) and interference peaks up to the 5th order were visible (using CuKα radiation). Apart from this non destructive test some samples were sputtered using Ar$^-$- or O$^+$-ions and an Auger electron spectroscopy or SIMS analysis of the Sn-signal vs sputter time was monitored. The abruptness of their inter-faces as deduced from the sputtering data is about 5 nm. For various reasons no better value can be expected using this technique (Maier 1985).

## Measurement and Effects of Strain on the Bandstructure

Due to the differences in the lattice constants between PbTe and $Pb_{1-x}Sn_xTe$ at room temperature there exists lateral biaxial strain, tensile for the $Pb_{1-x}Sn_xTe$ layers and com-pressive for the PbTe layers. For individual layer thicknesses up to 50 nm, no misfit dislocations are observed (Pongratz et al.1985) and the lattice mismatch is accomodated by lateral biaxial strain. At low temperatures, the differences in the expansion coefficients between the IV-VI films and the $BaF_2$ substrate cause additional strain, enhancing the tensile strain in the PbSnTe layers and diminishing the compressive one in the PbTe layers to zero or even to negative values. By X-ray diffraction techniques (Fantner et al.1984) these strains were measured quantitatively and independently for the PbSnTe and PbTe layers and typically $\Delta a/a$ amounts up to $3.5 \times 10^{-3}$ (PbSnTe) or $0.8 \times 10^{-3}$ (PbTe) at 20K (Pichler et al.1985).

For the [111] growth direction the strain present in the PbSnTe layer MQW's and SL's lift the degeneracy of the four equivalent L-extrema: the [111] valley oriented parallel to the surface normal is shifted downwards in the conduction as well as valence band. The three obliquely oriented valleys are slightly shifted upwards. If the thickness ratio between the PbSnTe and PbTe layers is 1:1, then at low temperatures the PbTe films are essentially unstrained, for a thickness ratio of 1:3, the PbTe films are also under tensile but much smaller strain. For the latter situation, e.g. the splitting of the <111> valleys in the conduction band of PbSnTe is 13 meV whereas for PbTe it amounts to 3 meV under the assumption that the deformation potentials for both compounds are the same.

In order to calculate these energy shifts in the conduction (c) or valence (v) band in the $\ell$-th valley, the expression

$$\delta E_{c,v}^{\ell} = \sum_{ij} D_{ij}^{\ell,c,v} \, \varepsilon_{ij} \tag{1}$$

is used where $\varepsilon_{ij}$ are the components of the strain tensor in the crystal axis coordinate system. The deformation potential tensor $D_{ij}^{\ell,c,v}$ is given by:

$$D_{ij}^{\ell,c,v} = D_d^{c,v} \cdot \delta_{ij} + D_u^{c,v} \, u_i^{\ell} \, u_j^{\ell} \tag{2}$$

where $D_d^{\ell,c,v}$ and $D_u^{c,v}$ are the dilatational and uniaxial deformation potential constants for the conduction or valence band and the $u^{\ell}$ are the cosines of the angles between the $x_i$ crystal axis and the $\ell$th ellipsoidal major axis. The energy shifts are given by (Kriechbaum et al.1984):

$$\delta E_{c,v}^{111} = (D_d^{c,v} + \frac{1}{3} D_u^{c,v}) \, (2\varepsilon_{11} + \varepsilon_{33}) + \frac{2}{3} D_u^{c,v}(-\varepsilon_{11} + \varepsilon_{33})$$

$$\delta E_{c,v}^{11\overline{1}} = (D_d^{c,v} + \frac{1}{3} D_u^{c,v}) \, (2\varepsilon_{11} + \varepsilon_{33}) - \frac{2}{9} D_u^{c,v}(-\varepsilon_{11} + \varepsilon_{33}) \tag{3}$$

In Fig.4 the resulting level structure is shown for a $PbTe/Pb_{1-x}Sn_xTe$ system with x=0.15. For this calculation the deformation potentials $D_u^{c,v}$ and $D_d^{c,v}$ as given by Kriechbaum et al.(1984) were used. Recently, Singleton et al.(1986) made an extensive study on strain effects and derived slightly different values for $D_u^v$ and $D_u^c$ than those given by Kriechbaum et al. The character of the splitting of the levels however

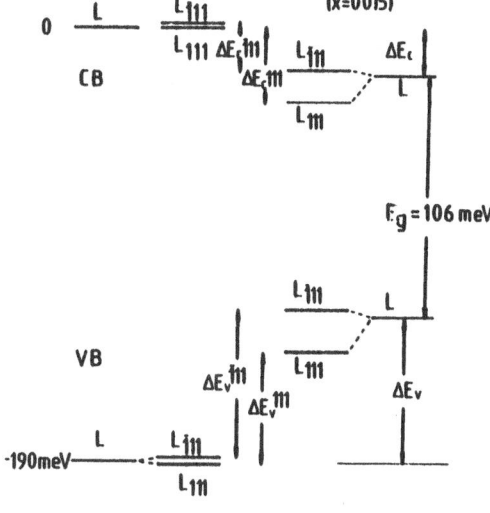

Fig.4: The effect of strain on the
conduction and valence band edges in
$PbTe/Pb_{1-x}Sn_xTe$.

does not change with these deformation potentials and we would
like to point out that the strains measured in the films always will shift the [111] (a) valleys of the $Pb_{1-x}Sn_xTe$ layers
downwards with respect to the corresponding [111] valleys of
PbTe. Thus the strain effects favor a type I SL model for the
a valleys. The corresponding energy shifts of the oblique <111>
(b) valleys are much smaller. However, the PbSnTe valleys are
shifted somewhat more upwards than the corresponding PbTe
ones and thus the barrier height $\Delta E_c^{111}$ will be decreased as
compared to the unstrained situation.

Whereas the energy shifts rely on measurements of the
strain, measurements of elastic constants and deformation
potentials as deduced from measurements on PbTe films, the
values $\Delta E_c$ and $\Delta E_v$ can only be obtained from measurements on
MQW or SL samples. For SnTe, the deformation potential tensor
components $D_u$ = 4.9 eV and $D_d/D_u$ = 1.2 were given by Katayama
and Mills (1980) as obtained from a fit to the temperature
dependence of the mobility of holes. Since even for PbTe the
uncertainty of the $D_u^{c,v}$, $D_d^{c,v}$ values is larger than the
difference between the PbTe and the SnTe values we used for
the pseudobinary compound $Pb_{1-x}Sn_xTe$ the same values as for
PbTe.

As far as the band alignment is concerned, and thus the
value of $\Delta E_c$, two conflicting models have been proposed for
the $PbTe/Pb_{1-x}Sn_xTe$ (x≤0.2) system. Kocevar (1986) has performed an empirical LCAO calculation for PbTe and SnTe. Using
a procedure which had been previously used successfully for
tetrahedral semiconductors (Vogl et al.,1983) the free atomic
energy values had to be scaled for the s and p valence states.
Atomic spin orbit couplings were used as described by Kriechbaum et al.(1984). The energy overlap integrals were considered as free parameters and were fitted to reproduce the experimental optical gaps, the position of secondary band extrema and the results of existing pseudopotential calculations.

The resulting absolute levels of conduction and valence band of both PbTe and SnTe are given by: -4.40, -4.59 eV and -4.20 and -4.53 eV, respectively (as measured from the vacuum level).

The other model (Murase et al.,1985) proposes a type I' band ordering from considerations of the energetic position of the In-level in PbTe and its pseudobinary alloy $Pb_{1-x}Sn_xTe$. For PbTe/SnTe SL's even a type II superlattice results according to Takaoka et al., (1986) i.e. the valence band edge of SnTe lies higher in energy than the conduction band edge of PbTe.

A problem related to the band edge discontinuities are electrostatic potentials at the interface. Usually n-PbTe and $p-Pb_{1-x}Sn_xTe$ layers are grown on each other. E.g. for a sample with $d^x$(PbTe) = 90 nm and d ($Pb_{1-x}Sn_xTe$, x=0.135) = 27 nm, n (PbTe) = $4\times10^{17}cm^{-3}$ ($\varepsilon_F$ = 16 meV) $\varepsilon^x$(PbTe) = 1300, p (PbSnTe) = $6\times10^{17}cm^{-3}$ ($\varepsilon_F$ = 22 meV), $\varepsilon_s$(PbSnTe) = 2500, the depletion lengths are 2240 Å and 3350 Å with potentials of 139 meV and 49 meV, respectively. For individual layer thicknesses below 100 nm, the electrostatic potentials amount to about 1meV and can thus not alter the shape of the potential well which is determined by the growth condition (abruptnes of interface, interdiffusion etc.).

MAGNETOOPTICAL INVESTIGATIONS

Since $Pb_{1-x}Sn_xTe$ with x = 0.135 - 0.175 has an energy gap which lies between 113 and 96 meV, and is thus about a factor of two smaller than that of PbTe, magnetooptical investigations can be useful to find out whether electrons are confined within PbSnTe or PbTe wells. The effective masses and g-factors associated with the narrow gap are quite different for $Pb_{1-x}Sn_xTe$ and PbTe. For this purpose magnetooptical intraband transitions were measured in the far infrared region using various FIR laser energies. These experiments were performed either in the configuration $B_\perp$ , i.e. B|| [111], the growth direction and B||k (Faraday geometry) or in $B_\parallel$ to the layers configuration (usually B|| [$\overline{1}$10] but $\vec{k}$|| [111]).

For an analysis of the data the dielectric function is neccessary. Since for the laser wavelength shown in Fig.5, the corresponding frequency lies between the TO and LO mode frequency the real part of the lattice dielectric function is negative. Cyclotron resonances of carriers within the wells are thus associated with dielectric anomalies. Two resonances, corresponding to carriers in the a and b pockets with circular or elliptic orbits are clearly observed. We would like to point out that the classical cyclotron mass of the oblique valley

resonance $m_c = m_t/3 \sqrt{1+ 8m_\ell/m_t}$ is considerably higher than for a three dimensional case and thus the resonance is shifted towards higher fields. For the analysis a single classical oscillator model has been used with

$$\varepsilon = 1 + \chi_\infty + \chi_{ph} + \chi_{f.c}$$

where $\chi_\infty$ is the high frequency susceptibility, $\chi_{ph}$ the optical phonon oscillator contribution and $\chi_{f.c}$ the free carrier contribution (Pichler et al., 1985).

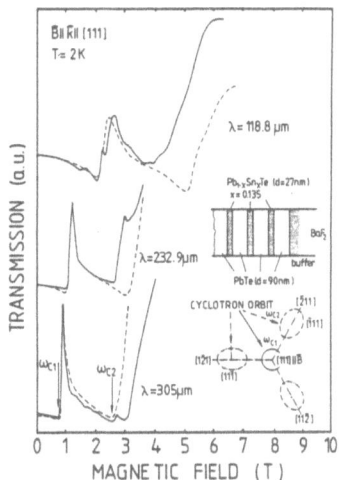

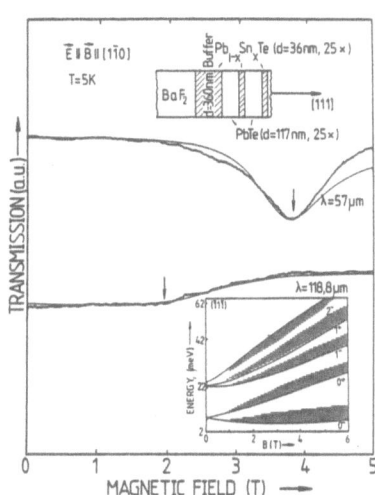

Fig.5:Magnetotransmission vs B, Faradaygeometry;——— exp., ---- classical oscillator fit using 2D orbits (insert),$m_t$= $0.02m_0$, $m_l/m_t$=10;n=$2\times10^{17}$cm$^{-3}$ $\omega_\tau$=3 cm$^{-1}$,phonon oscillator paramaters: Pichler et al., 1985

Fig.6: Magnetotransmission in $B_\parallel$ (Voigt), PbSnTe (x=0.175) parameters for fit.Insert: level energies vs B, hatched area: energy range for cyclotron center coordinate within PbSnTe well. ($\Delta E_c$(11$\bar{1}$)=22meV).

$$\chi^+_{\bar{f}.c} = \frac{-\omega^2_p}{4\omega} \frac{m_0}{m_t} (\omega \mp \omega_{c1} + i\omega_\tau)^{-1}$$

$$- \frac{\omega^2_p}{\omega} \frac{m_0 (\omega+i\omega_\tau) \frac{3}{8}(\frac{1}{m_t}+\frac{9}{m_t+8m_l}) \pm \omega_{c1} \frac{3}{8}(\frac{18}{m_t+8m_l})}{(\omega^2-\omega^2_{c2}-\omega^2_\tau + 2i\omega\omega_\tau)} \qquad (4)$$

where $\omega_{c1} = \frac{e}{m_t} B$ and $\omega_{c2} = \frac{e}{m_c} B$ ; $\omega_p$ denotes the plasma frequency.

For the actual calculation of the transmission interference effects and multiple reflection were taken into account assuming that the PbTe layers are not occupied by carriers but contributing with their lattice dielectric function to the total properties of the sandwich structure. A transfer matrix formalism (Harbecke 1986) was used for this calculation by Pichler (1986). The resonance positions are indicative for carriers being confined in the Pb$_{1-x}$Sn$_x$Te layers.

Experiments in the $B_\parallel$ geometry are shown in Fig.6. For sufficiently high fields in the $\vec{E}\parallel\vec{B}$ configuration a three dimensional like resonance is expected if the magnetic length is already small compared to the well width (d= 36 nm). For decreasing magnetic field the level structure changes its character and for vanishing field finally the Landau levels with different Landau quantum number n end in different electric subbands. The resonance positions result from model oscillator fits (the resonances are oblique valley resonances

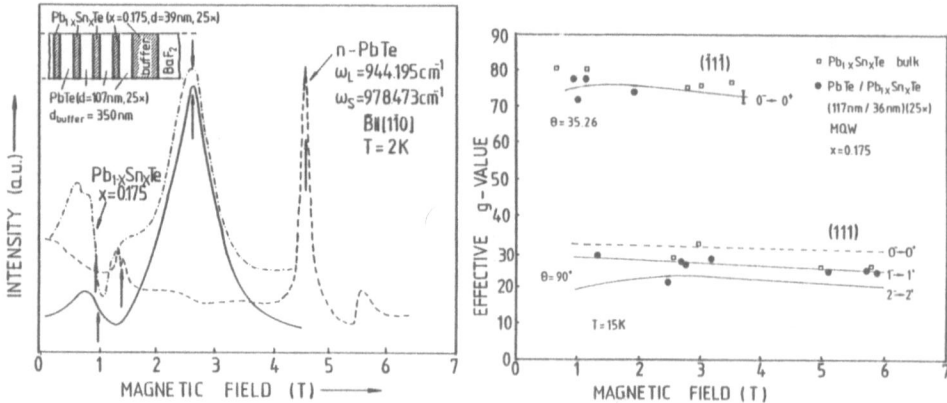

Fig.7a:Four wave mixing intensity as a function of magnetic
field.Full line: PbTe/PbSnTe
MQW structure, dashed line:
n-PbTe bulk sample and dash-
dotted line PbSnTe sample
for comparison. Arrows indicate the resonance positions
in the 35,26° and 90° valleys.

Fig.7b:Analysis of the data
of Fig.7a for various fields.
o: PbTe/PbSnTe MQW structure,
□: PbSnTe bulk sample. Broken
and full lines indicate spin
flip transition energies
according to EFA calculation
for center coordinate within
center of PbSnTe wells.

in the two valleys oriented by 35,36° to the magnetic field)
and a transverse cyclotron mass $m_t$=0.018 $m_o$ assuming K=$m_1$/$m_t$=10
is obtained. This value is again indicative for electrons
being confined in PbSnTe wells. The electronic g-factors were
measured directly in PbTe/Pb$_{1-x}$Sn$_x$Te (x=0.175) MQW's by Pascher
et al., (1986) using a coherent anti-Stokes Raman scattering
type of experiment. In this third order non linear susceptibility experiment two laser beams with $\omega_1$ and $\omega_2$ are superimposed in the MQW structure. The intensity of the mixing fre-
quency $2\omega_2-\omega_1$ has the same type of resonances as $\omega_1+\Delta\omega=2\omega_1-\omega_2$.
It resonates whenever $\hbar(\omega_1-\omega_2)$ is equal to the energy of a
Raman allowed transition in the scattering medium. In the B$_{||}$
configuration with $\vec{B}||\vec{E}$, $\vec{k}||$ [111], $\vec{B}||[1\bar{1}0]$, the spin flip re-
sonance g* $\mu_B$×B=$\hbar(\omega_1-\omega_2)$ is the strongest one. In Fig.7 a comparison between the resonances of an n-PbTe sample, an n-
Pb$_{1-x}$Sn$_x$Te sample and an n-PbTe/Pb$_{1-x}$Sn$_x$Te MQW sample with the
same composition x(=0.175) is shown. The transition at lower
fields is due to the valleys oriented by 35,26° with respect
to B, the one at higher fields to the carrier pockets oriented
by 90° with respect to B. This direct comparison shows that
the g-factors of the MQW sample are close to the corresponding
ones of Pb$_{1-x}$Sn$_x$Te of similar composition. Since in this geometry, in contrast to cyclotron resonance transitions the
spin flip transitions do not depend on the position of the
center of the cyclotron orbit with respect to the center coordinate of the wells, we consider these experiments as a
definite proof for a type I SL model for the structures under
investigation. Results of experimental data on interband magnetooptical transitions are shown in Fig.8. For a sample with
d(PbTe)=43 nm and d(PbSnTe)=47 nm, the extrapolations of the
transition within the [111] Landau ladders associated to several subbands within the Pb$_{1-x}$Sn$_x$Te conduction and valence band
wells yield information on subband energies. In the inset, these

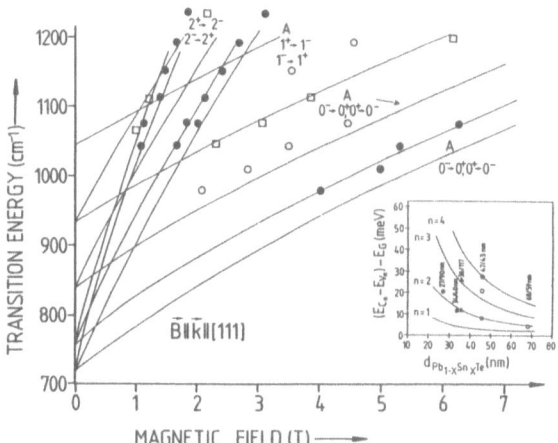

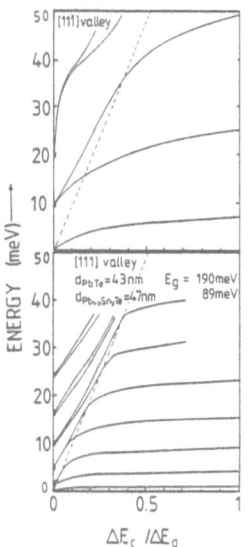

Fig.8a:Magnetooptical interband transition energies of PbTe/PbSnTe (x=0.18) in Faraday geometry.Solid lines:EFA calculation.Insert: Energy difference of mirror electric subbands vs PbSnTe well width;symbols:exp.data,——— EFA calculation.

Fig.8b:Energy of electric subbands vs conduction band offset normalized to $\Delta E_g$ for the two types of valleys.As in Fig. 8a:$d_{PbTe}$=43,$d_{PbSnTe}$=47nm.

energy differences of mirror electric subbands in the conduction and valence band for several samples with different well width to barrier width ratio are shown, too. The full curves represent calculations based on an envelope function treatment of the MQW structures. The influence of the conduction band offset $\Delta E_c$ on the energetic position of the various subbands in the two types of valleys is also shown.

## ANALYSIS IN TERMS OF ENVELOPE FUNCTION APPROXIMATION

In a semiconductor the properties of the electrons and their response to certain perturbations of the periodic lattice can readily be described by the envelope function approximation (EFA). The key assumption in deriving the EFA is the expansion of the Schrödinger function for the electron

$$\psi = \sum_b u_b(r)f_b(r) \qquad (5)$$

in a sum over all bands of products of Blochfunctions $u_b$ and envelope functions $f_b$. By assuming the perturbation $V_p$ of the crystal potential $V_c$ to be sufficiently slowly varying on the scale of the lattice periodicity the Schrödinger equation

$$H\psi = \{\frac{p^2}{2m} + U(r) + \frac{h^2}{4m^2c^2}[\nabla U \times p.\sigma]\}\psi = E\psi \qquad (6)$$

with $U = V_c + V_p$ can be transformed to a set of coupled equations for the $f_b$ alone. $V_c$ and the $u_b$ are hidden in matrix elements, which are treated as material parameters, like the differences of band energies, interband momentum matrix elements and deformation potentials. The number of coupled equations is determined by the number of close lying bands. In practice few bands are treated exactly and the far bands in perturbation theory. A magnetic field is taken into account by replacing p by p-eA. Although in general EFA is certainly not applicable to semiconductor heterojunctions, due to the rapid change of the crystal potential in the two layers, it may lead to reliable results for a heterojunction of very similar materials (Bastard, 1982, Altarelli, 1984). In the sense of Schrödinger perturbation theory one assumes that the Bloch functions $u_b$ are the same in both layers. The change in crystal potential causes then only a change in the material parameters across the interace. At an interface the solutions of the EFA equation Hf = Ef in both layers have to be joined. Due to the continuity of the Schrödinger function $\psi = \Sigma$ uf and due to the orthogonality of u every envelope function $f_b$ has to be continuous at the boundary

$$f_b(z^-) = f_b(z^+) \quad . \tag{7}$$

Although the equation Hf = Ef is a second order differential equation, no matching conditions are required for the derivatives of f (Kriechbaum, 1986). For lead salt superlattices, the conduction and valence band with symmetries $L_6^+$ and $L_6^-$ are treated exactly and all other bands in perturbation. The approximate 4x4 EFA Hamiltonian reads (Bauer, 1980)

$$H = \begin{bmatrix} h_{cc} & h_{cv} \\ h_{vc} & h_{vv} \end{bmatrix} \tag{8}$$

with

$$h_{cc} = E_c + A_c k_\perp^2 + B_c k_3^2 + D_d^c tr\varepsilon + D_u^c \varepsilon_{33} + \frac{1}{2} g_\ell^c \mu_B B_3 \sigma_3 + \frac{1}{2} g_t^c \mu_B (B_1 \sigma_1 + B_2 \sigma_2)$$

$$h_{vc} = h_{cv} = P_\perp (\sigma_1 k_1 + \sigma_2 k_2) + P_\| \sigma_3 k_3 \quad . \tag{9}$$

$h_{vv}$ is the same as $h_{cc}$ with the material parameters appropriate for the valence band. $P_\|$ and $P_\perp$ are the interband momentum matrix elements, $A_c, B_c, A_v, B_v$ the far band contributions to the mass and $g_\ell$, $g_t$ to the Landé factors. The indices 1,2,3 denote a Cartesian coordinate axis system with 3 being the main valley axis and $\sigma_i$ are the Pauli matrices. At a continuous compositional change these parameters are position dependent and Eq.(8) have to be replaced by their symmetrized forms, i.e. $Ak_\perp^2 \to k_\perp A k_\perp$ and $Pk \to (Pk+kP)/2$. In addition to the effective mass parameters the band set up has to be specified. The conduction and valence band edges are due to negligible band bending flat in each layer and the energy dispersion for the superlattice may be obtained by joining oscillating and evanescent plane wave states. In addition to the continuity condition (7) there holds a superlattice Bloch condition

$$f_b(z+D) = \exp(iKD) f_b(z) \qquad b = 1,2,3,4 \tag{10}$$

with $-\pi/D < K \leq \pi/D$ and D the superlattice periodicity. Eqs.(7) and (10) form eight conditions for the amplitudes of four degenerate plane wave states (two values $k_z$, spin up and down) to a given energy in each layer. The energy dispersion $E(K,k_x,k_y)$ with arbitrary in plane momenta $k_x,k_y$ may thus be calculated by

126

the zeroes of an 8x8 determinant (Kriechbaum et al., 1984).

For a magnetic field parallel to the SL growth direction, z, the solutions are still characterized by a SL Bloch vector because the magnetic field does not destroy the periodicity of the problem. If the magnetic field is not parallel to the main valley axis (cf. Fig. 1) coordinate transformations are required to obtain a form of the EFA Hamiltonian manageable for calculation. The new coordinates

$$
\begin{bmatrix} \rho_1 \\ \rho_2 \\ \rho_3 \end{bmatrix} = \begin{bmatrix} 1/P_{\parallel} & 0 & 0 \\ 0 & 1/w & 0 \\ 0 & \dfrac{sc}{w}\dfrac{P_{\parallel}^2 - P_{\perp}^2}{P_{\parallel} P_{\perp}} & \dfrac{w}{P_{\parallel} P_{\perp}} \end{bmatrix} \begin{bmatrix} x \\ y \\ z \end{bmatrix} \tag{11}
$$

and momenta

$$
\begin{bmatrix} \lambda_1 \\ \lambda_2 \\ \lambda_3 \end{bmatrix} = \begin{bmatrix} P_{\perp} & 0 & 0 \\ 0 & w & \dfrac{sc}{w}(P_{\perp}^2 - P_{\parallel}^2) \\ 0 & 0 & \dfrac{P_{\parallel} P_{\perp}}{w} \end{bmatrix} \begin{bmatrix} k_x \\ k_y \\ k_z \end{bmatrix} \tag{12}
$$

with $w = \sqrt{c^2 P_{\perp}^2 + s^2 P_{\parallel}^2}$, $c = \cos(3,z)$, $s = \sin(3,z)$ obey the commutation relations

$$
\left[\lambda_i, \rho_j\right] = -i\delta_{ij}, \qquad \left[\lambda_1, \lambda_2\right] = wP_{\perp}/i\ell^2 . \tag{13}
$$

$\ell$ is the cyclotron orbit radius ($\ell = 25.656/\sqrt{B[T]}\,nm$). By rotating the spin quantization axis by the transformation

$$
\sigma \rightarrow \sigma' = \begin{bmatrix} 1 & 0 & 0 \\ 0 & cP_{\perp}/w & sP_{\parallel}/w \\ 0 & -sP_{\parallel}/w & cP_{\perp}/w \end{bmatrix} \begin{bmatrix} \sigma_1 \\ \sigma_2 \\ \sigma_3 \end{bmatrix} \tag{14}
$$

and having $\sigma_3$ diagonal and introducing the Landau ladder operators $b, b^+$ with $[b, b^+] = 1$

$$
b = (\lambda_1 - i\lambda_2)/\sqrt{2wP_{\perp}} \tag{15}
$$

the Hamiltonian reads

$$
h_{cc} = E_c + A_c'(2b^+ b + 1) + B_c' \lambda_3^2
$$

$$
h_{vc} = h_{cv} = \begin{bmatrix} \lambda_3 & & b\sqrt{2wP_{\perp}}/\ell \\ & & \\ b^+\sqrt{2wP_{\perp}}/\ell & & -\lambda_3 \end{bmatrix} \tag{16}
$$

with

$$A' = A(wP_\perp + c^2 P_\perp^3/w)/2\ell^2 P_\perp^2 + Bs^2 P_\parallel^2 P_\perp /2\ell^2 P_\parallel^2 w$$
$$B' = As^2 P_\parallel^2/P_\perp^2 w + Bc^2 P_\perp^2/P_\parallel^2 w \quad . \tag{17}$$

In the diagonal elements terms proportional to b and $b^2$ have been neglected as also the higher band contributions to the g-factor. Note that for B parallel to the valley axis Eq.16 is exact. The solutions to the Hamiltonian (16) for each layer for flat band edges are given by

$$f = \begin{bmatrix} \psi_{n-1} \\ \psi_n \\ \psi_{n-1} \\ \psi_n \end{bmatrix} \exp(i\bar{\lambda}_3 \rho_3) \tag{18}$$

Here $\psi_n = 1/\sqrt{n!}(b^+)^n \psi_0(\rho_1 \rho_2)$ with $b\psi_0 = 0$ and $\bar{\lambda}$ a quantum number (n Landauquantum number). By choosing a proper gauge we may write the Landaufunctions as

$$\psi_n = \phi_n(\rho_1 - \lambda_2 \ell^2/wP_\perp) \exp(i\bar{\lambda}_2 \rho_2) \exp(i\bar{\lambda}_3 \rho_3) \tag{19}$$

$\phi_n$ is an harmonic oscillator function. Inserting back for the coordinates this is

$$\psi_n = \phi_n(x/P_\perp - \bar{\lambda}_2 \ell^2/wP_\perp) \exp(i\bar{\lambda}_2 y/w) \exp(i\bar{\lambda}_3 [sc(P_\parallel^2 - P_\perp^2)y + wz]/P_\parallel P_\perp). \tag{20}$$

In order to join continuously the solutions at the interface we must have the wavenumber

$$\bar{\Lambda} = \bar{\lambda}_2 + \bar{\lambda}_3 \, sc(P_\parallel^2 - P_\perp^2)/P_\parallel P_\perp \tag{21}$$

of the plane wave in y direction to be the same in both layers. This, however, implies that the oscillator functions are centered differently in both layers at

$$\bar{\Lambda} - \bar{\lambda}_3 sc(P_\parallel^2 - P_\perp^2)/(P_\parallel P_\perp)\ell^2/wP_\perp \quad \text{with} \quad \bar{\lambda}_3 = \bar{\lambda}_3^A \text{ or } \bar{\lambda}_3^B . \tag{22}$$

This difference must be small compared to the range of the oscillator function yielding a lower bound for the magnetic field:

$$\ell \ll \frac{1}{\Delta\bar{\lambda}_3}\sqrt{n+1/2} \; P_\parallel P_\perp\sqrt{wP_\perp}/sc(P_\parallel^2 - P_\perp^2) = \frac{1}{\Delta k_z}\sqrt{n+1/2}\,w^{3/2} P_\perp/sc(P_\parallel^2 - P_\perp^2) \quad . \tag{23}$$

For growth direction $[111]$ and valley axis $[11\bar{1}]$ one obtains $\ell \ll 15/\Delta k_z$ for n = 0. The eigenenergies E(n,K) may again be calculated by the zeroes of an 8x8 determinant thus yielding the amplitudes of the 4 possible states Eq.(18) to a given energy E in each layer. The oscillator strength for electric dipole transitions are calculated from the operators obtained after rotating the spin system (the operators are understood to be repeated once along the skew diagonal):

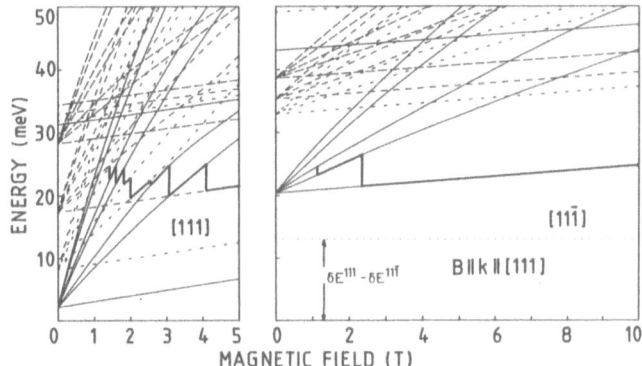

Fig.9: Calculated Landau levels vs B for $[111]$ and $<\bar{1}11>$ valleys for PbTe/PbSnTe (x = 0.135, $d_{PbTe}$ = 90 nm, $d_{PbSnTe}$ = 27 nm). Shift of CB zero due to strain effects. Position of Fermi energy as a function of B is indicated by the full line. Landau levels, $|n,+>$ and $|n+1,->$ repel each other.

$$j_x = \begin{bmatrix} 0 & P_\perp \\ P_\perp & 0 \end{bmatrix} \quad j_y = \begin{bmatrix} 0 & -iw \\ iw & 0 \end{bmatrix}$$

$$j_z = \begin{bmatrix} P_\| P_\perp/w & -ics(P_\perp^2 - P_\|^2)/w \\ ics(P_\perp^2 - P_\|^2)/w & -P_\| P_\perp/w \end{bmatrix}$$

(24)

Fig.9 shows fan charts for a $[111]$ and a $[11\bar{1}]$ valley in a $[111]$ SL. It is mainly an overlay of a bulk fan chart on every electric subband. Note however the peculiar anticrossing behaviour of the magnetic states (Eq.18) with the same n originating from different electric subbands. The selection rules for dipole transitions follow closely the bulk selection rules within Landau states originating from the same electric subband whereby the noncrossing behaviour is ignored. Only at the noncrossing points are the selection rules disturbed. Transitions between states originating from different electric subbands are forbidden.

For a valley with a main axis coinciding with B the above treatment is exact. For an oblique valley it is only correct for sufficiently high magnetic fields. For magnetic fields so small that the electric level spacing is larger than the magnetic one the cyclotron transition energies are for the lowest electric subband according to a one-band calculation for inversion layers by Stern and Howard (1967) governed by a "two dimensional" mass

$$\frac{1}{m_{2D}} = \sqrt{\frac{1}{m_t}} \sqrt{\frac{1}{m_\ell s^2 + m_t c^2}}$$

(25)

For high fields the cyclotron transition energy is proportional to the "three dimensional" mass (cf. Eq. 13)

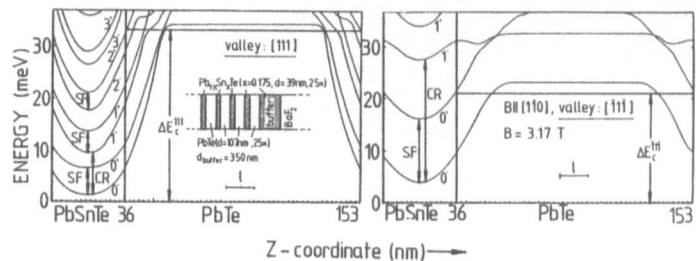

Fig.10: Landau levels for $B_\parallel$-geometry: energy vs cyclotron center coordinate for the $[\bar{1}11]$ (right) and $\langle 111\rangle$ valley (left). $\Delta E_c$ denote the conduction band offsets, SF and CR spin-flip and cyclotron resonance transitions, $\ell$ the cyclotron length.

$$\frac{1}{m_{3D}} = \sqrt{\frac{c^2}{m_t} + \frac{s^2}{m_\ell}} \ \sqrt{\frac{1}{m_t}} \ \sim \frac{2}{E_g} \ wP_\perp \qquad . \qquad (26)$$

This behaviour is clearly seen in Fig. 9b. If the band edge energies are not perfectly flat, either due to a considerable charge transfer or due to an inhomogeneous Sn concentration, there are no plane wave solutions along the growth direction. The wavefunction has then to be determined by solving the four coupled differential equations in z. The method of finite elements as also difference methods turned out to be quite easily manageable, as they lead to a hermitean eigenvalue problem.

For the $B_\parallel$ case a coordinate system is chosen with the x-axis parallel to B and z again parallel to the growth direction. Assuming a gauge $A = (0,-Bz,0)$ the momentum operators $k_i$ in the Hamiltonian (8) are given with the help of the coordinate rotation matrix T by

$$\begin{bmatrix} k_1 \\ k_2 \\ k_3 \end{bmatrix} = T \begin{bmatrix} -i\partial/\partial x \\ -i\partial/\partial y + z/\ell^2 \\ -i\partial/\partial z \end{bmatrix} \qquad (27)$$

As the parameters in the Hamiltonian (8) are constant along directions y and z the solution to Hf = Ef is found in the form

$$f_b = g_b(z-z_o)\exp(ik_y y)\exp(ik_z z) \qquad (28)$$

with the center coordinate $z_o = k_y \ell^2$. The $g_b$ have to be determined by coupled differential equations where the magnetic potential is superposed to the z-dependence of the material parameters. As seen from Fig.10 the degeneracy with respect to the center coordinate $z_o$ is lifted as it sweeps across the superlattice. For high magnetic fields the calculation is done most efficiently by expanding every $g_b$ in a set of harmonic oscillator functions and determining their amplitudes by a Hermitean eigenvalue problem. For lower fields again the methods of finite elements or finite differences are the easiest. Fig.6 shows results. It is seen that for high fields ($\ell \ll d_{PbSnTe}$) the fan chart is like the bulk fan chart for PbSnTe. This is

## Table 1. Material Parameters

| | $P_\parallel$ $P_\perp$ meVnm | | $A_C$ | $B_C$ meV nm$^2$ | $A_{2V}$ | $B_V$ | $g^C_l$ | $g^C_t$ | $g^V_l$ | $g^V_t$ |
|---|---|---|---|---|---|---|---|---|---|---|
| PbTe [a] | 141 | 486 | 544 | 70.7 | -370 | -26.4 | 1.3 | -2.0 | -0.3 | 5.5 |
| PbSnTe [b] | 137 | 464 | 605 | 92.9 | -428 | -23.1 | -2.4 | -1.5 | 0.2 | 0 |

[a]Singleton et al. (1986)
[b]Bauer (1980)

of course expected, as the magnetic confinement prevents the electron to "see" the superlattice.

DISCUSSION

The EFA method has been used successfully to analyse optical and magnetooptical properties of III-V compounds (Bastard 1982; Altarelli 1984). Quantizing magnetic fields could be included and for the $B_\parallel$ case Maan (1984) has derived the dispersion of the Landau states. Kriechbaum (1984, 1986) has extended the EFA method to the case of ellipsoidal surfaces of constant energy in a many valley band structure.

In this paper we have summarized several magnetooptical investigations which lend support to our previous claim that PbTe/PbSnTe (x $\lesssim$ 0.18) MQW and SL structures are of type I, i.e. electrons being confined within the PbSnTe wells. The conduction band offset is of the order of 40 meV and thus rather small dipole layer effects at the interface will influence considerably this value (Pascher et al. 1986).

In the $B_\perp$ geometry ($B \parallel [11\bar{1}]$), magnetooptical transitions within the oblique valleys pose a difficult problem. For high fields, the fan-charts as given in Fig.9b apply, which are, apart from a noncrossing behaviour essentially bulk like, i.e. "3D". For small fields, when the cyclotron energy becomes smaller than subband spacings, this fan chart is no longer valid and a treatment according to Stern and Howard (1967) should be used instead. This effect is experimentally observed in Fig.5 where the 2D cyclotron mass is observed just for small fields (B<< 3T). The transition from a "2D" dominated behaviour to a "3D" one shifts to higher magnetic fields for smaller well widths. Schaber and Doezema (1979) gave B < 1T for this limit, however for a considerably larger binding length (z $\sim$ 50 nm) than found in our samples.

In $B_\parallel$ geometry, the energy gap of the material which forms the well can be obtained from an extrapolation of the interband magnetooptical transitions at high fields (magnetic length $\ell$ < $d_{well}$) towards zero, considering the nonparabolicity of the three dimensional bulk fan chart. As far as intraband transitions are concerned, Fig.10 shows that the spin-flip transition energies do not depend on the center coordinate (within the well) in contrast to cyclotron like transitions. The width of the latter is indicated in the inset of Fig.6. For small fields therefore a rather broad band of cyclotron transitions is possible whereas for fields with $\ell$ << $d_{well}$ the transition energies are the same for a broad range of center coordinates.

Since in IV-VI compounds the control of the growth conditions and the control of the interace has not yet reached the status of III-V compounds it is not astonishing that the magnitude of the band offsets is not yet definitively settled (Ishida et al. 1985, Valenko et al. 1986, Pascher et al. 1986).

## ACKNOWLEDGMENTS

We thank P.Pichler, H.Clemens, H.Pascher, M.von Ortenberg and P.Kocevar for helpful discussions. Work supported by Fonds zur Foerderung der wissenschaftlichen Forschung (P5321), Vienna, Austria.

## REFERENCES

M.Altarelli, 1982, Lecture Notes in Physics 177, 174.
G.Bastard, 1982, Phys.Rev.B25, 7584.
G.Bauer, 1980, Lecture Notes in Physics, 133, 427.
G.Bauer, 1986, Surface Science, 168, 462.
H.Clemens, E.J.Fantner and G.Bauer, 1983, Rev.Sci.Instr.
    54, 685
E.J.Fantner, H.Clemens and G.Bauer, 1984, Advances in X-ray
    analysis, 27, 171.
E.J.Fantner, 1985, Appl.Phys.Lett., 47, 803.
W.Goltsos, J. Nakahara, A.V.Nurmikko and D.L.Partin, 1985,
    Appl.Phys.Lett., 46, 1173.
Th.Grandke and M.Cardona, 1980, Surface Science, 92, 385.
B.Harbecke, 1986, Appl.Phys. B39, 165.
A.Ishida and H.Fujiyasu 1985, Jap.J.Appl.Phys. 24, L956.
A.Ishida, M.Aoki and H.Fujiyasu, 1985, J.Appl.Phys. 58, 1901.
A.Ishida, H.Fujiyasu, H.Ebe, K.Shinohara, 1986, J.Appl.Phys.
    59, 3023.
S.Katayama and D.L.Mills, 1980, Phys.Rev. B22, 336.
P.Kocevar, 1986, to be published, see also Kriechbaum et
    al. 1984.
M.Kriechbaum, K.E.Ambrosch, E.J.Fantner, H.Clemens and
    G.Bauer, 1984, Phys.Rev. B30, 3394.
M.Kriechbaum, 1986, Springer Series in Solid State Sciences,
    67, eds.G.Bauer, F.Kuchar, H.Heinrich, p.120.
J.C.Maan, 1984, Springer Series in Solid State Sciences,53,183
M.Maier, 1985, private communication.
K.Murase, S.Shimomura, S.Takaoka, A.Ishida and H.Fujiyasu,
    1985, Superlattices and Microstructures, 1, 177.
G.Nimtz and B.Schlicht, 1983, Springer Tracts in Modern
    Physics, 98, p.1.
P.M.Northrop, 1971, J.Electrochem.Soc., Solid State Sci. 118,
    1365.
D.L.Partin, 1981, J.Electronic Materials, 10, 313.
D.L.Partin, 1984, Appl.Phys.Lett., 45, 487.
D.L.Partin, 1985, Superlattices and Microstructures, 1, 131.
D.L.Partin, 1985a, Optical Engineering 24, 367.
H.Pascher, G.Bauer and H.Clemens, 1985, Solid State Commun.
    55, 765
H.Pascher, P.Pichler, G.Bauer, H.Clemens, E.J.Fantner,
    M.Kriechbaum, 1986, Surface Science 170, 657.
H.Preier, 1979, Appl.Phys., 20, 189.
P.Pichler, 1986, unpublished.

P.Pichler, E.J.Fantner, G.Bauer, H.Clemens, H.Pascher, M.von
　　Ortenberg and M.Kriechbaum, 1985, Superlattices and
　　Microstructures, $\underline{1}$, 1.
P.Pongratz, H.Clemens, E.J.Fantner and G.Bauer, 1985, Inst.
　　Phys.Conf.Series No. $\underline{76}$, Section 7, 313.
H.Schaber and R.Doezema, 1979, Phys.Rev., $\underline{B20}$, 5257.
J.Singleton, E.Kress-Rogers, A.V.Lewis, R.J.Nicholas, E.J.
　　Fantner, G.Bauer and A.Lopez-Otero, 1986, J.Phys.C $\underline{19}$,77.
F.Stern and W.E.Howard, 1967, Phys.Rev., $\underline{163}$, 816.
S.Takaoka, T.Okomura, K.Murase, A.Ishida and H.Fujiyasu, 1986,
　　Solid State Commun. $\underline{58}$, 637.
M.V. Valenko, I.I.Zasavitskii, A.V.Matchenko, B.N.Mashonasvili,
　　1986, J.Exp.Theor.Phys. (in Russian), $\underline{43}$, (140), 1940.
P.Vogl, H.P.Hjalmarson, and J.D.Dow, 1983, J.Phys.Chem.Solids,
　　$\underline{44}$, 365.

QUANTUM WELLS AND SUPERLATTICES OF

DILUTED MAGNETIC SEMICONDUCTORS

J. K. Furdyna, J. Kossut* and A. K. Ramdas

Department of Physics
Purdue University
West Lafayette, IN 47907

## I.  INTRODUCTION

Diluted magnetic semiconductors (DMS) are semiconducting alloys containing substitutional transition metal ions.[1]  In this paper we shall concentrate on the most thoroughly understood group of these materials, i.e., on II-VI compounds containing substitutional $Mn^{2+}$.  Of those, $Hg_{1-x}Mn_xTe$ and $Hg_{1-x}Mn_xSe$ are narrow gap semiconductors, with the band structure analogous to that of $Hg_{1-x}Cd_xTe$, whereas $Cd_{1-x}Mn_xTe$ and $Zn_{1-x}Mn_xSe$ are examples of wide-gap DMS.  In Table 1. we give a list of all the $A_{1-x}^{II} Mn_x B^{VI}$ DMS materials, together with their crystal structures and ranges of compositions within which growth of single phase alloys has been successful.

One of the attractive features of the $A_{1-x}^{II} Mn_x B^{VI}$ family is the tunability of the energy gap by variation of the crystal composition (see Fig. 1). Also, the fact that the lattice parameters depend on the Mn molar fraction x makes DMS of interest in the area of heterostructures and superlattices, where lattice matching is an extremely important factor.  Furthermore, the presence of magnetic ions in DMS leads to an exchange interaction between band electrons and localized magnetic moments which, in turn, results in profound modification of various aspects of semiconductor band structure in the presence of an external magnetic field.  The effect of this interaction (the sp-d interaction) has been clearly demonstrated theoretically as well as in experimental studies of magnetooptical[2,3] and electric transport[4] properties, and has been largely responsible for the extensive interest generated by DMS materials in the scientific community.

The most important consequence of the sp-d interaction is a strong enhancement of the spin splitting of the electronic levels observed in DMS materials.  The origin of this "amplification" of spin properties can be best illustrated by the expression describing the electronic g-factor in the parabolic approximation (particularly suitable for the conduction electrons in wide gap DMS), where we can write

---

*On leave from The Institute of Physics, Polish Academy of Sciences, Warsaw, Poland.

Table I. $A^{II}_{1-x}Mn_xB^{VI}$ diluted magnetic semiconductors.

| Material | Crystal structure | Range of composition |
|---|---|---|
| $Zn_{1-x}Mn_xS$ | zinc blende | $0 < x \lesssim 0.10$ |
| | wurtzite | $0.10 < x \lesssim 0.45$ |
| $Zn_{1-x}Mn_xSe$ | zinc blende | $0 \cdot < x \lesssim 0.30$ |
| | wurtzite | $0.30 < x \lesssim 0.57$ |
| $Zn_{1-x}Mn_xTe$ | zinc blende | $0 < x \lesssim 0.86$ |
| $Cd_{1-x}Mn_xS$ | wurtzite | $0 < x \lesssim 0.45$ |
| $Cd_{1-x}Mn_xSe$ | wurtzite | $0 < x \lesssim 0.50$ |
| $Cd_{1-x}Mn_xTe$ | zinc blende | $0 < x \lesssim 0.77$ |
| $Hg_{1-x}Mn_xSe$ | zinc blende | $0 < x \lesssim 0.38$ |
| $Hg_{1-x}Mn_xTe$ | zinc blende | $0 < x \lesssim 0.75$ |

$$g_{eff} = g^* - \frac{\alpha M}{g_{Mn}\mu_B^2 B} = g^* - \frac{\alpha \chi}{g_{Mn}\mu_B^2} \quad . \tag{1}$$

Here $g^*$ is the g-factor determined solely by the band parameters, $\alpha$ is the sp-d exchange constant for the conduction band, M is the magnetization, $\mu_B$ is the Bohr magneton, $g_{Mn}$ is the g-factor of $Mn^{2+}$ ions, B is the external

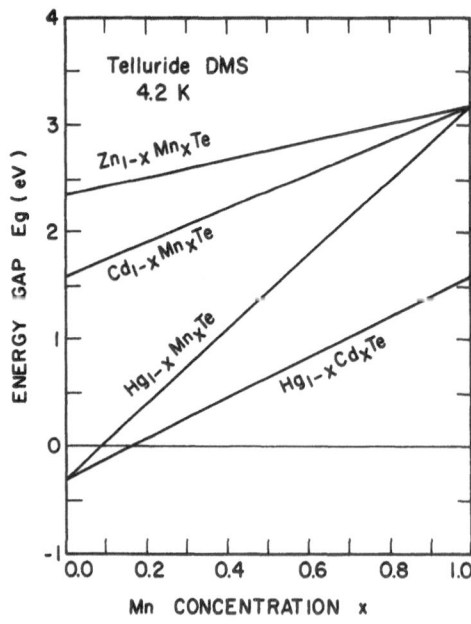

Fig. 1. Energy gap $E_g$ vs. Mn concentration x for telluride DMS at low temperatures, obtained by linear approximation. The behavior of $Hg_{1-x}Cd_xTe$ is shown for comparison.

magnetic field, and $\chi$ is the magnetic susceptibility. At low temperatures and for $x \gtrsim 0.02$, the magnitude of $g^*$ is typically smaller than that of the exchange-induced term in Eq. (1), even in narrow-gap DMS where $g^*$ itself can be quite sizable. Since the magnetic susceptibility, which appears explicitly in Eq. (1), is a function of both temperature and magnetic field, the effective g-factor--and the resulting spin splitting--may vary over a broad range depending on the values of these quantities, thus rendering the band structure of DMS much more sensitive to B and T than in ordinary (i.e., non-magnetic) semiconductors. The enhancement of the g-factor due to sp-d interaction, being directly proportional to the magnetization, is most pronounced in moderately concentrated DMS ($0.05 < x < 0.15$), because anti-ferromagnetic interactions between Mn ions tend to reduce the value of M at a given temperature in samples with higher values of x.

The magnetic properties of DMS themselves are determined by the Mn-Mn exchange interaction. For low values of x DMS behave as paramagnetic substances, while in more concentrated samples ($x \gtrsim 0.2$) a spin glass phase is observed at low temperatures,[5] and in the high-x limit ($x > 0.60$) there is evidence of long-range antiferromagnetic ordering.[6] Thus DMS constitute a rather unique family of materials where a whole spectrum of magnetic behavior can be encountered by varying the concentration of Mn ions.

In this review we concentrate on the properties of the two-dimensional (2D) electron gas in DMS. The paper is divided into two parts. First we describe those 2D DMS systems which have already been prepared and studied. These include both MIS structures on the surfaces of DMS crystals and MBE-grown quantum well and superlattice systems. Some of the most important experimental results found in these 2D DMS samples are also briefly reviewed. We then examine in some detail the novel electronic properties which are quite feasible in the 2D DMS systems, but which remain at present in the realm of speculation. Since the magnetic behavior plays a crucial role in determining the electronic properties of DMS, we finally consider--although very briefly--the possible modifications of the magnetic structure due to the two-dimensional character of the samples, and their consequences.


II. TWO-DIMENSIONAL DMS SYSTEMS

In this part we describe those quasi-2D electronic systems which have already been successfully prepared and, at least to some extent, studied experimentally. In the presentation we shall emphasize those features which are novel and/or unique to DMS. It should be mentioned at this point that the investigation of 2D DMS is still at a preliminary stage, although activity in this area is rapidly increasing.

## Accumulation and Inversion Layers in p-$Hg_{1-x}Mn_xTe$

Historically, the first 2D electronic systems in DMS involved MIS structures prepared on the surface of p-type $Hg_{1-x}Mn_xTe$, with x in the range 0.10 - 0.13, corresponding to an energy gap of the bulk material between ~0.08 to ~0.2 eV. This choice of crystal compositions was motivated by the requirement that the energy gap must be sufficiently large to limit the degree of tunneling of the conduction electrons from the surface layer to the valence band states in the bulk of the semiconductor. The oscillations of the electrical conductivity in the electron inversion layer, observed in the presence of an external magnetic field as a function of the gate voltage across the MIS structure,[7] showed a pronounced dependence on the temperature. This provided direct evidence that the strong temperature dependence of the electron energy levels characteristic for bulk DMS, is also present in 2D DMS systems.

Magnetooptical properties of both electron-inversion and hole-accumulation layers were also studied.[8] The effective mass of inversion electrons, determined from the position of cyclotron resonance, was found to depend on their density in a slightly more pronounced way than theoretically expected. On the other hand, similar calculations gave the correct populations of the surface subbands.

A very strong magnetic field dependence of the high frequency conductivity has been seen in accumulation layers of p-$Hg_{1-x}Mn_xTe$.[8] This can be--at least qualitatively--accounted for by the sp-d exchange-interaction-induced (therefore, magnetic-field-dependent) modifications of the hole effective masses,[9] thus constituting a unique feature of 2D DMS systems.

An interesting example of a 2D DMS system is the 2D electron gas confined to the vicinity of the boundary between two mono-crystalline grains often found in p-$Hg_{1-x}Mn_xTe$ samples grown by the Bridgman method.[10] The mobilities of those electrons were found to be higher than those exhibited by the inversion electrons in the MIS structures described above. In fact the quality of these grain-boundary n-inversion layers in $Hg_{1-x}Mn_xTe$ was sufficiently good to enable observation of the quantum Hall effect plateaus,[11] and the Shubnikov-de Haas oscillations at magnetic fields as low as 0.2 T. Similar 2D systems were previously studied in Ge[12,13] and InSb[14] containing grain boundaries. The Shubnikov-de Haas effect studies of the n-inversion layers at the grain boundaries in p-$Hg_{1-x}Mn_xTe$[11] ($x \approx 0.1$) did not reveal resolved spin splittings of the Landau levels. This fact is, at least partially, due to the fact that, because of the antiferromagnetic interactions between the $Mn^{2+}$ ions at this value of x, the sp-d exchange-induced contribution to the g-factor is relatively small (see Eq. 1). For this reason, the 2D inversion electrons at the grain boundaries were also studied[15] in the quaternary $Hg_{0.75}Cd_{0.23}Mn_{0.02}Te$, where the energy gap was primarily determined by the amount of Cd, and the smaller number of $Mn^{2+}$ ions resulted in the values of magnetization greater than in the previous case. These quaternary samples did indeed reveal the spin splitting of the Landau levels of 2D electrons. The magnitude of the splitting was in agreement with calculations assuming that the values of the sp-d exchange constants are the same as in the bulk. The calculated values for the effective masses and populations of the electron subbands associated with the grain boundaries also agreed with the experimental data.[15,16]

## Quantum Wells and Superlattices of Wide-Gap DMS

The first reports of successful growth of quantum wells and superlattices of $Cd_{1-x}Mn_xTe/Cd_{1-y}Mn_yTe$ by MBE date back to 1984.[17,18] More recently, superlattices of $Zn_{1-x}Mn_xSe/ZnSe$ have also been grown.[19] Because the energy gap in both cases is an increasing function of the Mn molar fraction, the material with the larger value of x constitutes the barrier layer. Even the very first measurements[18] on these systems indicated that the electronic properties of these systems are strikingly different from those of 3D samples, e.g., the intensity of the luminescence peak observed in $Cd_{1-x}Mn_xTe$ superlattice was found to be ~1500 times stronger than in the bulk material! The Raman scattering measurements clearly showed the zone-folding of the acoustic phonon dispersion curves,[20] as well as the existence of confined optical phonons.[21]

The fact that the lattice constants in both $Cd_{1-x}Mn_xTe$ (a =6.487 - 0.148 x [Å]) and $Zn_{1-x}Mn_xSe$ (a = 5.666 + 0.234 x [Å]) depend on the crystal composition results in a substantial lattice mismatch between the barriers and the wells in the superlattices involving these materials. For sufficiently thin layers the lattice mismatch can be accommodated by internal strains. These, in turn, modify the band offsets, thus complicating the subsequent analysis of the data. In particular, strain-related shifts of the band edges lift the degeneracy of the heavy and light hole states, leading to two distinct excitonic optical transitions[22-24] related to the two resulting hole species. The splitting was shown[23] to be approximately proportional to the strain. The strain also changes the energies of the across-the-gap transitions: in $ZnSe/Zn_{1-x}Mn_xSe$ superlattices one observes[23] a red shift of the transition energies instead of the usual blue shift due to the quantization of energy levels of electrons confined in the quantum well. The opposite variation of the lattice constant with x in $Cd_{1-x}Mn_xTe$ compared to $Zn_{1-x}Mn_xSe$ (leading to the opposite sign of strains in the $CdTe/Cd_{1-x}Mn_xTe$ superlattices) results in an enhancement of the confinement-induced blue shift of the excitonic transition energies (see, e.g., Ref. 22). As could be expected, strain effects are also responsible for large differences in the optical properties shown by the superlattices grown along the (100) and the (111) crystallographic directions.[25]

The photoluminescence measurements on both $CdTe/Cd_{1-x}Mn_xTe$[26,27] and $ZnSe/Zn_{1-x}Mn_xSe$[22] superlattices in the presence of a magnetic field showed that the spin splitting of the relevant energy levels is surprisingly large, although the wells in both cases consist of non-magnetic material, with small values of the g-factors in the bulk. Although such an enhancement is, in principle, possible due to the "leakage" of the wave functions of quantum-well-confined electrons into the DMS barrier layers, where the exchange interaction with localized moments of Mn can take place (see, e.g., Ref. 28 and later in this review), the magnitude of the observed effect is unexpectedly large and cannot be quantitatively accounted for by this simple mechanism. This disagreement led to the conclusion that the excitons in the quantum wells and superlattices are trapped at the heterointerfaces.[26,27,29] This conclusion is further supported by time resolved exciton recombination studies[30] and by the strong polarization of light emitted from these quantum wells in the presence of an external magnetic field.[31] The trapping of excitons at the interfaces is viewed now[32-34] as being due to the sp-d exchange interaction of the hole spins in the wells with Mn ions in the barriers - a phenomenon somewhat analogous to formation of magnetic polarons in bulk DMS.

To complete this brief review of results obtained on $Cd_{1-x}Mn_xTe$ and $Zn_{1-x}Mn_xSe$ superlattices, we finally quote the observation of stimulated emission[35-37] from these structures. This is of considerable interest from the point of view of possible applications because of the possibility of tuning of the emitted radiation by an external magnetic field[36] and also-- owing to the large energies of the transitions occurring in $Zn_{1-x}Mn_xSe/ZnSe$--because of the eventual possibility of stimulated emission in the blue region of the spectrum.

## Quantum Wells and Superlattices of Narrow Gap DMS

At this time considerably less experimental information is available on quantum wells and superlattices of narrow-gap DMS. However, even preliminary (and, as yet, unpublished) results[38] of far-infrared magneto-absorption exhibit unusual and interesting features. Without attempting to give their physical explanation, we shall enumerate some of the findings in this area.

The measurements were carried out on $Hg_{1-x}Mn_xTe/HgTe$ superlattices with x=0.03 and x=0.10, both grown on $Cd_{1-x}Zn_xTe$ substrates, with layers in the former 48 Å thick, and those in the latter 54 Å thick. Because of the value of x, the first system constitutes a "zero-gap/zero-gap" superlattice, with possibly a very small conduction band offset but with spatial modulation of the electron effective mass, while the second is an "open-gap/zero-gap" system (type III superlattice).

The magnetoabsorption spectra in both cases are very rich. The resonances observed exhibit a strong temperature dependence both in their position and intensity (see Fig. 2). This latter dependence suggests that the electron densities in the adjacent layers vary with temperature, which may tentatively be ascribed to the existence of exchange-interaction-induced (and thus temperature dependent) band offsets. Although an unambiguous identification of the observed resonance has not yet been made, their qualitative behavior suggests that they are due both to inter- and intraband transitions. It is interesting to note in Fig. 3 that some of the lines either do not extrapolate to zero energy in vanishing magnetic field, or that they exhibit a very strong "nonparabolic" behavior. The effective masses given by the position of cyclotron resonance (or, rather, by the most prominent absorption line seen in the cyclotron-resonance-active polarization) are quite different from bulk values corresponding to either the barrier or the well layers. Moreover, by tilting the magnetic field with respect to the superlattice, a clearly defined shift of the magnetoabsorption lines was found, indicating strong anisotropy of the electron effective mass even in the case of the x=0.03 superlattice, where the band offset is probably very small.

## III. NEW OPPORTUNITIES IN 2D-DMS STRUCTURES

Here we describe some of the effects which have not yet been realized in DMS quantum wells

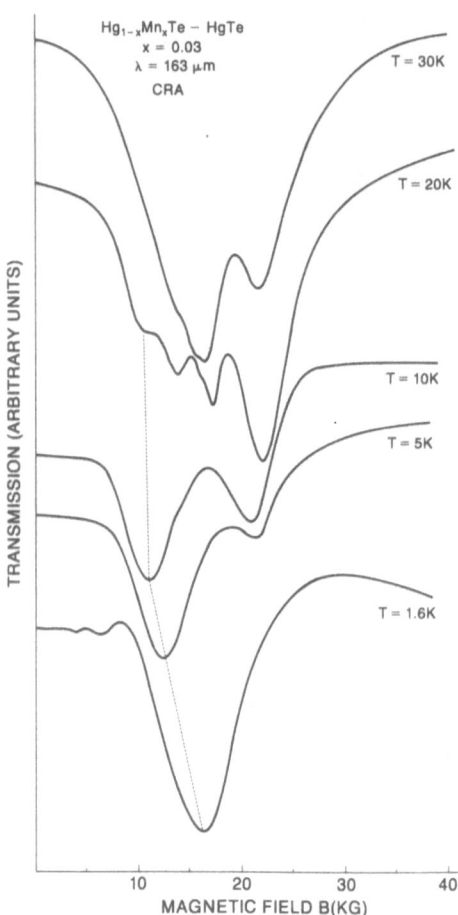

Fig. 2. The magneto-transmission spectra in $Hg_{1-x}Mn_xTe/HgTe$ (x=0.03) superlattice in the cyclotron-resonance-active configuration (with magnetic field parallel to the superlattice growth axis) for several temperatures. The broken line shows the temperature shift of one of the features. [38]

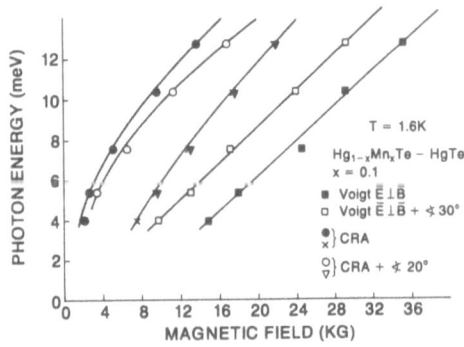

Fig. 3. The "fan chart" of the magneto-transmission features in $Hg_{1-x}Mn_xTe/HgTe$ (x=0.1) superlattice observed in various configurations of the field with respect to the superlattice axis. The lines are drawn merely to guide the eye. [38]

and superlattices, but which appear to be both feasible and highly promising. We begin with electronic properties, which will be followed by a discussion of the modifications of magnetic properties brought about by the quasi-2D nature of the DMS layers.

## Exchange Effects in DMS Superlattices and Quantum Wells

There are three basic facts which underlie the phenomena to be described in this section: (i) the spin splitting in DMS layers is considerably greater than in adjacent non-DMS layers (easily by two orders of magnitude in the case of wide-gap DMS at low temperatures); (ii) the spin splitting in DMS can be comparable to the ionization energies of shallow impurities; (iii) the spin splitting in wide-gap DMS far exceeds the energy difference between consecutive Landau levels.

One of the structures making use of the unique properties of DMS is the idea of a "spin superlattice" which was put forward by von Ortenberg.[39] This structure was to consist of layers of $Hg_{1-x}Mn_xSe$ and $Hg_{1-y}Cd_ySe$, with x and y chosen in such a manner that the energy gaps (and, therefore, the electron effective masses) in both materials were the same. The layers would be quite different, however, as far as their g-factors are concerned. In this situation an electron traveling along the axis of the superlattice would experience an approximately homogeneous potential in the absence of an external magnetic field, but when the field was present the potential "seen" by the electron would have the periodicity of the superlattice. This periodicity of the potential would lead to the formation of "minibands" separated by "minigaps", their magnitudes depending on the difference between the g-factors in the adjacent layers (thus being magnetic field and/or temperature dependent). The calculations performed for the specific case of $Hg_{0.99}Mn_{0.01}Se/Hg_{0.976}Cd_{0.024}Se$ superlattice (both layers 97 Å thick) showed that the minigap varies between 0.6 meV at 25 K and 3 meV at 1.8 K, corresponding to energies characteristic for the realm of submillimeter spectroscopy.[39]

Turning to wide-gap materials, consider now a single quantum well of a modulation-doped superlattice depicted in Fig. 4, with the wells consisting of non-DMS material sandwiched between n-type DMS barriers, e.g., $Cd_{1-x}Mn_xSe/CdSe/Cd_{1-x}Mn_xSe$. By an appropriate choice of the Mn mole fraction x, the energy gap of the barrier material may be so adjusted that the donor level in $Cd_{1-x}Mn_xSe$ is slightly above the ground state $E_1$ of the electron confined in the well, $E_1$ being determined primarily by the well width. At low temperatures the donor electrons will then "spill" from the DMS layer onto the $E_1$ states, rendering the system conducting in 2D. When a magnetic field is applied to this system, one of the spin components of the donor state will shift strongly down in energy, while the $E_1$ level in the non-magnetic well remains practically unaffected by the field. The electrons will then transfer back to the donor states in the DMS layers, with a corresponding decrease of the conductivity (the freeze-out effect). Figure 5 shows the results of a model calculation for the population of the $E_1$ state in a non-magnetic well as a function of the field at two temperatures.

A reverse effect (i.e., electron boil-off) is easily conceivable in a complementary system, consisting of DMS wells and n-type non-magnetic barriers, with parameters so chosen that in zero magnetic field the donor levels are now slightly below the $E_1$ state in the well. In this configuration, application of the field forces the donor electrons to transfer to the well states (owing to the down shift of the lower spin level within the well), where 2D conductivity can take place.

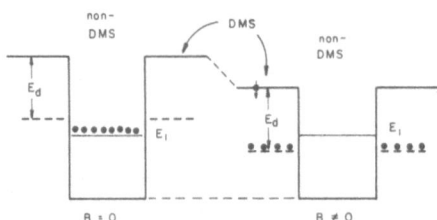

Fig. 4. A non-DMS quantum well between DMS barriers, illustrating exchange-induced
freeze-out. Only the ground
state (spin down) is shown
for B≠0.

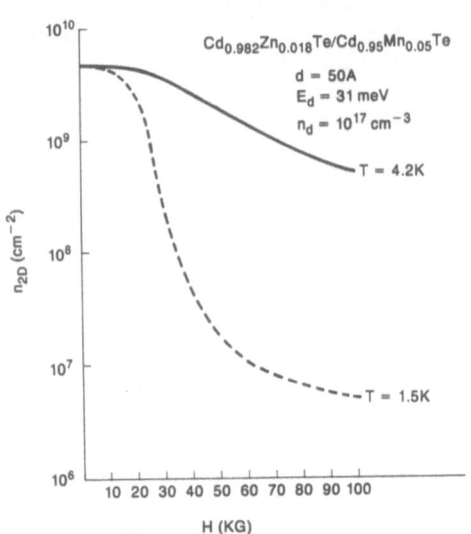

Fig. 5. Calculated density of 2D
electrons confined in
the 50 Å thick quantum
wells formed by the
$Cd_{0.982}Zn_{0.018}Te/$
$Cd_{0.95}Mn_{0.05}Te$ layers, showing the freeze-out effect
and its dependence on the
temperature. The ionization
energy of donors in the non-
DMS barriers was assumed to
be $E_d=31$ meV, and their concentration $n_d=10^{17}$ cm$^{-3}$.

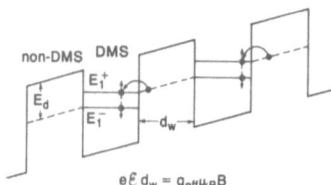

Fig. 6. DMS quantum wells between
non-DMS barriers in the
presence of an external
electric and magnetic fields,
showing sequential resonant
tunneling between spin split
states in the well via the
donor levels in the barriers.

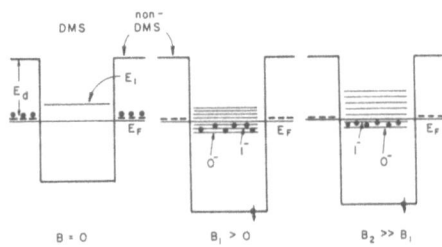

Fig. 7. A DMS quantum well between non-DMS barriers, illustrating the mechanism for the unusual sequence
of the quantum oscillations. $B_1$ and $B_2$ denote
two values of the applied magnetic field. Note
that the energy of the spin-down states in DMS
does not continue to decrease indefinitely with
increasing field because of saturation of the
magnetization, as illustrated by the diagram on
the right-hand side of the figure.

The fact that the relative position of the donor levels in the barriers and $E_1$ states in the wells can be adjusted suggests a possibility of resonant tunneling across the barriers tuned by the magnetic field and/or the temperature. A sequential tunneling (originally demonstrated[40] in non-magnetic superlattices for the case involving resonance of the ground state $E_1$ and the first excited state $E_2$ in the successive wells) is in the present case conceivable when the two spin components of the $E_1$ state coincide with the donor levels in the adjacent barriers, as shown in Fig. 6. This (or similar configurations involving various spin components of the excited states in the well) may ultimately lead to the construction of efficient (and tunable) infrared emitters, with specific polarization of the out-going light.

As pointed out elsewhere,[28] the quantum well structures of DMS may show novel features in quantum oscillation phenomena, of which the most unusual is shown in Fig. 7. It should be noted that the sequence of the oscillations (each oscillation occurring as Fermi level passes one of the Landau levels) takes place in the order reverse to that usually observed in semiconductors. That is, in the case of non-magnetic semiconductors, the sequence of quantum oscillations is determined by the fact that the index of Landau levels crossing the Fermi level <u>decreases</u> as the magnetic field is increased. In the case shown in Fig. 7, on the other hand, one may expect series of the oscillations to begin with the oscillation associated with the <u>lowest</u> Landau level $0^-$, followed by those associated with $1^-$, $2^-$, etc. as the field increases. This phenomenon is a consequence of the fact that in wide gap DMS spin splittings far exceed the spacing between consecutive Landau levels. This sequence will continue until the Mn spins saturate, and the sequence of oscillations should then reverse.

Finally, in the case of a <u>shallow</u> <u>non-DMS</u> quantum well sandwiched between DMS barriers one may expect the electron states in the well to acquire some of the sensitivity to the magnetic field and temperature characteristic for the DMS material. This is because the effective depth of the well, given by the difference between positions of the band edges, is itself strongly dependent on the field. This effect may be viewed as being due to the leakage of the wave function of the electron confined in the well into the DMS barriers, so that the electron interacts (at least in part) with the Mn ions via sp-d exchange.

The situations cited above are by no means an exhaustive list of the phenomena made possible by the sp-d exchange interaction in the context of quantum wells and superlattices involving DMS, but are rather meant as illustrative examples of the exciting opportunities which this interaction holds in store.[28,41]

The Magnetism of 2D Layers

As has been suggested earlier,[28] the 2D character of DMS layer samples should also affect the nature of the magnetism exhibited by these structures. Although direct measurements of magnetic properties of the 2D DMS have not yet been undertaken, some of the observations already made indicate indirectly that the type of magnetism of 2D DMS is indeed different from that of their bulk counterparts. The study of this problem is of vital importance not only because of its fundamental nature, but also because--as we have already argued--the <u>electronic</u> properties of DMS depend in a very sensitive way on the magnetization exhibited by these materials.

The first indication of differences between the magnetic behavior in 2D and in bulk DMS samples comes from Raman scattering measurements, where it was noted[20] that in $Cd_{0.5}Mn_{0.5}Te/Cd_{0.89}Mn_{0.11}Te$ superlattices the line associated with characteristic collective excitations (magnons) was conspicuously missing at low temperatures. Such a line was clearly observed in previous Raman studies on bulk $Cd_{0.5}Mn_{0.5}Te$. Instead, in the case of the superlattices, the low temperature Raman spectrum shows only well resolved paramagnetic resonance lines, such as those observed in bulk $Cd_{0.5}Mn_{0.5}Te$ at high temperatures, as well as at low Mn concentrations in general, i.e., in the absence of any short range magnetic order. Thus it was concluded[20] that, in contrast with the bulk material, the 2D samples studied exhibit a paramagnetic rather than magnetically ordered behavior.

The most recent results also indicating that the character of magnetic behavior is altered when the dimensionality of the samples is reduced, was made possible because of the successful growth of very thin (8 Å) layers ("spin sheets") of zinc-blende MnSe[42] sandwiched between thicker (45 Å) layers of non-magnetic ZnSe. It may be expected that zinc-blende MnSe, similarly to β-MnS, will show an antiferromagnetic ordering. Surprisingly, the luminescence line observed at low temperature exhibited a vigorous shift toward longer wavelengths in the presence of an external magnetic field. The large magnitude of this shift is inconsistent with antiferromagnetic ordering and can be taken as a signature of paramagnetic behavior of the "spin sheets". On the other hand the thicker layers (~30 Å) of zinc-blende MnSe do not reveal any strong dependence of the luminescence line, a behavior which agrees with the existence of antiferromagnetism in this material.

While the above results still await a full explanation, let us note here that the spin glass phase observed in bulk DMS, where the Mn-Mn interactions are of short range character, is most probably due to the fcc or hcp lattice frustration within the Mn spin subsystem. Therefore, by changing the magnetic environment of the Mn ions, one can alter substantially the degree of the frustration. The "spin sheet" mentioned above can serve as an example of such modification, since the intercalating ZnSe layers certainly result in a reduction of the coupling of Mn ions belonging to different sheets, thus altering the number of possible magnetic nearest neighbors of a given Mn ion. Less drastic changes of the number of nearest neighbors can be achieved by growing DMS films on various appropriately chosen substrates, with each substrate "forcing" its own lattice structure on the DMS film. For instance, the films of zinc-blende MnSe mentioned above have been grown on ZnSe substrates, although MnSe normally crystallizes in the NaCl structure. Similarly, $Zn_{0.34}Mn_{0.66}Se$ (which normally crystallizes in the wurtzite structure) has been grown in the form of zinc-blende epilayers on ZnSe.[42] Let us note in passing that, because of the possibility of "tuning" the lattice constants of ternary compounds by varying the composition, one can judiciously choose substrates which are perfectly lattice matched to specific crystalline phases, thus making it possible to grow epitaxially a wider variety of crystal structures, at the same time avoiding problems related to internal strains.[43]

Finally--although here we part company with the $A_{1-x}^{II}Mn_xB^{VI}$ DMS--we wish to raise the possibility of an effective <u>long-range</u> coupling between Mn ions via mobile band carriers (RKKY), which might be possible, e.g., in $Pb_{1-x-y}Sn_xMn_yTe$.[44] Recent theoretical calculations show[45] that such coupling may be very sensitive to the 2D nature of the mediating electrons, leading to a strong dependence of the coupling constant on the film thickness. Furthermore, since the strength of the RKKY interaction depends on the density of the mediating electrons, a unique possibility arises of controlling the Mn-Mn coupling constant by means of the gate voltage across, say, an MIS device. These possibilities are, however, still to be demonstrated.

# VI. CONCLUDING REMARKS

2D structures, quantum wells and superlattices have attracted such a great deal of interest largely because they enable, by their proper design, to "tailor" various band parameters, such as an effective energy gap and an effective electron mass. 2D systems of DMS further extend this possibility by adding an independent control over the spin properties by means of an external magnetic field and temperature. Application of the required fields may be achieved not only by means of conventional magnets, but also --on a local microscopic scale--by depositing on the DMS layers thin monolithic films of a ferromagnetic material in appropriate configurations. Recent growth of ferromagnetic Fe on semiconductor surfaces[46] by MBE is an important step in this last direction.

Realization of many of the ideas presented in this paper depends critically on the development of effective epitaxial techniques in the area of controlled doping of DMS materials. Recent achievements in the area of understanding impurities and defects in $A^{II}B^{VI}$ compounds[47] make the study of 2D DMS systems such as those described above not only tempting but also quite realistic.

## Acknowledgments

The authors are grateful to the National Science Foundation for support under Grants DMR-8316988 and DMR-8600014.

## REFERENCES

1. For a recent review of DMS see: J. K. Furdyna, J. Appl. Phys. <u>53</u>, 7637 (1982); N. B. Brandt and V. V. Moshchalkov, Adv. Phys. <u>33</u>, 193 (1984).
2. G. Bastard, C. Rigaux, Y. Guldner, J. Mycielski, and A. Mycielski, J. Phys. (Paris) <u>39</u>, 87 (1978.
3. J. Gaj, R. R. Galazka, and M. Nawrocki, Solid State Commun. <u>25</u>, 193 (1978).
4. M. Jaczynski, J. Kossut, and R. R. Galazka, phys. stat. sol. (b) <u>88</u>, 73 (1978).
5. S. Nagata, R. R. Galazka, D. P. Mullin, H. Akbarzadeh, G. D. Khattak, J. K. Furdyna, and P. H. Keesom, Phys. Rev. B22, 3331 (1980).
6. T. Giebultowicz, W. Minor, H. Kepa, J. Ginter, and R. R. Galazka, J. Magn. Magn. Mater., <u>31-34</u>, 1373 (1982).
7. G. Grabecki, T. Dietl, J. Kossut and W. Zawadzki, Surface Sci. <u>142</u>, 588 (1984).
8. M. Chmielowski, T. Dietl, F. Koch, P. Sobkowicz, and J. Kossut, Acta Phys. Polon., in press.
9. J. Gaj, J. Ginter, and R. R. Galazka, phys stat. sol. (b) <u>89</u>, 655 (1978).
10. G. Grabecki, T. Dietl, P. Sobkowicz, J. Kossut, and W. Zawadzki, Appl. Phys. Lett. <u>45</u>, 1214 (1984).
11. G. Grabecki, T. Suski, T. Dietl, T. Skoškiewicz, and T. Przeor, Acta Phys. Polon., to be published.
12. B. M. Vul and E. T. Zavaritskaya, Zh. Eksp. Teor. Fiz. <u>76</u>, 1089 (1979).
13. S. Uchida, G. Landwehr, and E. Bangert, Solid State Commun. <u>45</u>, 869 (1983).
14. R. Herrmann, W. Kraak, G. Nachtwei, and G. Worm, Solid State Commun. <u>52</u>, 843 (1984).
15. G. Grabecki, T. Dietl, P. Sobkowicz, J. Kossut, and W. Zawadzki, Acta Phys. Polon. <u>A67</u>, 297 (1985).
16. G. Gobsch, J.-P. Zöllner, and G. Pausch, phys. stat. sol. (b) <u>134</u>, K149 (1986).

17. R. N. Bicknell, R. W. Yanka, N. C. Giles-Taylor, D. K. Blanks, E. L. Buckland, and J. F. Schetzina, Appl. Phys. Lett. 45, 92 (1984).

18. L. A. Kolodziejski, T. C. Bonsett, R. L. Gunshor, S. Datta, R. B. Bylsma, W. M. Becker, and N. Otsuka, Appl. Phys. Lett. 45, 441 (1984).

19. L. A. Kolodziejski, R. L. Gunshor, T. C. Bonsett, R. Venkatasubramanian, S. Datta, R. B. Bylsma, W. M. Becker, N. Otsuka, Appl. Phys. Lett. 47, 169 (1985).

20. S. Venugopalan, L. A. Kolodziejski, R. L. Gunshor, A. K. Ramdas, Appl. Phys. Lett. 45, 974 (1984).

21. D. U. Bartholomew, E-K. Suh, L. A. Kolodziejski, R. L. Gunshor, and A. K. Ramdas, Bull. Am. Phys. Soc. 31, 349 (1986), and in preparation.

22. D. K. Blanks, R. N. Bicknell, N. C. Giles-Taylor, J. F. Schetzina, A. Petrou, and J. Warnock, J. Vac. Sci. Technol., in press.

23. R. B. Bylsma, R. Frohne, J. Kossut, W. M. Becker, L. A. Kolodziejski, and R. L. Gunshor, Proc. of the Material Research Society Symposium, Boston, 1985, vol. 51, in press.

24. Y. Hefetz, J. Nakahara, A. V. Nurmikko, L. A. Kolodziejski, R. L. Gunshor, and S. Datta, Appl. Phys. Lett. 47, 989 (1985).

25. S.-K. Chang, A. V. Nurmikko, L. A. Kolodziejski, and R. L. Gunshor, Phys. Rev. B33, 2589 (1986).

26. X.-C. Zhang, S.-K. Chang, A. V. Nurmikko, D. Heiman, L. A. Kolodziejski, R. L. Gunshor, and S. Datta, Solid State Commun. 56, 255 (1985).

27. A. V. Nurmikko, X.-C. Zhang, S.-K. Chang, L. A. Kolodziejski, R. L. Gunshor, and S. Datta, J. Lumines. 34, 89 (1985).

28. S. Datta, J. K. Furdyna, and R. L. Gunshor, Superlatt. Microstruct. 1, 327 (1985).

29. X.-C. Zhang, S.-K. Chang, A. V. Nurmikko, L. A. Kolodziejski, R. L. Gunshor, and S. Datta, Phys. Rev. B31, 4056 (1985).

30. X.-C. Zhang, S.-K. Chang, A. V. Nurmikko, L. A. Kolodziejski, R. L. Gunshor, and S. Datta, Appl. Phys. Lett. 47, 59 (1985).

31. A. Petrou, J. Warnock, R. N. Bicknell, N. C. Giles-Taylor, and J. F. Schetzina, Appl. Phys. Lett. 46, 692 (1985).

32. C. E. T. Goncalves da Silva, Phys. Rev. B32, 6962 (1985).

33. J.-W. Wu, A. V. Nurmikko, and J. J. Quinn, Solid State Commun. 57, 853 (1986).

34. C. E. T. Goncalves da Silva, Phys. Rev. B33, 2923 (1986).

35. R. B. Bylsma, W. M. Becker, T. C. Bonsett, L. A. Kolodziejski, R. L. Gunshor, M. Yamanishi, and S. Datta, Appl. Phys. Lett. 47, 1039 (1985).

36. R. N. Bicknell, N. C. Giles-Taylor, J. F. Schetzina, N. G. Anderson, and W. D. Leidig, J. Vac. Sci. Technol., in press.

37. E. D. Isaacs, D. Heiman, J. J. Zayhowski, R. N. Bicknell, and J. F. Schetzina, Appl. Phys. Lett. 48, 275 (1986).

38. M. Dobrowolska, Z. Yang, H. Luo, J. K. Furdyna, K. A. Harris, J. W. Cook, Jr., and J. F. Schetzina, J. Vac. Sci. Technol. [to be published (1987)].

39. M. von Ortenberg, Phys. Rev. Lett. 49, 1041 (1982).

40. F. Capasso, K. Mohammed, and A. Y. Cho, Appl. Phys. Lett. 48, 478 (1986).

41. J. K. Furdyna, J. Vac. Sci. Technol., (1986) in press.

42. L. A. Kolodziejski, R. L. Gunshor, N. Otsuka, B. P. Gu, Y. Hefetz, and A. V. Nurmikko, Appl. Phys. Lett. 48, 1482 (1986).

43. J. K. Furdyna and J. Kossut, Superlattices and Microstructures, 2, 89 (1986).

44. T. Story, R. R. Galazka, R. B. Frankel, P. A. Wolff, Phys. Rev. Lett. 56, 777 (1986).

45. A. E. Kuchma, Sov. Phys. Solid State 27, 1537 (1985).

46. C. Vittoria, F. J. Rachford, J. J. Krebs, and G. A. Prinz, Phys. Rev. B30, 3903 (1984).

47.  R. N. Bhargava, in Proceedings of the 17th International Conference on the Physics of Semiconductors, San Francisco, 1984 (J. D. Chadi and W. A. Harrison, eds.) Springer Verlag, p. 1531.

# OPTICAL NONLINEARITIES IN NARROW-GAP SEMICONDUCTORS

Alan Miller and Duncan Craig

Royal Signals and Radar Establishment
Great Malvern
Worcs. WR14 3PS, UK

## SUMMARY

We discuss the origin of band gap resonant refractive non-linearities in narrow gap semiconductors and describe the excess carrier dynamics. The band gap dependence and transient nature of these non-linearities are illustrated by nonlinear etalon, optical bistability and self-defocusing results in CdHgTe and InSb at room temperature.

## INTRODUCTION

Giant optical nonlinearities in semiconductors[1] were first discovered in InSb[2] using band gap resonant laser excitation and were subsequently used to achieve optical bistability[3] at milliwatt optical power levels in thin polished etalons of this material. The band gap of low temperature InSb is coincident with the CO laser output at a wavelength of about 5μm, while two-photon excitation can be employed at room temperature[4] using a $CO_2$ laser at 10μm. This material has allowed a thorough study of optical bistability to assess the potential of all-optical devices for digital optical processing and computing[5]. Optical bistability has now been demonstrated in semiconductors over a large range of band gaps and wavelengths through either electronically induced nonlinearities, as in InSb, or by making use of thermally induced changes in refractive index[6]. Theory and experiment have shown however, that the largest band gap resonant nonlinear refractive phenomena of electronic origin occur in semiconductors with the smallest energy gaps[7,8]. Other semiconductors with similar electronic properties which exhibit large nonlinear optical effects are InAs and the alloy semiconductor, $Cd_xHg_{1-x}Te$. The low temperature band gap of InAs is compatible with the HF laser at 3μm and optical bistability has been achieved at 7mW incident power in this semiconductor[9]. A strong incentive for considering CdHgTe is due to the coincidence which can be achieved between its band gap and the output of the $CO_2$ laser at around 10μm given a suitable alloy composition. Indeed, the ability to vary the band gap of CdHgTe with composition and/or temperature gives the opportunity to examine the band gap dependence of resonant nonlinearities[7] in this well researched infrared detector material. We have demonstrated extremely large nonlinear effects in low temperature CdHgTe under band gap resonant conditions using a low power $CO_2$

laser[10,11].   In  this  paper,  we limit our dicussion to our studies  of
room  temperature nonlinear phenomena in CdHgTe (with various band gaps)
and  InSb  in  order  to  illustrate  the general  features  of  the  giant
refractive nonlinearities in narrow gap semiconductors.


THEORY OF NONLINEAR REFRACTION IN NARROW GAP SEMICONDUCTORS

Even  moderate densities of optically generated electron-hole pairs
can  cause  large changes in the refractive index of  narrow  gap  semiconductors.   This  nonlinearity  can  be quite complex because  of  the
dynamics of the electrons and the different time constants involved.   In
figure  1 we illustrate the expected electron motion for different  band
gap energies at a fixed wavelength of excitation covering the conditions
of  our  experiments using a $CO_2$ laser.   For $Cd_xHg_{1-x}Te$  the  band  gap
dependence with composition, $x$,  and temperature, $T$, is consistent with
the expression[12],

$$Eg(x,T) = -0.302 + 1.93x + 5.35 \times 10^{-4} \, T \, (1-2x) - 0.810x^2 + 0.832x^3 \quad (1)$$

Thus,  for  x=.18 (Eg=.120eV),  10.6μm wavelength excitation  gives
single photon absorption into the band tail (fig.  1a).   These carriers
are  rapidly scattered into the band (within a few picoseconds) and will
recombine on longer timescales (a few nanoseconds or less).  Up to x=.26
(Eg=.234eV),  two photon absorption excites electrons directly into  the
conduction  band  with  some excess kinetic  energy  (fig.  1b).   These
carriers  drop to the bottom of the band by phonon emission on  a  picosecond  timescale  before recombining in a time on the order of tens  of
nanoseconds.   This  is  also the situation pertaining to InSb at  room
temperature.   In  the range x=.26 to .35,  three photon absorption  is
expected  with carrier dynamics similar to the two photon case but  here
the carrier lifetime is longer at a few μs (fig.  1c).  Although  multiphoton  excitation  is  typically  not as  efficient  as  single  photon
absorption for the generation of carriers, small gap semiconductors have
relatively large two and three photon absorption coefficients[13].  Multiphoton  absorption  is  of course a nonlinear mechanism which  gives  an
additional intensity dependence to be considered in the analysis of  any
results.   Background  absorption can be much smaller in the multiphoton
case  resulting in a much more even distribution of  generated  carriers
through the material.   To complicate matters further however,  both the
absorption  coefficients and carrier lifetimes are density dependent due
to free carrier absorption and Auger recombination respectively.

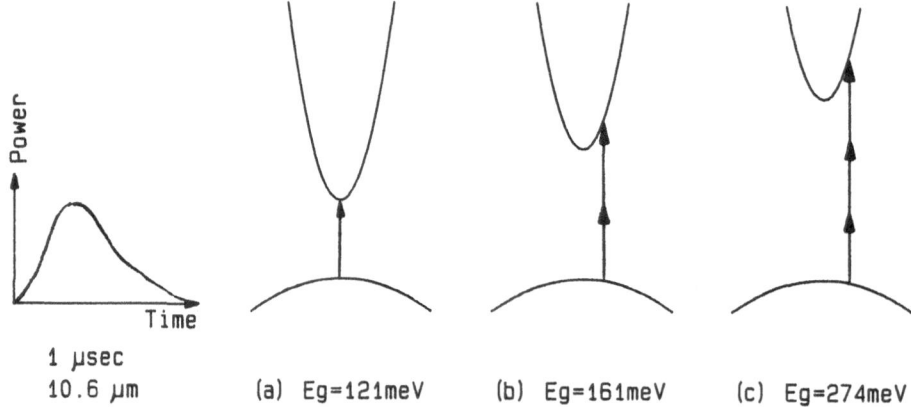

1 μsec
10.6 μm            (a) Eg=121meV        (b) Eg=161meV        (c) Eg=274meV

Fig. 1.  Excitation conditions at 10.6μm wavelength for different alloy
         compositions of $Cd_xHg_{1-x}Te$.

The refractive index change, $\Delta n$, for photon energies below the band gap energy, due to an excess carrier concentration, $\Delta N$, can be considered to arise from two contributions. Expressing the nonlinearity as a change of refractive index per electron-hole pair per unit volume, $\sigma$, then

$$\sigma = - \frac{e^2}{2\epsilon_0 m^* n \omega^2} \left[ 1 + \frac{2}{\sqrt{3}\pi} \frac{m^*}{m} \left( \frac{mP^2}{\hbar^2} \right) \frac{1}{kT} J \left( \frac{\hbar\omega - Eg}{kT} \right) \right] \qquad (2)$$

where,

$$J(a) = \int_0^\infty \frac{dx \, x^{\frac{1}{2}} \, e^{-x}}{(x-a)}$$

Here, $e$ is the electronic charge, $\epsilon_0$ the free space dielectric constant, $m^*$ the conduction band effective mass, $m$ is the electron mass, $\hbar\omega$ is the photon energy, $n$ the linear refractive index and $P$ the momentum matrix element (the quantity $mP^2/\hbar^2$ is in Joules). The first term describes the plasma contribution to the nonlinear refraction and is the Drude expression derived from standard dispersion theory modified by the electron effective mass[14]. The second term is the result of the excess carriers blocking virtual transitions between the valence and conduction bands and shows a resonance for photon energies close to the band gap, embodied in the thermodynamic integral, $J(a)$.

Thus, free carrier plasma and band filling contributions to the nonlinear refraction result in a direct proportionality between the change of refractive index and the density of generated carriers to a first order of approximation,

$$\Delta n = \sigma \Delta N \qquad (3)$$

Density dependent carrier recombination rates due to the dominance of Auger processes in narrow gap semiconductors complicate the dynamics of the nonlinearity. The very different thermal populations for different band gap energies gives carrier lifetimes from 1ns for $Eg=100meV$ to 1$\mu$s for $Eg=250meV$. These lifetimes decrease as excess carriers are optically excited. The net result is that fixed values of the effective third order nonlinear coefficient, $\chi^{(3)}$, cannot be quoted for materials having density dependent carrier recombination since the nonlinearity is dependent on the precise excitation conditions. We can describe the excess carrier population in terms of the rate equation,

$$\frac{d\Delta N}{dt} = G - r \Delta N \qquad (4)$$

where for single photon absorption, $G = \alpha I/\hbar\omega$; two-photon absorption, $G = \beta I^2/2\hbar\omega$; and three-photon absorption, $G = \gamma I^3/3\hbar\omega$; and $\alpha, \beta, \gamma$ are one-, two- and three-photon absorption coefficients respectively. The Auger recombination rate, $r$, is given by,[15]

$$r = \frac{(2 + \Delta N/N_i)(1 + \Delta N/N_i)}{2\tau_i} \qquad (5)$$

where $N_i$ is the intrinsic carrier concentration and $\tau_i$ is the intrinsic Auger recombination time. For steady state conditions, $d\Delta N/dt = 0$, and from equations (3) to (5), we see that,

$$\begin{aligned} &\Delta n \, \alpha \, I^q \quad \text{for} \quad \Delta N \ll N_i \\ &\Delta n \, \alpha \, I^{q/3} \quad \text{for} \quad \Delta N \gg N_i \end{aligned} \qquad (6)$$

where q denotes the number of photons in the carrier generation process.

Nonlinear absorption should also be considered. In particular, generated electron-hole pairs give rise to free carrier absorption which can be significant under multiphoton conditions. This leads to optical loss without generation of carriers; the dominant free carrier absorption arises from intervalence band transitions since a phonon is not required for this process. Values of absorption cross-sections at 10.6µm for electrons and holes in $Cd_xHg_{1-x}Te$, x=.23 have been measured as[16], $\sigma_e = 8.2\times10^{-17}cm^2$ and $\sigma_h = 3.8\times10^{-16}cm^2$ and for InSb[17] $\sigma_e = 3.3\times10^{-17}cm^2$ and $\sigma_h = 8.7\times10^{-16}cm^2$.

Absorption of the radiation inevitably leads to heating of the material and therefore also results in refractive index changes but with a slower time constant than for electronic processes. Whereas electronic nonlinearities give negative contributions for both CdHgTe and InSb, the band gap shift with temperature for CdHgTe gives a negative refractive index change with increasing temperature, of opposite sense to the thermal contribution in InSb[11].

## THEORY OF NONLINEAR OPTICAL ETALONS

The Fabry-Perot equations for an absorbing etalon relating transmission, T, to the front and back mirror reflectivities, $R_F$ and $R_B$, the intermediate layer absorption coefficient, $\alpha$, and the thickness, L, are[18-20],

$$T = \frac{A}{1 + F \sin^2\delta} \tag{7}$$

where,

$$A = e^{-\alpha l}(1-R_F)(1-R_B)/(1-R_\alpha)^2$$

$$F = 4R_\alpha/(1-R_\alpha)^2$$

$$R_\alpha = (R_F R_B)^{\frac{1}{2}}e^{-\alpha L}$$

and the single pass phase is

$$\delta = 2\pi nL/\lambda$$

The cavity finesse is defined as $\mathscr{F} = \pi F^{\frac{1}{2}}/2$. If we initially assume a refractive index change, $\Delta n$ proportional to the average cavity intensity, $I_c$, and linear absorption, then,

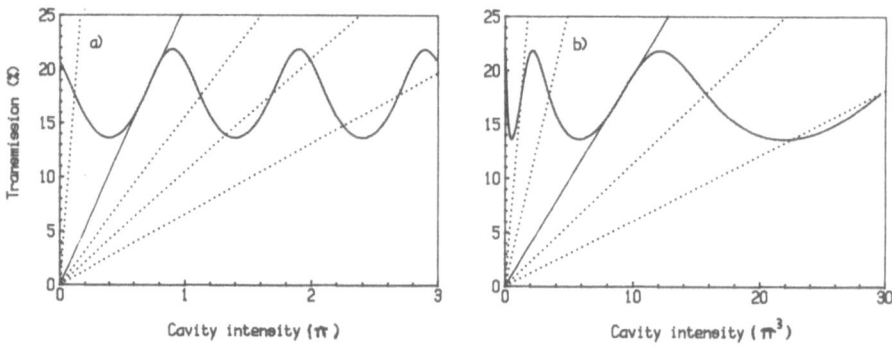

Fig. 2. Nonlinear Fabry-Perot transmission versus mean cavity intensity for $R_F = R_B = 32\%$ and $\alpha L = 1$ for (a) $\Delta n \propto I$ and (b) $\Delta n \propto I^{1/3}$.

$$\Delta n = n_2 \, I_c \qquad\qquad (8)$$

and,

$$\delta = \delta_0 + \gamma \, I_c \qquad\qquad (9)$$

An example of the oscillatory Airy function of equation (7) as a function of phase, $\gamma I_c$, is shown for a low finesse case in fig. 2a for $R_F = R_B = 32\%$ and $\alpha l = 1.0$ typical of a polished semiconductor under band gap resonant excitation. The transmission at low intensities is determined by the initial phase, $\delta_0$. The input-output characteristics of the nonlinear etalon are sensitive to the value of this initial detuning.

The relationship between cavity intensity and transmitted intensity leads to the additional criteria:

$$\frac{T}{I_c} = \frac{C_T}{I_I} \; ; \; C_T = \frac{\alpha L e^{-\alpha L}(1-R_B)}{(1-e^{-\alpha L})(1+R_B e^{-\alpha L})} \qquad\qquad (10)$$

Equations (7) and (10) solved simultaneously to eliminate $I_c$ describe the nonlinear Fabry-Perot transmission as a function of incident intensity[21]. Equation (10) describes straight lines through the origin of fig. 2a; the slopes represent different values of incident intensity, the shallower slopes corresponding to higher intensities. Bistability occurs at multiple crossings of the periodic and straight lines. In fig. 2a, the initial phase has been adjusted to give the critical conditions for optical bistability.

Figure 2b shows the effect of single photon absorption in a semiconductor exhibiting Auger recombination[20]. The saturation of the nonlinearity as the third power (eq.6) has the effect of causing an increased spacing of the etalon resonances at higher powers altering the conditions for optical bistability. If nonlinear absorption was also included, the effect would be to cause a change in the transmission in the oscillatory function and to bend the lines representing the incident intensities.

TRANSIENT NONLINEAR ETALON EXPERIMENTS

Samples of CdHgTe and InSb (Table 1) were polished plane parallel. The output from a hybrid $CO_2$ laser was focused onto the samples (1/e² radius $\simeq 675\mu m$) after passage through suitable attenuators. The incident and transmitted pulses were measured by fast, room temperature CdHgTe detectors (1ns response time) and recorded on transient digitizers. Pinholes (200μm diameter) were placed on the back surface of the samples to select only the central portion of the incident

TABLE 1                 Material parameters

| Sample | #1 | #2a | #2b | #3 | InSb |
|---|---|---|---|---|---|
| x | .18 | .20 | .23 | .29 | – |
| $E_g$(meV) | 121 | 143 | 193 | 274 | 161 |
| $N_i$($\times 10^{16}$cm$^{-3}$) | 4.8 | 3.2 | 1.8 | .45 | 2 |
| $\tau_i$(ns) | 4 | 16 | 90 | 3500 | 50 |
| L (μm) | 171 | 208 | 211 | 240 | 502 |
| I (kW/cm²) | 20 | 90 | 235 | 1130 | 115 |
| $n_2$($\times 10^{-6}$cm²/W) | 2.21 | .31 | .106 | .019 | .116 |

transverse profile and so approximate plane-wave illumination. The samples were initially examined with the low-power c.w. section of the hybrid laser to characterize the Fabry-Perot transmission fringes at single-photon energies as the samples were rotated in the beam.

The samples were then all set to the same initial detuning (a transmission maximum) and the attenuation of the incident laser pulse adjusted to give some modulation of the transmitted temporal pulse profile. Measurements were then made of the incident power required to tune the etalon to the first transmission minimum and hence obtain a comparison of the strength of the nonlinearity over this bandgap range[22].

Figure 3 shows incident and transmitted pulse profiles for the InSb sample for a peak incident intensity of 232kW/cm². The modulation on the transmitted pulse profile is due to the tuning of the etalon through its transmission fringes. We see a sharp switch up to high transmission as two photon absorption creates carriers which cause a change in refractive index and hence tuning of the etalon, and then recovery to low transmission in the trailing edge as the carriers recombine. Figure 4 shows transmission plotted against time by dividing the transmitted pulse by the output pulse for InSb and CdHgTe samples #1, #2b & #3, (the sample numbers indicate the number of photons contributing to the dominant absorption process for each CdHgTe sample). Each sample shows a modulation in transmission due to the induced etalon tuning and the intensity required to reach the first transmission noted in table 1. Samples #1, #2a, #2b and InSb all exhibit quasi-steady state nonlinear behaviour because of the short carrier lifetimes compared to the pulse length. On the other hand, sample #3 does not recover on the timescale of the pulse since $\tau_i$ is long. This results in hysteretic behaviour which is not bistable. Two-photon samples can give quasi-steady-state bistability (see below) because of the relatively fast response. In the band gap resonant sample, #1, the nonlinear response is very fast, but because of the larger absorption and hence low finesse, optical bistability is not expected in first order. Since the nonlinearity saturates at higher power as $I^{1/3}$, the etalon response will be similar to that shown in figure 2.

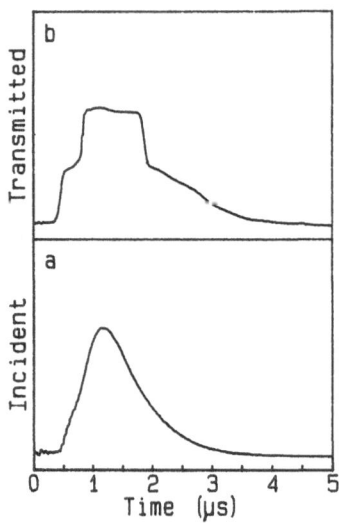

Fig. 3. Incident and transmitted pulses for InSb. Peak incident intensity 232kW/cm².

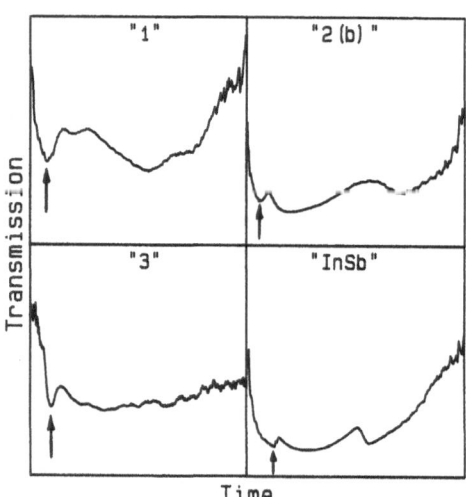

Fig. 4. Transmission versus time for CdHgTe samples and InSb. Arrows show the first etalon minima.

Comparison of the magnitudes of the nonlinearities in the different band gaps can be misleading because of the different time constants and power dependences involved. However, we have calculated an average, effective third order nonlinearity, $\chi^{(3)}$, over the intensity required to tune each etalon from a transmission maximum to a minimum. Although strictly speaking this gives magnitudes only valid for the conditions of the experiments used here, it does provide some guide for comparison with other nonlinear materials. The result is a $\chi^{(3)}$ which varies from $5\times10^{-6}$ e.s.u. for the three photon case, sample #3, up to $\chi^{(3)} = 5\times10^{-4}$ e.s.u. for band gap resonance, #1. These values are large for room temperature conditions. The reduction of nonlinear refraction in the multiphoton cases compared to single photon excitation is less than might be expected because of the much longer carrier lifetimes in the higher bandgap alloys.

## TWO-PHOTON INDUCED OPTICAL SWITCHING AND BISTABILITY

Optical bistability in a room temperature, 360µm thick, polished etalon of $Cd_xHg_{1-x}Te$, x = .23, was studied at 10.6µm under two photon excitation conditions[23,24]. The pulsed (1.75µs FWHM) output from a hybrid $CO_2$ laser was focused to a spot size of 250µm. A 100µm diameter pinhole located on the rear surface of the sample limited the detected output to the central region of the illuminated area in order to approximate to plane wave conditions and also to reduce the effects of sample inhomogeneities. Clear Fabry-Perot fringes could be observed on rotation of the sample using low power radiation from the c.w. source of the hybrid laser. Optical bistability was achieved at incident intensities above 100kW/cm². Figure 5 shows results obtained by plotting the transmitted power during a pulse against the simultaneously recorded incident power (maintaining careful time synchronization of both recordings) for four angles of incidence. Since the pulse length is longer than the carrier recombination time ($\tau_i$=90ns) essentially steady-state conditions prevail. At 18°, the etalon is initially tuned near a transmission maximum as indicated by the slope of the characteristic at low intensities and bistability is observed as the

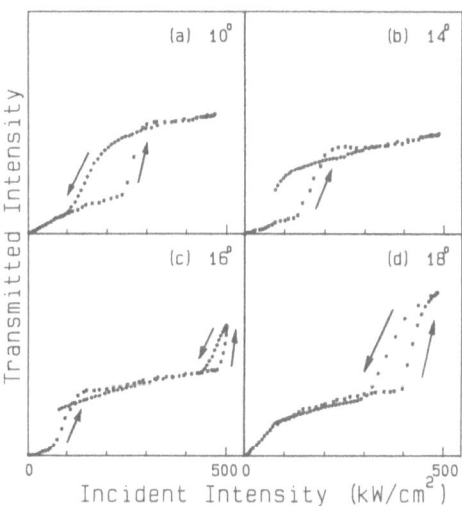

Fig. 5. Input-output characteristics at 10.6µm
wavelength for a 360µm thick CdHgTe
etalon for different angles of incidence.

etalon tunes through the next etalon resonance. Going to smaller angles, the initial etalon detuning moves progressively towards higher order fringes, thus, higher powers are required to reach the bistable feature observed at 18° because the refractive index change with carrier density is negative. The next (higher) etalon order is clearly bistable at 14°.

Although close to quasi-steady state conditions were achieved in these experiments, the dynamics of the bistability is very complex[24,25]. This is partly because the two photon absorption process is itself non-linear but also because both the background absorption and carrier recombination rates are density dependent. The net result of these effects is depicted in fig. 6 where the conventional graphical construction for optical bistability[21] is calculated[23] using the experimental etalon and material parameters and plotted for each of the conditions of fig. 5. Increasing absorption by the optically generated free carriers causes the peak transmission to decrease with increasing intensity and lines of constant incident intensity to be slightly curved. The carrier lifetime decreases at higher densities due to Auger recombination. This results in saturation of the nonlinearity ($I^{1/3}$) and thus an increase of etalon resonance spacing would be expected as a function of cavity intensity. However, the two-photon nonlinearity ($I^2$) has the opposite effect so that the actual fringe spacing remains fairly uniform. Multiple intersections give the conditions for bistability as

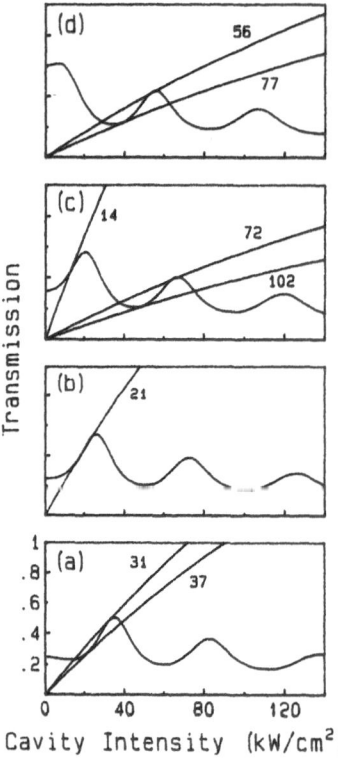

Fig. 6. Calculated periodic curves of transmission versus cavity intensity for a CdHgTe etalon for the four initial detunings corresponding to those used in fig. 5. Lines of constant incident intensity (in kW/cm²) are drawn to illustrate bistable conditions.

usual and we see that this construction predicts the conditions for optical bistability very well when compared with the results of fig. 5. A full theoretical dynamical modelling of optical bistability in CdHgTe under these conditions is given elsewhere[23].

THEORY OF SELF-DEFOCUSING

A material possessing an intensity dependent refractive index can give rise to self-focusing or defocusing of a laser beam through distortion of the wavefront phase profile as the beam passes through the material[2,10,26]. For a negative refractive index change with intensity, the central more intense part of the beam experiences a phase advance relative to the wings resulting in self-defocusing[26]. For a thin slice of material at a beam waist, nonlinear refraction produces only a phase advance without any alteration of the beam width within the sample but this results in an increased angle of refraction from the sample and distortion of the beam in the far field. Experiments described in the next section, time resolved far field beam distortions of initially Gaussian transverse profile $CO_2$ laser pulses after passage through InSb and CdHgTe under one-, two- and three-photon excitation conditions[27].

Knowledge of the induced carrier population is required throughout the pulse length in order to correctly model the phase change at any instant in the sample caused by the electronically induced nonlinearity. This is again directly influenced by the Auger recombination process (see eq. 6) and determines the phase distortion across the transverse beam profile at the exit face of the sample. For a nonlinearity of the form $\Delta n = \gamma_s I^s$, the instantaneous on-axis phase shift is given by[10],

$$\Delta\theta = \frac{2\pi\gamma_s}{\lambda} I_0^s \frac{[1 - \exp(-s\alpha L)]}{s\alpha} \tag{11}$$

where $I_0$ is the intensity just inside the front face of the sample. Thus, at the exit face, the field amplitude is given by,

$$E(r,0) = E(0,0) \exp\left[-\frac{r^2}{\omega_0^2} + i\Delta\theta \exp\left(-\frac{2sr^2}{\omega_0^2}\right)\right] \tag{12}$$

Using the theory of Weaire et al[26], this radial field profile at the exit face can be expressed as a sum of Gaussian profiles of decreasing spot size radius[27,28],

$$E(r,0) = E(0,0) \sum_{m=0}^{\infty} \frac{(i\Delta\theta)^m}{m!} \exp\left(-\frac{r^2}{\omega_m^2}\right) \tag{13}$$

In this way, the initial beam profile is expressed as a sum of Gaussian beams of increasing radius,

$$\omega_m^2 = \frac{\omega_0^2}{2ms + 1} \qquad m = 0, 1, 2... \tag{14}$$

and the propagation of such beams is described straightforwardly to any point, z, by the usual linear propagation formula,

$$E(r,z) = E(0,0) \sum_{m=0}^{\infty} \frac{(i\Delta\theta)^m}{m!} \left[1 + \left(\frac{z}{d_m}\right)^2\right]^{-\frac{1}{2}} \exp\left[-\frac{r^2}{\omega_m^2(z)} - i\left(\frac{kr^2}{2R_m(z)} - P_m(z)\right)\right] \tag{15}$$

157

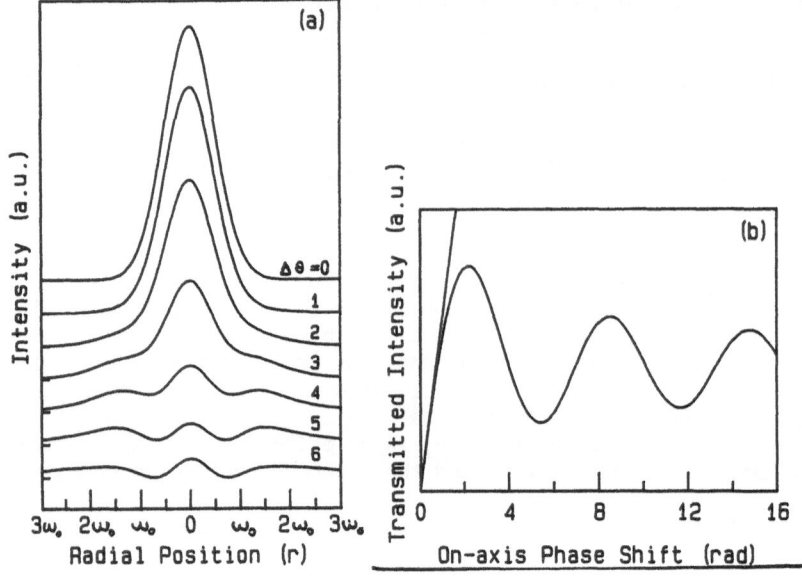

Fig. 7. a) Calculated far field beam profiles for different
on-axis phase shifts, $\Delta\theta$.
b) On-axis intensity as a function of on-axis shift.

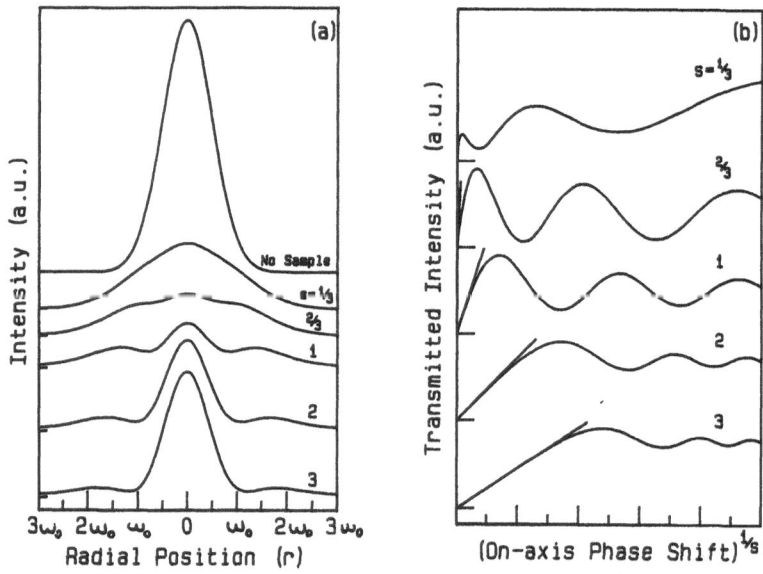

Fig. 8. Comparison of a) far field beam profiles, $\Delta\theta = 4$rad.
and b) on-axis intensity as a function of
intensity for different power dependences, s.

For the mth Gaussian beam, the beam radius at distance, z, is given by, $\omega_m(z) = \omega_m(0) [1 + (z^2/d_m^2)]^{\frac{1}{2}}$, the radius of curvature is, $R_m(z) = z [1 + (d_m^2/z^2)]$, the phase parameter is $P_m(z) = -\tan^{-1}(z/d_m)$ and the Rayleigh range is $d_m = \pi\omega_m(0)^2/\lambda$.

To illustrate the type of laser beam profiles that can be expected after transmission through a thin nonlinear medium located at a beam waist, we have used eq.15 to calculate some examples under steady state conditions. Figure 7(a) shows a sequence of profiles at a (far field) distance of 10cm beyond a sample for the same wavelength and focusing conditions used in our experiments. Results for values of on-axis phase shift, $\Delta\theta$, from 0 to 6 radians are plotted (i.e. equivalent to increasing intensity) assuming first a nonlinearity in which the refractive index change is proportional to intensity. Considerable distortion of the beam is seen to occur with much of the energy being spread into the wings of the profile for $\Delta\theta>3$. Figure 7(b) shows the on-axis transmitted beam intensity at up to an on-axis shift in the sample of 16 rad. giving an oscillatory dependence.

For the present case of single and multiphoton excitation of semiconductors with Auger carrier recombination we know that different power dependencies result. Figure 8 illustrates how important it is to take proper account of these dependences by making a comparison of far field profiles for $\Delta\theta=4$. We can expect other effects to influence the results[27]. Additional absorption from the generated free carriers will alter the transmission. Heating of the crystal lattice because of absorption has been found to give significant nonlinear refractive contributions. It should be remembered that the refractive index change on heating is negative for CdHgTe compared to positive for InSb because of the opposite band gap temperature coefficient.

TRANSIENT SELF-DEFOCUSING EXPERIMENTS

We have studied CdHgTe and InSb samples[27] with the same band gap energies as those listed in table 1. A hybrid $Co_2$ laser operating at 10.6μm was focused to a beam waist of, $\omega_0 = 150$μm at the sample plane, the pulses being monitored before and after the sample with fast detectors to time resolve the temporal pulse profiles. The detector monitoring transmission was placed on-axis 21mm beyond the sample, the 1mm detector element only intercepting the central portion of the transmitted transverse profile. The CdHgTe samples were 200μm thick while the InSb was 500μm thick. In this case, the samples were not highly polished to minimise etalon effects caused by the surface reflectivity. The samples did show residual etalon fringes on rotation and so for this experiment they were rotated to an etalon minimum so that any initial decrease in transmission could not be attributed to nonlinear etalon action.

Figure 9 shows an incident temporal pulse profile and examples of transmitted pulses under two photon excitation conditions for CdHgTe, (the same band gap as sample #2b, table 1) and InSb. At low power, the pulse profiles remain undistorted but as the peak incident intensity is increased the profile flattens and then extreme modulation develops, as can be seen for InSb in fig. 9 (b to d). This modulation occurs due to the transverse profile expanding into a series of rings, each trough appearing at the instant when a dark ring appears on axis (see figure 7(b)). Because the low excitation intrinsic carrier lifetimes of both materials are much shorter than the pulse length of 1μs, (and these lifetimes decrease further during the pulse because of the density dependent Auger lifetime, eq.5), the materials can respond in a quasi-

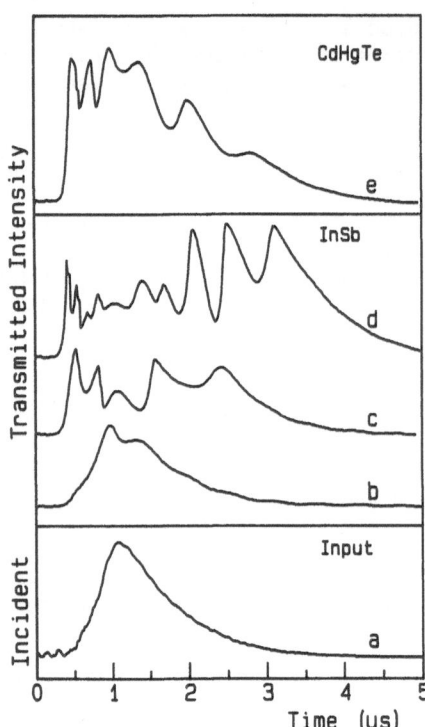

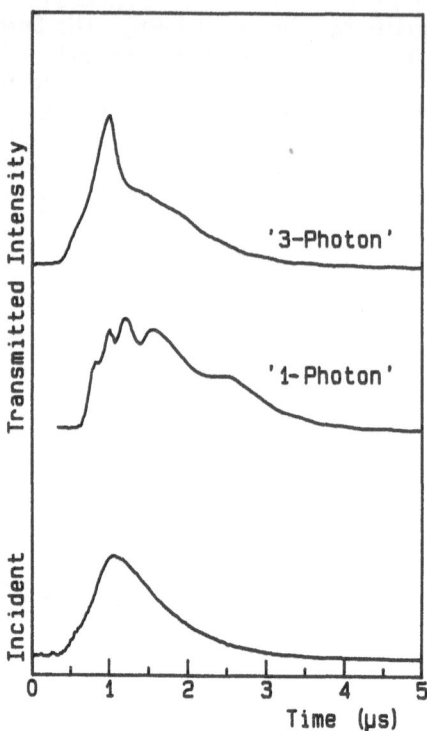

Fig. 9. Laser input pulse (a) and transmitted pulses for InSb (b) 58, (c) 337, (d) 2300kW/cm² and (e) Cd$_{.23}$Hg$_{.77}$Te at 2100kW/cm² peak intensity.

Fig. 10. Incident pulse shape and transmitted pulses for Cd$_{.18}$Hg$_{.82}$Te at 430 and Cd$_{.29}$Hg$_{.71}$Te at 1570kW/cm²

steady state fashion to changes in incident intensity. The modulation is repeated on the falling edge of the pulse due to the collapse of the ring structure. We estimate on-axis phase changes of up to $\Delta\theta = 40$rad. in the 500µm thick sample of InSb.

Figure 10 shows time resolved, on axis transmission results for CdHgTe with band gaps equal to samples #1(band tail absorption) and sample #3 (three photon absorption) respectively. For the band gap resonant sample, beam distortion becomes apparent at relatively low intensities (<50kW/cm²) with a fast response time (<4ns). In contrast, the three photon case requires a much higher intensity to produce an observable effect and the long carrier lifetime means that the modulation is not repeated in the later part of the pulse.

Two photon excitation conditions in CdHgTe were modelled by first computing the dynamics of the induced carrier population and then calculating the propagation of the wavefronts beyond the sample using equations 4, 5, and 11 to 15. These equations are solved self-consistently by a predictor-corrector type computer model to account for any dynamic effects assuming no spot size change in the sample and a spatially uniform absorption. It was found that for the majority of the pulse duration the excited carrier population obeys a $\Delta N \propto I^{2/3}$ dependence. An s value of 2/3 is therefore used in the propagation calculation (eq. 11 to 15). Figure 11 shows experimental and corresponding theoretical transmitted pulse shapes. The calculation uses estimated material parameter values obtained from the literature[29], two-photon absorption coefficient, $\beta = 5cm/MW$, $\tau_i = 98ns$, $N_i = 1.67 \times 10^{16}cm^{-3}$, $\alpha = 7.5cm^{-1}$ and a calculated $\sigma = -1.2 \times 10^{-18}cm^3$. It can be seen that the calculation reproduces the main features of the experimental result implying that the model contains the essential physics of the phenomenon and that the parameters used are approximately correct. The depth of modulation observed in the experiment is shallower than the predicted result because the detector is of finite size and hence there is some integration over the transmitted transverse pulse profile, whereas the calculation is for a single on-axis point. The peak intensity of the pulse used in the calculation is $600kW/cm^2$, in good agreement with the experimental value of $610kW/cm^2$.

CONCLUSIONS

Nonlinear optical phenomena in semiconductors are currently being studied to determine whether they can provide optical signal processing and computing functions which could out-perform electronics in appropriate applications. Optical bistability and phase conjugation are under very active examination for this reason. Narrow gap semiconductors exhibit extremely large band gap resonant optical non-linearities which could prove suitable for infrared image processing, for instance. Most studies of narrow gap materials have been carried out at low temperature, partly because the smaller thermal carrier density results in more sensitive effects and partly because of compatibility of low temperature band gaps in InSb and InAs with available laser sources. Low temperatures are inconvenient for device applications and in this paper, we have described room temperature optical nonlinearities at 10.6µm in narrow gap semiconductors through experiments which have exploited the variation of band gap energy with composition in the alloy $Cd_xHg_{1-x}Te$. Etalon effects and self-defocusing cause dramatic changes in the transmission of $CO_2$ laser radiation through samples of CdHgTe and InSb and the values of nonlinear refraction, $n_2$, under the conditions employed for the different band gap energies are listed in table 1. These are usefully large values of nonlinear refraction. We find a drop of two orders of magnitude in the value of $n_2$ over the range from band gap resonance to three photon absorption conditions. The nonlinear refraction has a very fast response under band gap resonant conditions, while multi-photon excitations still provide relatively large nonlinearities with low background absorption because of the longer carrier lifetimes in higher gap materials. Narrow gap, low dimensional semiconductors should give further scope for novel nonlinear optical phenomena[30].

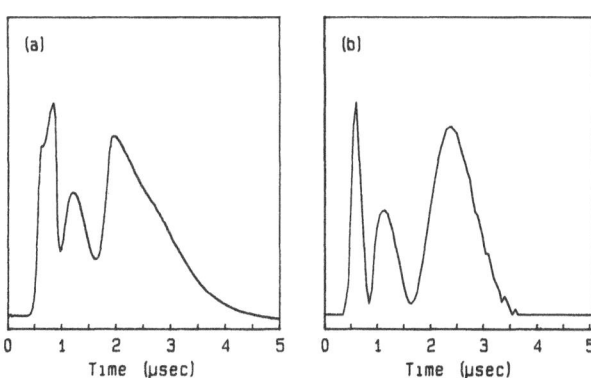

Fig. 11. (a) Measured transmitted pulse for $Cd_{.23}Hg_{.77}Te$ at $610kW/cm^2$.
(b) Theoretically modelled pulse at $600kW/cm^2$ peak intensity.

ACKNOWLEDGEMENTS

We thank J.G.H. Mathew, A.K. Kar and M.J. Soilleau for contributions to this work and Mullard Ltd for supplying the CdHgTe samples.

REFERENCES

1. A. Miller, D.A.B. Miller and S.D. Smith, Dynamic nonlinear optical processes in semiconductors, Adv.Phys. 30:697 (1981).
2. D.A.B. Miller, M. Mozolowski, A. Miller and S.D. Smith, Nonlinear optical effects in InSb with a cw CO laser, Opt.Comm. 27:133 (1978).
3. D.A.B. Miller, S.D. Smith and A. Johnston, Optical bistability and signal amplification in a semiconductor crystal: applications of new low power nonlinear effects in InSb, Appl.Phys.Lett. 35:658 (1979).
4. A.K. Kar, J.G.H. Mathew, S.D. Smith, B. Davis and W. Prettl, Optical bistability in InSb at room temperature with two photon excitation, Appl.Phys.Lett. 42:334 (1983).
5. A.C. Walker, F.A.P. Tooley, M.E. Prise, J.G.H. Mathew, A.K. Kar, M.R. Taghizadeh and S.D. Smith, InSb devices: transphasors with high gain, bistable switches and sequential logic gates, Phil.Trans.R.Soc.Lond. A313:249 (1984).
6. H.M. Gibbs, "Optical bistability: Controlling light with light," Academic Press, Orlando, (1985).
7. D.A.B. Miller, C.T. Seaton, M.E. Prise and S.D. Smith, Band gap resonant nonlinear refraction in III-V semiconductors, Phys.Rev. Lett. 47:197 (1981).
8. B.S. Wherrett and N.A. Higgins, Theory of nonlinear refraction near the band edge of a semiconductor, Proc.R.Soc.Lond. A379:67 (1982).
9. C. Poole and E. Garmire, Bandgap resonant optical nonlinearities in InAs and their use in optical bistability, IEEE J.Quantum Electron. QE-21:1370 (1985).
10. J.R. Hill, G. Parry and A. Miller, Nonlinear refractive index changes in CdHgTe at 175K with 10.6$\mu$m radiation, Opt.Commun. 43:151 (1982).
11. A. Miller, G. Parry and R. Daley, Low power nonlinear Fabry-Perot reflection in CdHgTe at 10$\mu$m, IEEE J.Quantum Electron. QE-20:710 (1984).
12. G.L. Hansen, J.L. Schmit and T.N. Casselman, Energy gap versus alloy composition and temperature in $Hg_{1-x}Cd_xTe$, J.Appl.Phys. 53:7099 (1982).
13. B.S. Wherrett, Scaling rules for multiphoton interband absorption in semiconductors, J.Opt.Soc.Am. B1:67 (1984).
14. J.P. Woerdman, Diffraction of light by laser generated free carriers in Si: dispersion or absorption?, Phys.Lett. A32:305 (1970).
15. J. Blakemore, Ch.6, "Semiconductor statistics," Pergamon, Oxford, (1962).
16. J.A. Mroczkowski and D.A. Nelson, Optical absorption below the absorption edge in $Hg_{1-x}Cd_xTe$, J.Appl.Phys. 54:2041 (1983).
17. S.W. Kurnick and J.M. Powell, Optical absorption in pure single crystal InSb at 298° and 78°K, Phys.Rev. 116:597 (1959).
18. D.A.B. Miller, Refractive Fabry-Perot bistability with linear absorption: theory of operation and cavity optimization, IEEE J. Quantum Electron. QE-17:306 (1981).

19. B.S. Wherrett, Fabry-Perot bistable cavity optimization on reflection, IEEE J. Quantum Electron. QE-17:306 (1981).

20. A. Miller and G. Parry, Optical bistability in semiconductors with density dependent carrier lifetimes, Opt.Quantum Electron. 16:339 (1984).

21. J.H. Marburger and F.S. Felber, Theory of lossless nonlinear Fabry-Perot interferometer, Phys.Rev. A17:335 (1978).

22. D. Craig and A. Miller, Room-temperature optical nonlinearities in CdHgTe, Optica Acta 33:397 (1986).

23. D. Craig, A.K. Kar, J.G.H. Mathew and A. Miller, Two photon induced optical bistability in CdHgTe at room temperature, IEEE J. Quantum Electron. QE-21:1363 (1985).

24. D. Craig, A. Miller, J.G.H. Mathew and A.K. Kar, Fast optical switching and bistability in room temperature CdHgTe at 10.6μm, Infrared Phys. 25:289 (1985).

25. J.G.H. Mathew, D. Craig and A. Miller, Optical switching in a CdHgTe etalon at room temperature, Appl.Phys.Lett. 46:128 (1985).

26. D. Weaire, B.S. Wherrett, D.A.B. Miller and S.D. Smith, Effect of low-power nonlinear refraction on laser-beam propagation in InSb, Opt.Lett. 4:331 (1979).

27. D. Craig, A. Miller and M.J. Soileau, Time resolved self-defocusing in $Cd_{0.23}Hg_{0.77}Te$ and InSb, Opt.Lett. (Dec. 1986).

28. J.G.H. Mathew, A.K. Kar, N.R. Heckenberg and I. Galbraith, Time resolved self defocusing in InSb at room temperature, IEEE J. Quantum Electron. QE-21:94 (1985).

29. A. Miller, D. Craig, G. Parry, J.G.H. Mathew and A.K. Kar, Optical bistability in $Cd_xHg_{1-x}Te$, in: "Digital optical circuit technology," AGARD conference proceedings, No. 362, NATO, Neuilly sur Seine (1985).

30. D.A.B. Miller, Optical nonlinearities in low dimensional structures, this volume.

OPTICAL NONLINEARITIES IN LOW-DIMENSIONAL STRUCTURES

D.A.B. Miller

AT & T Bell Laboratories
Holmdel, NJ

Fabrication of semiconductor microstructures opens up new opportunities in optics. In quantum wells, the resulting particle-in-a-box behaviour leads to new optical properties near the optical absorption edge that are applicable at room temperature in GaAs/GaAlAs and other materials systems.

One application is to semiconductor diode lasers, where the different density of states has several beneficial consequences [1]. Recently, the nonlinear-optical [2] and electro-optical [3] properties of the so-called exciton absorption resonances have been examined. Unlike bulk semi-conductors, these sharp resonances are seen at room temperature. Their absorption can be saturated relatively easily [2], and this has been applied as a modelocker to generate picosecond pulses from diode lasers [4]. This absorption also shows interesting dynamic effects on a subpicosecond timescale [5]. The physics of this nonlinear absorption differs from bulk material first of all in that exciton effects are stronger. Also, direct Coulomb screening effects appear to be relatively weaker in the quantum wells [6,7], but state-filling and exchange remain strong nonlinear absorption mechanisms for excitonic absorption.

Electric fields parallel to the layers destroy the exciton resonances just as in bulk materials [3]. These large electroabsorptive effects persist to $\lesssim$ 500 fs [8], and may be applicable to sensitive optical diagnostics of very high speed electrical devices. For electric fields perpendicular to the layers, a new effect arises, called the Quantum-Confined Stark Effect (QCSE), that shifts the exciton resonances with field [3]. The relation between the QCSE and the bulk Franz-Keldysh electroabsorption has recently been established [9], with a smooth transition predicted with increasing layer thickness. The QCSE is applicable to small, high-speed electrically-driven optical modulators [10], and to so-called Self Electro-optic Effect Devices (SEED's) [11]. The SEED uses only optical inputs and outputs, and is a potential low energy switch for optical processing applications. Two-dimensional arrays of optically-bistable SEED's have recently been demonstrated [12].

In general, it is clear that the ability to fabricate semiconductor microstructures enables us to investigate new physical mechanisms and engineer them for practical applications. The possible consequences of confinement in other dimensions are tantalizing, with, for example, further enhanced nonlinear optical response recently predicted for fully-confined "quantum dots" [13]. Fabrication of such uniform and small micro-crystallites represents both a major challenge and a significant opportunity.

REFERENCES

1. W.T. Tsang , IEEE J. Quantum Electron. QE20:1119 (1984).
2. D.S. Chemla and D.A.B. Miller, J.Opt.Soc.Am. B2:1155 (1985).
3. D.A.B. Miller, D.S. Chemla, T.C. Damen, A.C. Gossard, W. Wiegmann, T.H. Wood and C.A. Burrus, Phys.Rev.Lett. 53:2173 (1984); Phys.Rev. B32:1043 (1985).
4. P.W. Smith, Y.S. Silberberg and D.A.B. Miller, J.Opt.Soc.Am. B2:1228 (1985).
5. See e.g. W.H. Knox, C. Hirlimann, D.A.B. Miller, J. Shah, D.S. Chemla and C.V. Shank, Phys.Rev.Lett. 56:1191 (1986).
6. S. Schmitt-Rink, D.S. Chemla and D.A.B. Miller, Phys.Rev. B32:6601 (1985).
7. W.H. Knox, C. Hirlimann, D.A.B. Miller, J. Shah, D.S. Chemla and C.V. Shank, Phys.Rev.Lett. 56:1191 (1986).
8. W.H. Knox, D.A.B. Miller, T.C. Damen, D.S. Chemla, C.V. Shank and A.C. Gossard, Appl.Phys.Lett. 48:864 (1986).
9. D.A.B. Miller, D.S. Chemla and S. Schmitt-Rink, Phys.Rev. B33:6976 (1986).
10. T.H. Wood, C.A. Burrus, R.S. Tucker, J.S. Weiner, D.A.B. Miller, D.S. Chemla, T.C. Damen, A.C. Gossard and W. Wiegmann, Electron. Lett. 21:693 (1985).
11. D.A.B. Miller, D.S. Chemla, T.C. Damen, T.H. Wood, C.A. Burrus, A.C. Gossard and W. Wiegmann, IEEE J. Quantum Electron. QE21:1462 (1985).
12. D.A.B. Miller, J.E. Henry, A.C. Gossard and J.H. English, to be published.
13. S. Schmitt-Rink, D.A.B. Miller and D.S. Chemla, to be published.

# ENERGY RELAXATION PHENOMENA IN GaAs/GaAlAs STRUCTURES

Erich Gornik

Institut für Experimentalphysik, Univ. Innsbruck
A-6020 Innsbruck, Austria

The energy relaxation of 2D electrons in GaAs/GaAlAs structures has been investigated by analysing the electric field dependence of Shubnikov-de Haas oscillations, the far infrared emission and photoluminescence spectra. A quite general behavior of the electron heating $\Delta T = T_e - T_L$ as a function of the input power per electron $P_e$ is found: $T \propto \sqrt{P_e}$. The corresponding energy relaxation times in the range of nsec are independent of the electron temperature up to 30 K and inversly proportional to the electron density. At higher electron temperatures the energy relaxation is governed by optical phonon emission. However, the onset depends on electron concentration and is different for heterostructures and quantum wells. From intensity dependent cyclotron resonance transmission experiments Landau level lifetimes between 0.2 ns and 1 ns depending on the electron density are found in agreement with data from time-resolved photoluminescence.

## INTRODUCTION

The mechanism of energy relaxation is of fundamental interest for the understanding of electric field effects in two-dimensional (2D) electron systems. At low temperatures, where lattice scattering is weak, hot electron phenomena can be produced by fields as low as V/cm in Si as well as GaAs inversion layers.

The first investigations of energy relaxation in 2D systems were performed by Hess et al.[1]. The energy loss was determined from Shubnikov-de Haas (SdH) experiments as a function of electric field for temperatures up to 20 K in p-Si inversion layers. Similar investigations were performed by Englert and Landwehr[2] for Si-n-inversion layers. Sakakai et al.[3] investigated with the same technique n-GaAs inversion layers in heterostructures. For temperatures below 30 K a dominance of acoustic phonon relaxation is found for both materials.

Emission techniques including the analysis of the broadband hot electron emission[4,5], subband emission[6] and the plasmon emission[7,8] have been used to determine the energy loss rate as a function of input power. It was found that this rate is sample-independent as long as acoustic phonon scattering is dominant. A theoretical description of this phenomenon was given by Vass[9,10] and Price[11,12].

Very fundamental information on the hot electron distribution and energy loss rate is obtained from time resolved photoluminescence experiments performed by Shah et al.[13,14]. Over a wide range of electric fields a hot electron temperature-like distribution function was found. For temperatures above 50 K the energy loss rate is dominated by the emission of longitudinal optical (LO) phonons. Evidence for screening of the electron LO phonon interaction[15,16] and for hot phonon effects[17,18] is found.

The energy relaxation in the presence of a strong magnetic field has been studied only recently. The saturation behavior of the cyclotron resonance (CR) absorption as a function of laser power yields an electron density dependent energy relaxation time[19]. From time dependent luminescence experiments in GaAs/GaAlAs quantum wells a decrease in the relaxation rate with magnetic field is found for fields below 10 T[20] and an increase for fields of 20 T[21].

In the present paper a summary on experimental techniques to determine the energy relaxation in GaAs/GaAlAs heterostructures in the temperature range up to 100 K is given. The results will be compared with recent theoretical results. In the presence of a magnetic field a qualitative description of the experimental findings will be presented due to the lack of theoretical work.

ENERGY RELAXATION AT ZERO AND LOW MAGNETIC FIELDS

The electron heating and the energy relaxation in 2D GaAs was first investigated by analysing Shubnikov-de Haas (SdH) oscillations at low magnetic fields (B < 3 T) as a function of the input power. Optical techniques as photoluminescence and far infrared (FIR) emission at zero magnetic field followed. The classical technique evaluating the time dependence of the hot electron current following a small step variation in electric field[22] has not been applied to a 2D system so far.

a) Electric field dependence of SdH oscillations

A damping of SdH oscillations in magneto-resistance is observed when the 2D electron gas (2DEG) is heated in an electric field. To determine the electron temperature $T_e$ from the temperature and electric field dependent oscillations, the Dingle temperature must be constant[23]. If that is the case the change in amplitude can be described by one parameter $T_e$. Fig. 1 shows a typical result of SdH measurements after Sakaki et al.[3] at 4.2 K. The decay in SdH amplitude is clearly observed with increasing current (corresponding to increasing electric field). Significant differences in the heating behavior are found as a function of zero field mobility. However, a very general behavior is obtained when $T_e$ is plotted against the input power per electron. The weakness of this technique lies in the fact that $P_e = e\mu E^2$ is calculated using the electric field dependent mobility at zero magnetic field while the $T_e$ values are determined from SdH oscillations at finite B.

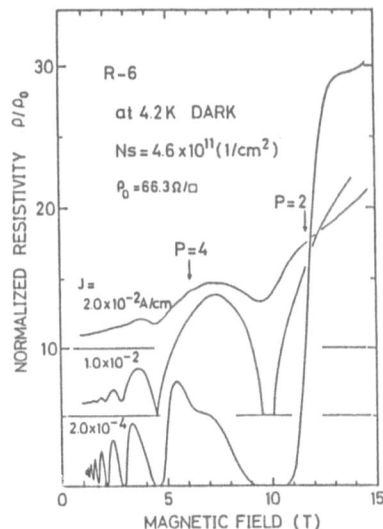

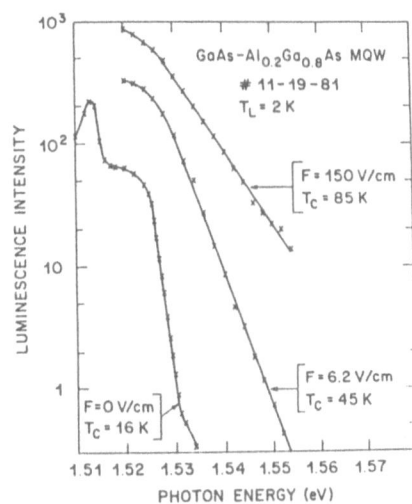

Fig. 1: Shubnikov-de Haas oscillations of Sample R-6 after Ref. 3.
The sample has a mobility of
$2.05 \times 10^5 \text{cm}^2/\text{Vs}$.

Fig. 2: Photoluminescence spectra
of GaAs/AlGaAs heterostructures
for three in-plane electric fields.
The electron temperatures deduced
from the high-energy slope are indicated. Relative intensities of curves
are arbitrary.(after Ref. 13).

 

The results from Sakaki et al.[3] will be shown in Fig. 4
together with the data from the other techniques. Similar
results were obtained by other authors studying the mobility
and electron temperature as a function of electric field[24,25,26].

## b) Hot electron luminescence analysis

Luminescence experiments on GaAs/GaAlAs multiple quantum
well (MQW) structures were performed by Shah et al.[13,14]. A
cw infrared dye laser (7700 Å $< \lambda <$ 7900 Å, P = 2 mW) was used
to weakly excite carriers in the GaAs across the gap. Luminescence spectra as a function of electric field were analysed
with a double monochromator. Typical spectra after Shah et al.[14]
are shown in Fig. 2 for three different electric fields. The
high energy tail is attributed to electron hole recombination.
This tail can be well fitted by a single exponential (see Fig.2)
at all fields up to the highest field (150 V/cm). The spectra
are fitted with a Maxwell-Boltzmann type distribution with a
temperature $T_e$ higher than $T_L$.

The obtained electron heating $\Delta T$ is also plotted in Fig.4
over the input power derived from I-V measurements. In
a steady state situation the power input must be equal to the
power loss to the lattice.This technique gives direct means
to determine energy loss rates at zero magnetic field. It is
found that the electron heating shows a somewhat weaker dependence on input power than the previous technique and that there
is also practically no dependence on the carrier concentration.

The disadvantage of this technique is the simultaneous excitation of holes which can influence the total energy loss rate. In addition we have to be aware of the fact, that the experiments were performed on MQW which may give a somewhat different behavior than heterostructures. A comparison will only be meaningful for rather wide wells (as in the present case: well thickness 250 Å).

## c) Far infrared (FIR) emission

The basic idea of this experiment is to heat up the 2D carrier gas by electric field pulses and to measure the spectrum and intensity of the emitted broadband FIR light and to correlate it with an electron temperature.

FIR broadband emission has first been used in Si-MOSFETs to determine the energy loss of 2D carriers[4]. The electron temperature $T_e$ was determined from the absolute value of the emitted power. For 2D electrons in GaAs a somewhat different experimental technique based on a relative intensity measurement was applied. The signals of two narrowband FIR detectors (n-GaAs at 35 cm$^{-1}$, Ga doped Ge at 100 cm$^{-1}$) are measured as a function of the applied electric field and calculated for each detector according to

$$U = \int_0^\infty R(\omega)A(\omega)I_{BB}(\omega,T_e) \cdot d\omega \tag{1}$$

where $R(\omega) = R_o r(\omega)$ and $A(\omega) = A_o a(\omega)$ are the frequency dependent and known detector response and the absorptivity of the electron system respectively. In equilibrium the spectral emission intensity of a system with absorptivity $A(\omega)$ is given by

$$I(\omega,T) = I_{BB}(\omega,T) \cdot A(\omega) \tag{2}$$

with $I_{BB}$ the black body emission intensity (Planck-function). Assuming a quasi equilibrium of the electron system at a temperature $T_e$, we can express the broadband emission from the 2D carrier gas by $I(\omega,T_e)$.

The absorptivity can be expressed for normal incidence according to

$$A(\omega) = \frac{4ReF}{(\sqrt{\varepsilon_s} + 1 + F)^2} \tag{3}$$

with $F = \sigma(\omega)/\varepsilon_o \cdot c$ and $\varepsilon_s$ being the dielectric constant of the substrate. For $\sigma(\omega)$ a Drude type frequency behavior is assumed.

The unknown parameters in equ. (1) are determined by measuring with two detectors at two different electric fields giving four equations for four unknown factors. Fig. 3 shows the observed detector signal in Volts as a function of electric field for two different samples: a low mobility sample with $n_s = 8.7 \times 10^{11}$cm$^{-2}$ and $\mu = 1.6 \times 10^3$cm$^2$/Vs (Sample 1) and a high mobility sample with $n_s = 2.4 \times 10^{11}$cm$^{-2}$ and $\mu = 1.0 \times 10^6$cm$^2$/Vs (Sample 2, dark). It is directly evident that the lower mobility sample shows a considerably higher emission signal.

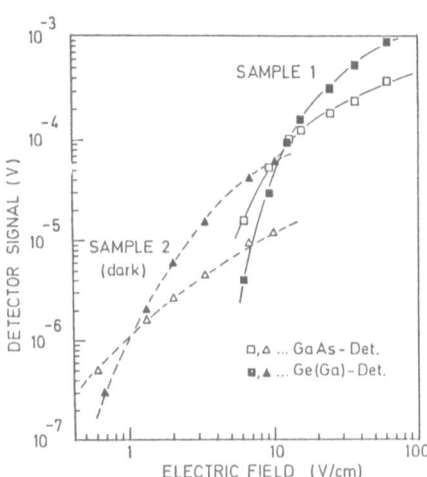

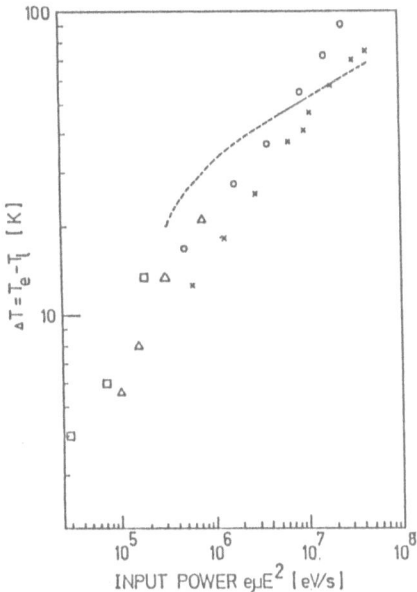

Fig. 3: Detector signals as a function of applied electric field for two different detectors after Ref. 5. The sample and detector properties are given in the text.

Fig. 4: Electron heating versus input power per electron obtained from the different techniques:
Ref. 3: ($\Delta$) sample described in Fig. 1, ($\square$) $N_s = 3.5 \times 10^{11} cm^{-2}$, $\mu = 7.0 \times 10^4 cm^2/Vs$.
Ref. 5: (x) Sample 1, (o) Sample 2 (see text); dashed curve: data from Ref. 13.

Quantitative values for the electron temperature are determined from the fit of Fig. 3 according to equ. (1) and also plotted as a function of the input power in Fig. 4. The basic behavior of $\Delta T$ is the same as determined with the other techniques. The emission data have a somewhat steeper slope than the luminescence data. There is also a weak dependence on density evident in the emission data.

The emission technique seems to be the most appropriate method to determine $T_e$. However, the analysis requires several assumptions which might introduce sample dependences.

A similar FIR emission technique is the analysis of the emission from 2D plasmon excitations as demonstrated first in Si[27,7] and later in GaAs[28]. In a recent paper Sambe et al.[29] have analyzed the emitted power from 2D plasmon excitation in GaAs and derived electron temperatures as a function of electric field. As mobility on electric field data were not published an inclusion of their data in Fig. 4 is not possible.

The summary of the experimental findings is shown in Fig. 4 as a plot of $\Delta T = T_e - T_L$ over the input power per electron $P_e$. It is clearly evident that the emission data continue very well the SdH results. The data can be described by a relation $\Delta T \propto \sqrt{e\mu E^2}$ up to electron temperatures of 30 K. For higher temperatures the emission data show a weak, while the luminescence data show a considerably stronger change in slope.

## d) Determination of the energy relaxation time

To derive an energy relaxation time from the data an energy balance equation for the average electron energy $\varepsilon(T)$ using Fermi-Dirac statistics with an electron temperature $T_e$ is used:

$$|\varepsilon(T_e) - \varepsilon(T_L)|/\tau_\varepsilon = \frac{\pi^2 k_B^2}{6\varepsilon_F}(T_e^2 - T_L^2)/\tau_\varepsilon = e\mu E^2 = P_e \qquad (4)$$

where $\varepsilon_F$ is the Fermi energy and $\mu$ the electric field dependent mobility. It is directly evident that the expression $\Delta T = T_e - T_L = \text{const.}\ \sqrt{P_e \cdot \varepsilon_F \cdot \tau_\varepsilon}$ describes the experiments quite well indicating that for a given $P_e$ the product $\varepsilon_F \cdot \tau$ has to be sample independent. As $N_s = D(\varepsilon) \cdot \varepsilon_F$ with $D(\varepsilon) = \text{const.}$ $\tau_\varepsilon$ is inversely proportional to the 2D carrier density.

The resulting energy relaxation times as a function of density as obtained from Fig. 4 for the range of electron temperatures, where the slope is constant, is plotted in Fig.5. As long as the slope of $\Delta T$ over $P_e$ is constant we obtain a $\tau_\varepsilon$-value independent of $T_e$ for a given sample. It is clearly evident that the energy relaxation time is a linear inverse function of the density as a consequence of the analysis with equ. (4). Both methods, the SdH oscillations and the emission technique, give within experimental accuracy the same $\tau_\varepsilon$-values so that we can be quite confident about the obtained results.

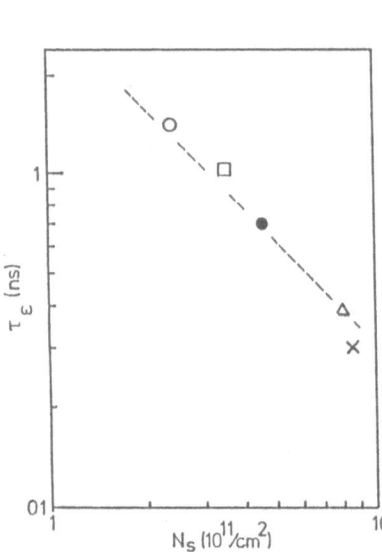

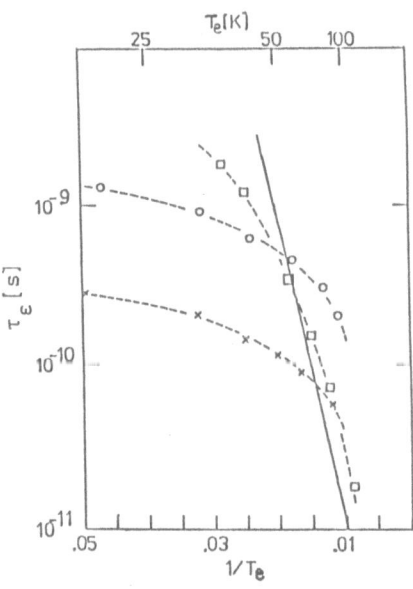

Fig. 5: Calculated energy relaxation time versus 2D electron density for three samples from Ref. 3 ($\square$, $\bullet$, $\triangle$) and two samples from Ref. 5 (o, x). The symbols represent the same samples as in Fig. 4.

Fig. 6: Energy relaxation times versus electron temperature for Sample 1 (x) and Sample 2 (o) from Ref. 5 and for a sample with $N_s = 3.9 \times 10^{11} cm^{-2}$ and $\mu = 7.9 \times 10^4 cm^2/Vs$ from Ref. 3. The full curve represents the loss due to optical phonons[14].

An evaluation of $\tau_\epsilon$ over $T_e$ is only meaningful for temperatures above 30 K, where the slope starts changing. But also in the higher temperature range we use equ. (4) to evaluate $\tau_\epsilon$ which means that we use degenerate Fermi-Dirac statistics with an electron temperature $T_e$. The use of the statistics has a strong influence on the results. A plot of $\tau_\epsilon$-values from the emission data for two different samples and for the photoluminsescence data is shown in Fig. 6 as a function of the inverse electron temperature (and electron temperature). The energy relaxation times derived from the emission data show a similar behavior for both concentrations: For temperatures above 40 K $\tau_\epsilon$ decreases significantly reaching values of a few hundred psec. The analysis of the photoluminescence data gives a considerably stronger dependence of $\tau_\epsilon$ on $T_e$ which is already evident in the weaker $\Delta T$ over $P_e$ curve. The full curve shows calculations of the energy relaxation rate due to optical phonons for a 3D case[14]. This curve has a slope proportional to the optical phonon energy (36 meV) in GaAs. It is evident from Fig. 6 that for both techniques there is a tendency toward the optical phonon line. However, there is a clear difference in the behavior of the results from the two techniques. The photoluminescence data seem to be more strongly influenced by the optical phonon emission. The reason for this is not clear but might be due to the fact that in the photoluminescence experiment electrons are always excited above the optical phonon energy by the laser excitation.

In Ref. 14 it was argued that the energy loss rate for electrons was considerably smaller than for holes. However, we do not find this result since the energy loss rates shown in Fig. 6 are comparable with loss rates of holes. The difference in the data comes from the analysis. Shah et al.[14] used non-degenerate statistics. The $\tau_\epsilon$-values obtained this way are longer by a factor $\varepsilon_F/\Delta T$. For samples with Fermi energies of the order of 20 meV the use of non-degenerate statistics seems not appropriate.

A critical analysis of the energy relaxation was performed by Tsubaki et al.[30] who analysed transport studies as a function of electric field from several authors. The influence of the used statistics in the evaluation was examined. It is shown that the application of degenerate statistics is meaningful down to $\varepsilon_F/k_B T$-values of 3. However, the obtained $\tau_\epsilon$-values are independent of $N_s$ and are nearly an order of magnitude shorter than the data shown in Fig. 6. At low temperatures the values are comparable, however, the slope of the $\tau_\epsilon$ versus $T_e$ plot is considerably steeper (in Ref. 30) than the optical phonon curve (see full curve in Fig. 6).

A theoretical description of the energy relaxation rate for 2D electrons in GaAs in the case of acoustic phonon scattering has been published by Vass[9]. Vass derived an expression for the energy loss rate: $\Delta T = $ const. $\sqrt{P_e}$ where the calculated value of the constant agrees well with the experimental findings in GaAs and Si[4]. The power loss is dominated by acoustic deformation potential scattering up to temperatures of 40 K. In a recent paper[10] values for the energy relaxation time as a function of electric field and electron concentration were reported including optical phonon emission and electron-electron scattering. The calculated values show the same electric field (or $T_e$) dependence as the experimental values but are an order of magnitude to high. The predicted carrier density dependence is also too weak.

Price has investigated the mobility and energy loss rate for piezo-electric and deformation potential acoustic phonon scattering[13]. For temperatures up to 10 K he predicts piezo-electric scattering to be dominant giving a considerably steeper dependence of $\Delta T$ on $P_e$ than from equ. (4). For higher temperatures he predicts a linear dependence of $\Delta T$ on $P_e$ which is also dependent on the sample concentration. These results are not in agreement with the experimental observations.

## ENERGY RELAXATION IN HIGH MAGNETIC FIELDS

A direct way to obtain the energy relaxation rate in the presence of a strong magnetic field is the measurement of the incoherent saturation of the CR transmission. This technique has been applied to bulk n-type InSb and GaAs[31,32]. It was found that an interplay between electron-electron scattering and optical phonon emission governs the energy relaxation. The same technique was applied to GaAs/GaAlAs heterostructures by Helm et al.[19]. A high power cw optically pumped FIR laser was used to perform the experiments. Typical transmission spectra for two different samples are shown in Fig. 7. It is evident that powers below 1 W/cm$^2$ reduce the transmission already significantly and that the lower density sample saturates stronger. In the analysis rate equations for three Landau levels and a constant relaxation time $\tau_B$ was used. Fig. 8 shows the obtained relaxation time $\tau_B$ plotted against the total 2D carrier concentration. A systematic decrease of $\tau_B$ with $N_s$ is observed even though there is considerable scatter in the data. The dependence of $\tau_B$ on $N_s$ indicates that the mechanism may be similar to that found in bulk semiconductors[31]. It is also interesting to note that in the zero magnetic field case a similar dependence is found.

Data on relaxation rates between Landau levels were also reported by Ryan et al.[20] and Hollering et al.[21] using psec photoluminescence techniques. Ryan et al. find relaxation times in the order of 0.3 ns between the first excited and lowest Landau level for MQW samples with $N_s = 5 \times 10^{11}$cm$^{-2}$. The relaxation times in magnetic fields of 7 and 8 T are longer than without magnetic field. On the other hand Hollering et al. find a considerably shorter time for fields of 20 T as compared to the zero magnetic field case.

The whole behavior seems to be consistent: In fields below 10 T the energy relaxation time derived is increased as compared to the field free case. The data from Fig. 8 have to be compared with $\tau_\varepsilon$-values in Fig. 6 at the highest temperatures. This comparison shows that they are somewhat longer in agreement with Ryan's observation on the same sample. The considerably shorter values for 20 T are somewhat surprising. In Ref. 21 they are briefly explained by a reduced screening of the electron LO phonon interaction.

Further experimental and theoretical studies in the magnetic field are necessary to get clear evidence for the relaxation mechanism. In a very recent paper Rodriguez et al.[33] report an intensity dependent CR transmission experiment at rather low

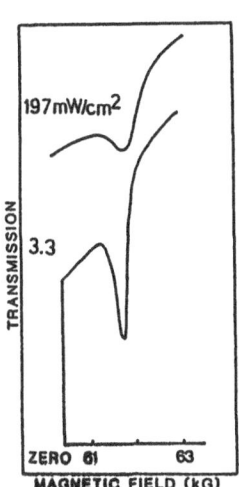 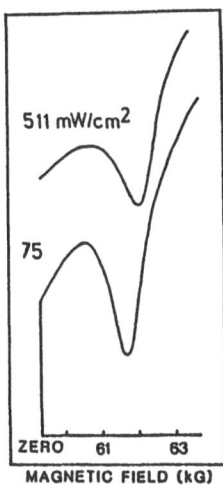 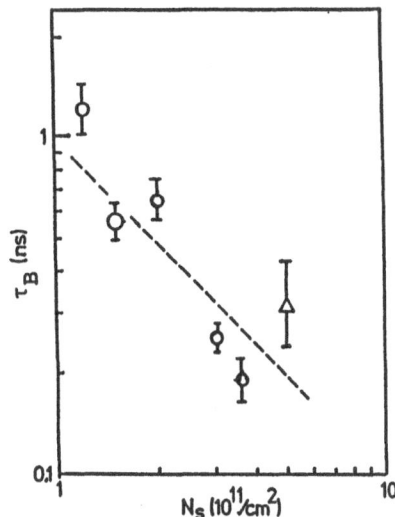

Fig. 7: Cylotron resonance absorption for two different GaAs/GaAlAs hetero-layers at 118 μm for different laser intensities: left sample with $N_S = 1.2 \times 10^{11} cm^{-2}$, right sample with $N_S = 3.0 \times 10^{11} cm^{-2}$.

Fig. 8: Calculated lifetimes from saturation behavior as a function of $N_S$: (o) cyclotron resonance, (Δ) photoluminescence. The dashed curve represent a linear dependence.

frequencies. Saturation is observed at considerably higher intensities than in Ref. 19 for magnetic fields below 4 T. This indicates a non-resonant heating process similar to direct electric field heating. The observed relaxation times in the order of 10 psec probably directly reveal the optical phonon relaxation time. The main reason for this behavior is the nearly equidistant Landau level spacing at low magnetic field. In this situation a resonant saturation of a transition between two levels is not possible. Only if a sufficient amount of nonparabolicity is present the level shifts are larger than the individual Landau level linewidths and a saturation of an individual transition becomes possible as demonstrated in Ref. 19.

ACKNOWLEDGEMENT

This work was partly supported by the European Research Office of the U.S. Army, London, and the Stiftung Volkswagen-werk, Projekt-Nr. I 61 840, Hannover.

## REFERENCES

1. K. Hess, T. Englert, T. Neugebauer, G. Landwehr and
   G. Dorda, Phys. Rev. B 16, 3652 (1977).
2. T. Englert and G. Landwehr, Phys. Rev. B 21, 702 (1980).
3. H. Sakaki, K. Hirakawa, J. Yoshino, S.P. Svensson,
   Y. Sekiguchi, T. Hotta, S. Nishii and N. Miura, Surf. Sci.
   142, 306 (1984).
4. R.A. Höpfel, E. Vass and E. Gornik, Solid State Commun.
   49, 501 (1984).
5. R.A. Höpfel, E. Gornik and G. Weimann, Proc. of the
   17th Int. Conf. on the Physics of Semiconductors, p. 579,
   Springer-Verlag (1984);
   R.A. Höpfel and G. Weimann, Appl. Phys. Lett. 46, 291 (1985).
6. E. Gornik and D.C. Tsui, Solid State Electron. 21, 139
   (1978).
7. R.A. Höpfel, E. Vass and E. Gornik, Phys. Rev. Lett. 49,
   1667 (1982).
8. R.A. Höpfel and E. Gornik, Surface Science 142, 412 (1984).
9. E. Vass, Solid State Commun. 55, 847 (1985).
10. E. Vass, Physica 134 B, 337 (1985).
11. P.J. Price, J. Appl. Phys. 53, 6863 (1982).
12. P.J. Price, Physica 134 B, 164 (1985).
13. J. Shah, A. Pinczuk, H.L. Störmer, A.C. Gossard and
    W. Wiegmann, Appl. Phys. Lett. 42, 55 (1983).
14. J. Shah, A. Pinczuk, A.C. Gossard and W. Wiegmann, Physica
    134 B, 174 (1985).
15. C.H. Yang and S.A. Lyon, Physica 134 B+C, 305 (1985).
16. S. Das Sarma and B.A. Mason, Physica 134 B+C, 301 (1985).
17. P. Kocevar, Physica Status Solidi (b) 84, 581 (1977).
18. P.J. Price, Phys. Rev. B 30, 2236 (1984).
19. M. Helm, E. Gornik, A. Black, G.R. Allan, C.R. Pidgeon
    and K. Mitchell, Physica 134 B, 323 (1985).
20. J.F. Ryan, Physica 134 B, 403 (1985);
    J.F. Ryan, R.A. Taylor, A.J. Turberfield and J.M. Worlock,
    Physica 134 B, 318 (1985).
21. R.W.J. Hollering, T.T.J.M. Berendshot, H.J.A. Blyssen,
    P. Wyder, M.R. Leys and J. Wolter, Physica 134 B, 422 (1985).
22. J.P. Maneval, A. Zilberstein and H.F. Budd, Phys. Rev. Lett.
    23, 848 (1969).
23. G. Bauer and H. Kahlert, Phys. Rev. 5, 556 (1972);
    H. Kahlert and G. Bauer, Phys. Rev. B 7, 2670 (1973).
24. T.J. Drummond et al., Electron. Lett. 17, 545 (1981).
25. S. Hiyamizu, T. Fujii, T. Mimura, K. Nanbu, J. Saito and
    H. Hae, Japan. J. Appl. Phys. 20, 455 (1981).
26. M. Inoue, M. Inayama, S. Hiyamizu and Y. Inuishi, Japan.
    J. Appl. Phys. 22, 357 (1983) Suppl. 22-1.
27. D. Tsui, E. Gornik and R.A. Logan, Solid State Commun. 35,
    875 (1980).
28. R.A. Höpfel, G. Lindemann, E. Gornik, G. Stangl,
    A.C. Gossard and W. Wiegmann, Surf. Sci. 113, 118 (1982).
29. Y. Sambe et al., Ext. Abstract 17th Conf. on Solid State
    Devices and Materials (Tokyo), 95 (1985).
30. K. Tsubaki, A. Sugimura and K. Kumabe, Appl. Phys. Lett.
    46, 764 (1985).
31. E. Gornik, T.Y. Chang, T.J. Bridges, V.T. Nguyen, I.D.
    Mc Gee and W. Müller, Phys. Rev. Lett. 40, 1151 (1978).
32. G.R. Allan, A. Black, C.R. Pidgeon, E. Gornik, W. Seiden-
    busch and P. Colter, Phys. Rev. B 31, 3560 (1985).
33. G.A. Rodriguez, R.M. Hart, A.J. Sievers, F. Keilmann, preprint.

THE RATE OF CAPTURE OF ELECTRONS INTO THE WELLS OF A SUPERLATTICE

B.K. Ridley

Department of Physics
University of Essex
Colchester, England

Factors affecting the rate of capture from states above the barrier into states localized in the well are discussed. An estimate of the rate is made on the basis of a simple model founded on the polar optical phonon scattering rate in an infinitely deep quantum well with the bulk phonon spectrum. The initial state is assumed to correspond to a well-transmission resonance. This model predicts a rate approximately proportional to the square of the well-width. Modifications to this simple model are made which take account of the actual superlattice bandstructure and eigenfunctions, and non-resonant initial states. In contradistinction to other work only weak resonances are predicted, and it turns out that the simple model remains useful.

INTRODUCTION

The rate at which carriers injected into a superlattice either electrically or optically relax energy is relevant to the speed of operation of a number of devices. In all materials this rate is usually determined by the rate at which optical phonons can be emitted, which means that the spectrum of optical phonons in the superlattice is directly relevant. In polar materials; in addition to the necessity of taking into account modifications of the bulk phonon spectrum wrought by the superlattice structure, it is necessary to consider the effects of screening. In all materials there are also complications associated with other effects of high carrier concentration, for example, the interaction with plasmons and the phenomenon of hot phonons resulting from a high phonon-emission rate. Not surprisingly, in view of the complexity of the problem and its youthfulness, a full description has yet to be made. Simple models always have a role to play in establishing insight, and in the present context they can be, in addition, valuable precursors to a full-blooded theory. It is the purpose of this paper to suggest a simple model of energy relaxation of hot carriers in a polar superlattice.

The problem can be conceived to be in two parts: that of describing relaxation at transverse energies greater than the well-depth i.e. $E_T > V$; and that of describing "capture" into the well, i.e. into states where $E_T < V$, where V is the well-depth. By transverse energy is meant the component of the total energy (measured from the bottom of the well) associated with motion transverse to the layers forming the superlattice: if $E_T < V$ the state tends to be localized in the well, whereas if $E_T > V$ the state tends to be delocalized. Splitting up the problem in this way is useful whenever transitions among delocalized states occur more rapidly than do those to localized states. This is often the case since there are often more delocalized states to scatter into than localised states. It will obviously not be the case if hot carriers are injected directly into a localized state, which is possible if their longitudinal momenta are large. Thus another aspect of the problem is the initial distribution of hot electrons and its evolution in time; an energy relaxation rate is therefore dependent, in general, on initial conditions and statistical processes.

In view of this we limit our attention to the rate at which an electron in the lowest allowed state with $E > V$ makes a transition to a state with $E < V$. Connection with reality implies the assumption that the hot electrons, however introduced, have relaxed energy rapidly and occupy the lowest of allowed delocalized states. (Strictly speaking, all states in a superlattice are delocalized, but it is useful to retain the nomenclature appropriate to the case of single quantum wells since some states are more delocalized than others, as long as it is understood that in the present context delocalized states refer to those with $E > V$, and localized states to those with $E < V$.)

The simplest model is one which identifies the well-capture rate with the intersubband scattering rate associated with the emission of a bulk polar optical phonon in an infinitely deep quantum well[1,2,3]. This is discussed in Section 2. Comparison with the superlattice situation is made in Section 3, and the large differences from the case of an infinitely deep well are pointed out. The model is modified in the light of this discussion but it is shown that the predictions of the unmodified model remain surprisingly valid. The major difference between the two models is that the modified version, which takes into account the actual superlattice energy band and eigenfunction structure, predicts weak resonances associated with the onset of localization, whereas the simple model does not. Nevertheless, such resonant capture as is predicted is very much weaker than has been predicted for the case of a single quantum well[4,5]. This discrepancy is discussed in Section 4.

## 2.  SIMPLE MODEL

It is well known[6] that in the case of a single quantum well transmission resonances occur when $k_a a = n\pi$, (n integer), where $k_a$ is the transverse component of the wavevector of the electron wave in the well and a is the well-width. Capture into the well is expected to be rapid when such a condition holds[4]. The condition $k_a a = n\pi$ is also the condition for a level to appear in an infinitely deep well. Thus a convenient estimate of a resonant capture rate can be made by equating it to the intersubband scattering rate due to bulk phonons in an infinitely deep well, for which an analytic expression exists[2], modified only by a normalization factor viz:

$$W_{nm} = W_o \frac{a}{a+b} \Gamma_{nm}$$

$$\Gamma_{nm} = \frac{x^{\frac{1}{2}}}{2} \left[ \frac{1}{2n(n-m)x-1} + \frac{1}{2n(n+m)x-1} \right]$$

(1)

where

$$W_o = (e^2/4\pi\varepsilon_p\hbar)(2m^*\omega/\hbar)^{\frac{1}{2}}, \quad \varepsilon_p^{-1} = \varepsilon_\infty^{-1} - \varepsilon_s^{-1}$$

$$x = E_o/\hbar\omega, \quad E_o = \hbar^2\pi^2/2m\ a^2$$

In these expressions b = barrier width, n and m are quantum numbers of initial and final states, $\varepsilon_\infty$ and $\varepsilon_s$ are high-frequency and static permittivities, $\hbar\omega$ is the phonon energy and $m^*$ is the effective mass of the electron. It has been assumed that parallel motion is zero in the initial state (but, of course, not in the final state). This analytic result is obtained in the momentum-conservation approximation (MCA) in which it is assumed that crystal momentum in the transverse direction is strictly conserved. The MCA has been shown to be a good one for intersubband transitions[3]. The normalization factor is simply the result of the charge density of the incident wave being divided between well and barrier, with only that in the well contributing to the transition matrix element. The spectrum of phonons has been taken to be that of the bulk, which is not likely to lead to serious error in the case of GaAs/AlGaAs superlattices[7,8].

The normalized rate, $\Gamma_{nm}$, is plotted in Fig. 1 as a function of x. The principal behaviour is that the rate decreases as the wells get narrower. This directly reflects the characteristic weakening of the polar interaction with increasing wavevector change. As the width lessens the subbands move apart and larger phonon wavevectors are required to effect the intersubband transition. Another feature is that, very roughly,

$$\sum_{m<n} \Gamma_{nm} \approx \Gamma_{21} \tag{2}$$

This means that $\Gamma_{21}$ is a reasonably good estimate of the total capture rate, independent of initial state. For example, if n = 4 corresponds to the lowest transmission resonance and m = 3, 2, 1 correspond to the localized states, the total capture rate, $\Gamma_{41} + \Gamma_{42} + \Gamma_{43}$, is approximately equal to $\Gamma_{21}$. The reciprocal of $\Gamma_{21}$, $T_{21}$, is plotted in Fig. 2 for the case of a GaAs/AlGaAs superlattice, where it is shown that $T_{21}$ obeys an inverse square law over the range of well-widths considered. Naturally, no resonances are predicted since the model already assumes capture is via resonant transmission.

## 3. CRITIQUE AND MODIFICATIONS OF THE SIMPLE MODEL

The principal defect of the simple model is that it assumes an energy-band structure (that of an infinitely deep well) which is quantitatively different from the actual band-structure of a superlattice. The electron band structure for a GaAs/AlGaAs superlattice (assuming parabolic bands) is shown in Fig. 3, and it can be immediately seen that there are several factors which will modifiy the estimate of capture rate.

1. Energy differences (and therefore transverse-wavevector differences) between subbands are smaller than for the infinitely deep well. This means that intersubband scattering rates will be larger than estimated.

2. The trend of subband energy differences with well-width is not as predicted by the simple model. For example, the difference in energy between first and second subbands does not increase monotonically with decreasing well-width but rather it goes through a maximum near a = 50Å. The inverse square law ought therefore to be invalid.

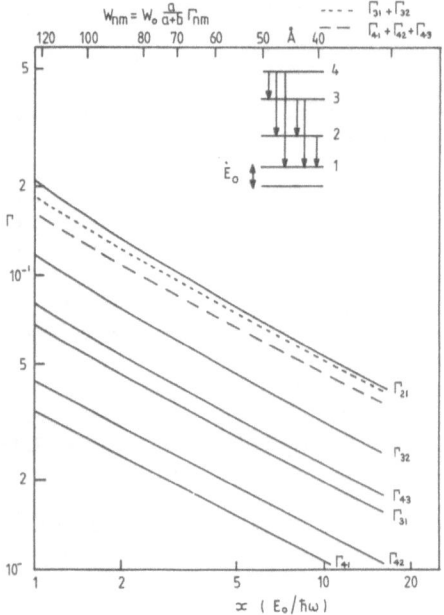

Fig. 1   Intersubband scattering rate
as a function of ground state
energy in an infinitively deep
well.   Corresponding well-
widths for GaAs are also shown.

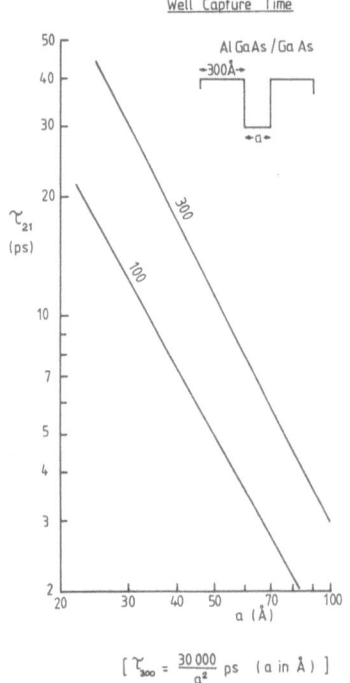

Fig. 2   Well-capture times as a function
of well-width in GaAs for 100Å
and 300Å barriers.

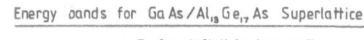

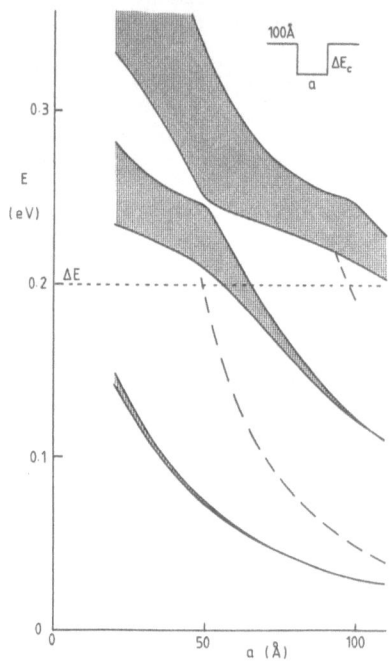

Fig. 3   Superlattice band structure in the parabolic
approximation for elec-
trons in GaAs/Al.$_3$Ga.$_7$
as a function of well-
width for 100Å barriers.

3.  In practice one is interested in the capture of electrons from the lowest delocalized energy states, but these do not, in general, correspond to transmission resonances. Rates will be therefore smaller than estimated.

The first two factors can be accommodated in the simple model very simply by replacing the quantum numbers in the expression for $\Gamma_{nm}$ (eq. 1) by energies and using actual superlattice band structure values. Thus

$$\Gamma_{nm} = \frac{x_1^{\frac{1}{2}}}{2} \left[ \frac{1}{2x_n^{\frac{1}{2}}(x_n^{\frac{1}{2}}-x_m^{\frac{1}{2}})-1} + \frac{1}{2x_n^{\frac{1}{2}}(x_n^{\frac{1}{2}}+x_m^{\frac{1}{2}})-1} \right] \tag{3}$$

The x outside the bracket in eq. (1) has been identified with $x_1$, Where $x_n = E_n/\hbar\omega$, and $E_n$ are the true eigenvalues. The third factor can be accommodated by introducing into eq (1) a "resonance" factor $\gamma < 1$. Thus the rate of capture takes the form

$$W_{nm} = W_0 \frac{a\gamma}{a+b} \Gamma_{nm} \tag{4}$$

with $\Gamma_{nm}$ given now by eq (3).

The resonance factor, $\gamma$, measures the degree to which the wavefunction of the delocalized state in the well corresponds to a transmission resonance. In the latter case the amplitude in the well will be comparable to that in the barrier. An estimate of the effect of off-resonance is therefore provided by the ratio of these amplitudes. Fig. 4 shows the wavefunction for the lowest energy delocalized state in a number of cases.[9] From these $\gamma$ can be found.

When the well-width is small, charge density in the lowest delocalized state piles up on the barrier. This is simply a consequence of the boundary conditions: $k_a \gg k_b$, and consequently slopes can be fitted only with a large amplitude in the barrier and a small amplitude in the well. As a $\rightarrow 58\overset{\circ}{A}$ a transmission resonance ($k_a a = \pi$) is approached $\gamma \rightarrow 1$, and the state paradoxically becomes localized. Beyond a = $68\overset{\circ}{A}$ the lowest delocalized state corresponds to n = 3, and once more the amplitude in the well is small. It rises steadily towards a = $116\overset{\circ}{A}$ when the next transmission resonance ($k_a a = 2\pi$) occurs.

The results for $\Gamma$ ($\Gamma_{21}$ for a < $68\overset{\circ}{A}$ and $\Gamma_{32} + \Gamma_{31}$ for a $68\overset{\circ}{A}$) and as a function of well-width for a GaAs/Al$_{.3}$Ga$_{.7}$As superlattice with $100\overset{\circ}{A}$ barriers is shown in Fig. 5. Although both terms vary rapidly their product is less volatile. Fig. 6 shows the capture time as a function of well-width and a comparison with the prediction of the unmodified model. A relatively weak resonance occurs when a state is being localized, but otherwise the overall behaviour accords reasonably well with the predictions of the simple model.

4.  DISCUSSION

Taking account of a more realistic energy-band and eigenvalue structure for the superlattice has had rather a small effect on the predictions of the simple model. The reason for this rather unexpected result lies with the characteristics of the polar optical phonon interaction. The smaller gaps between subbands in the superlattice compared with the situation for an infinitely deep well enhance the scattering rate for the polar interaction, but their effect is largely nullified by the off-resonant concentration of charge on the barrier sites. The latter effect is independent of scattering mechanism, and its influence on capture rate would appear more strongly in cases where energy

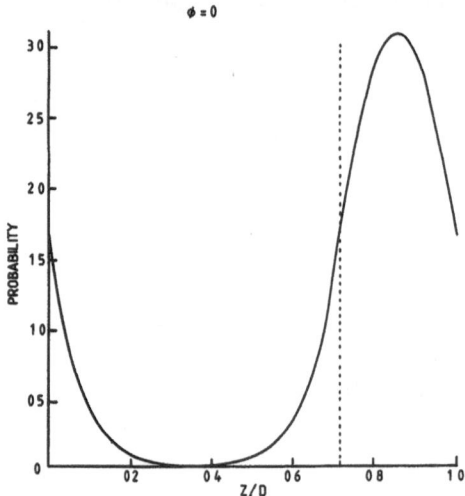

(a) n = 1, $\phi$ = 0, a = 40Å
("localised" state)

(b) n = 2, $\phi$ = $\pi$, a = 40Å
(lowest "delocalized" state)

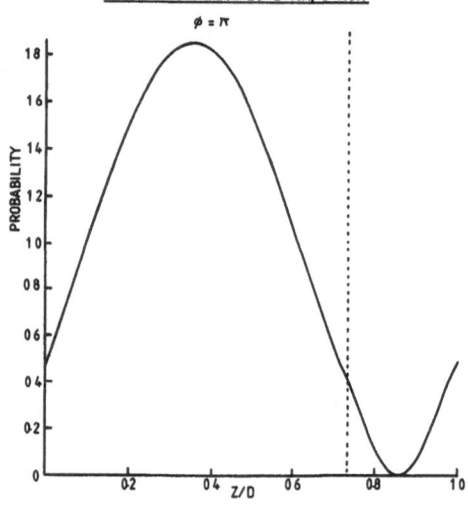

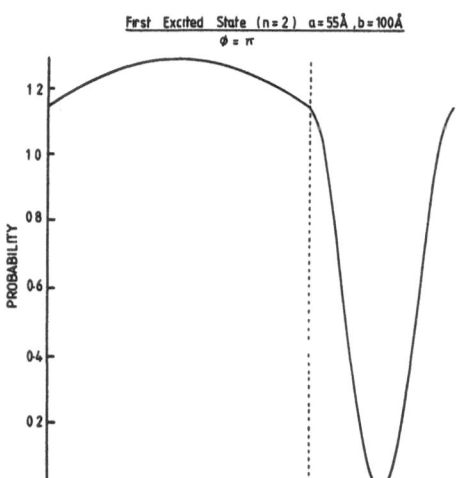

(c) n = 2, $\phi$ = $\pi$, a = 55Å
(near resonance)

Fig. 4.  Electron probability densities

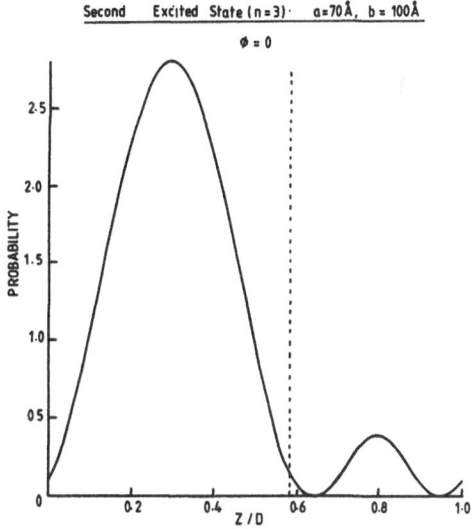

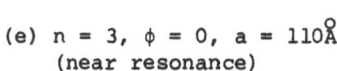

(d) n = 3, φ = π, a = 70Å
(lowest "delocalized" state)

(e) n = 3, φ = 0, a = 110Å
(near resonance)

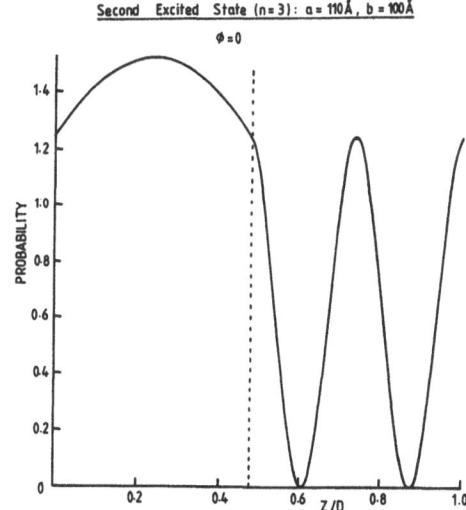

Fig. 4. Electron probability densities (cont'd)

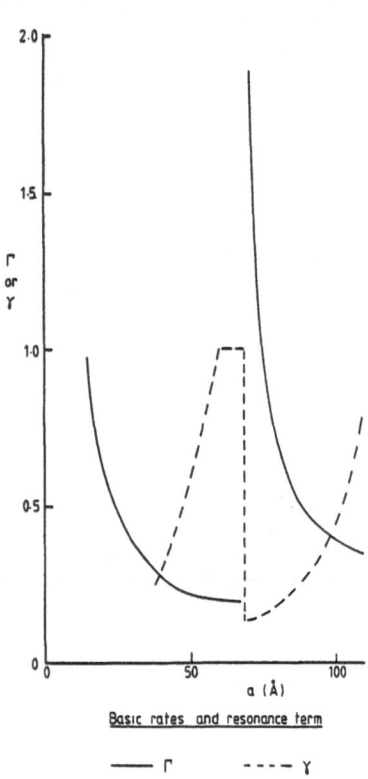

Basic rates and resonance term

———— Γ    - - - - γ

Fig. 5    Basic intersubband
          scattering rates and
          resonance factor as
          a function of well-
          width in GaAs/Al$_{.3}$Ga$_{.7}$As

Fig. 6    Capture time as a
          function of well-width.

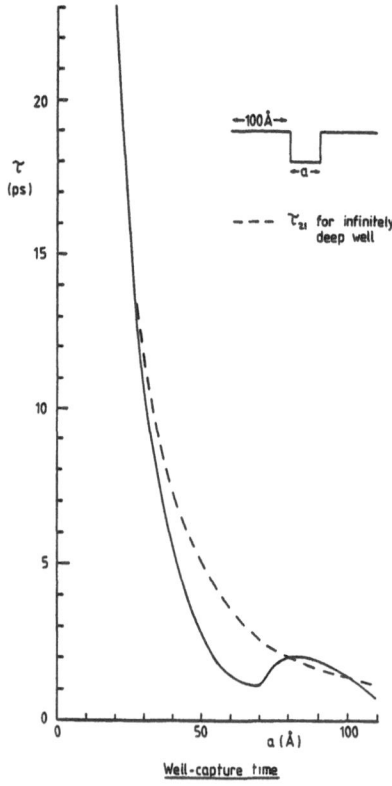

Well-capture time

relaxation was primarily via the interaction with non-polar phonons. In other words, resonant capture ought to be more evident in non-polar than in polar materials, but even with non-polar optical phonon scattering dominant the effect will be moderated by the proportionality of the scattering rate to density of final states, since the latter increases with diminishing well-width. As shown by the behaviour of $\gamma$ in Fig. 5, a factor of over 10 separating resonant from non-resonant capture is possible, but only if the basic transition rate is independent of well-width. As shown in Fig. 6, polar optical phonon scattering reduces this to a factor of about 5. Strongly resonant capture is therefore not predicted, especially in the case of polar materials. This conclusion is quite different from that obtained by Brum and Bastard[4], who predict strong oscillations. The cause of this discrepancy is not clear, though it may be associated with their treatment of phonon emission as being one-dimensional (zero momentum along the layers in the final state).

The foregoing analysis touches only lightly on the problem. A fuller discussion taking into account the true superlattice phonon spectrum and the effect of screening will be presented shortly[9]. It is clear, however, that the amplitude of the resonances is going to be profoundly affected by any change in the phonon spectrum. A bulk spectrum always allows the existence of phonons with a wavevector which can satisfy momentum conservation in the transverse direction, but this is generally no longer the case for superlattice modes. As a result, the coupling is weakened. In particular, we cannot expect the rate to rise as the gap between subbands diminishes, and consequently, the cancellation of the resonant effect due to the periodic bunching of electronic charge in the well (Fig. 9) will not occur. Strong resonances are then predicted, in contrast to the weak resonances obtained assuming a bulk phonon spectrum.

ACKNOWLEDGEMENTS

It is a pleasure to acknowledge the invaluable contribution made to this work by Dr. M. Babiker. The project was supported by the U.S. Office of Naval Research.

REFERENCES

1. P.J. Price, Ann. Phys. N.Y. 133 217 (1981)
2. B.K. Ridley, J. Phys. C: Solid State Phys. 15 5899 (1982)
3. F.A. Riddoch and B.K. Ridley, J. Phys. C: Solid State Phys. 16 6971 (1983)
4. J.A. Brum and G. Bastard, Phys. Rev. B33 1420 (1986)
5. S.V. Kozyrev and A.Ya. Shik, Sov. Phys. Semicond. 19 1024 (1986)
6. D. Bohm, "Quantum Theory" (Prentice Hall, 1951)
7. B.K. Ridley, Festkorperprobleme XXV 449 (1985)
8. F.A. Riddoch and B.K. Ridley, Physica 134B 342 (1985)
9. M. Babiker and B.K. Ridley (to be published)

# SUBBAND PHYSICS FOR $Hg_{0.8}Cd_{0.2}Te$ IN THE ELECTRIC QUANTUM LIMIT

Frederick Koch

Physik-Department, Technische Universität München
8046 Garching, Fed. Rep. of Germany

ABSTRACT

We argue that essential aspects of the electron surface bands for narrow-gap semiconductors are best studied on p-type material in the electric quantum limit. For the case of a single occupied subband in strongly doped $Hg_{0.8}Cd_{0.2}Te$ we consider the surface level energies in the presence of strong tunneling overlap of the subband and valence band wave functions. We discuss the electric-field induced spin-splitting of the electron subbands.

## I. INTRODUCTION

The unique and characteristic feature of electronic subbands in a surface layer on a narrow-gap semiconductor like $Hg_{0.8}Cd_{0.2}Te$ is that the level energies $E_n$ are comparable with the gap energy $E_g$. It follows that the electrostatic surface potential $V(z)$ will influence the mixing of the valence and conduction band levels. The surface band structure is to be solved by a priori inclusion of a self-consistent $V(z)$ in the Kane-model Hamiltonian.

A special case, one that still demands particular attention, is that for which surface level energies coincide with the continuum of filled states in the valence band. When the tunneling overlap is not negligible, the numerical evaluation of the surface bands becomes complicated. The difficulties have been pointed out over and over again in the calculation of such bands /1/, but have proved in practice not to be of much numerical significance. Practice is defined in this case from the experimental parameters used in earlier work on HgCdTe, where either n-accumulation or only light p-type doping ($N_A - N_D \sim 10^{15}$ cm$^{-3}$) was employed. Thus even for the approximately 40 meV gap energy in Fig. 1, the authors of Ref./1/ conclude that tunneling overlap of the $E_0$ state with the valence band can be ignored. Raising $N_s$ above $2 \times 10^{12}$ cm$^{-2}$ in Fig. 1 does not really change the situation very much because many more subbands are rapidly filled. The potential is dominated by the self-consistency aspect and is accumulation-layer-like. The Fermi energy $E_F$ rises only slowly. Subband electrons extend many hundreds of Å into the semiconductor. Especially the excited states are quasi-three-dimensional with bulk-like properties. It is no surprise that such electronic states show the polaron-coupling expected of bulk electrons. They do not really probe the surface region

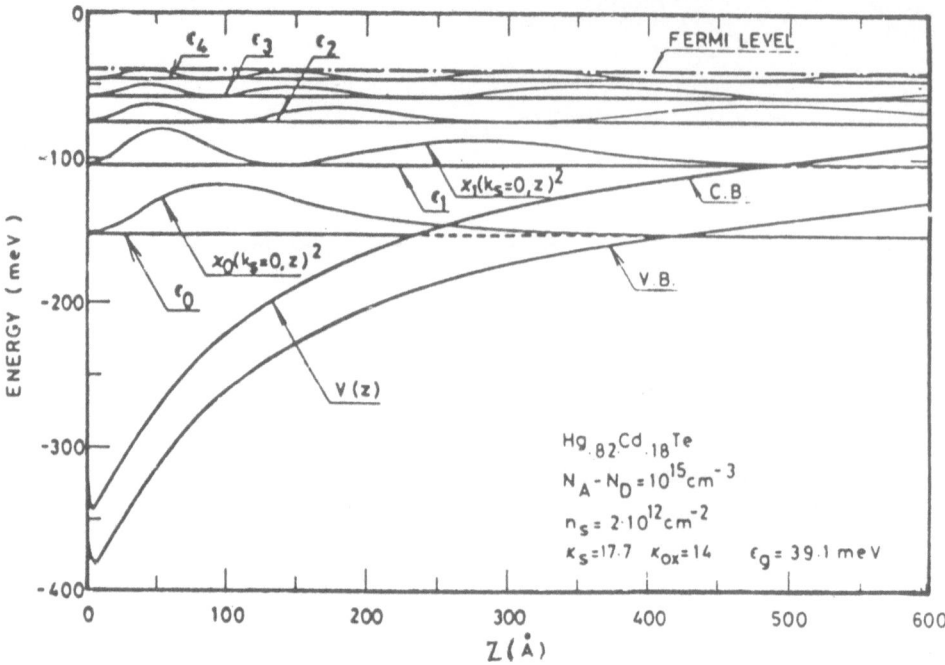

Fig. 1: Subbands and electronic charge densities for
p-HgCdTe ($N_A = 1 \times 10^{15}$ cm$^{-3}$) with $E_g \sim 40$ meV after
Ref. /1/. For a typical "experimental" charge density
of $N_s = 2 \times 10^{12}$ cm$^{-2}$ many bands are occupied.

and are insensitive to exact boundary conditions at the anodic oxide
interface or to the gap-grading that was proposed in Ref. /2/. Not only
is the multi-band calculation in the self-consistent potential rather complex, and thus not easily verified and compared with experiments, but also
such features as the electric-field induced spin-splitting and the coupling
to the valence band states cannot clearly be demonstrated.

II.   THE CASE FOR THE ELECTRIC QUANTUM LIMIT

The multi-subband aspect that so much dominates subband physics for
the narrow-gap materials also complicates the evaluation of Shubnikov-de
Haas and cyclotron resonance data, it makes  difficult the analysis of
subband resonance spectra. Challenging problems - i.e. exact band energies
in the presence of band coupling, spin-splitting, possible gap-grading in
the surface layer because of a change in the stoichiometric composition
etc. - still await exacting comparison of theory with experiment. This
makes one wish for a return to the simplicity of the electric quantum
limit, the case of a single, occupied band of carriers bound tightly
to the surface, as in Si (100).

What would it take to have such conditions in $Hg_{0.8}Cd_{0.2}Te$? The answer
is not difficult. One needs p-type material, doped sufficiently high to
provide a surface depletion-layer field like that in Si. Since the field
scales as $\sqrt{E_g \cdot N_A}$ , or more exactly as $\sqrt{(E_g - E_A) \cdot N_A}$ when the acceptor
ionization energy $E_A$ is not small, we need for the x = 0.2 material whose
$E_g = 60$ meV and $E_A = 20$ meV, about 30 times higher doping than the
$10^{15} - 10^{16}$ cm$^{-3}$ typical for Si. Current work in our laboratory is with
p-type $Hg_{0.8}Cd_{0.2}Te$ for $N_A$ ranging from $5 \times 10^{16}$ to $3 \times 10^{17}$ cm$^{-3}$. For
such crystals the electric quantum limit extends up to $N_s \cong 4 \times 10^{11}$ cm$^{-2}$
for the lower doping and almost to $1 \times 10^{12}$ cm$^{-2}$ for the upper limit.

## III. COUNTING THE CHARGES $N_s$

Examining the Shubnikov-de Haas oscillations for p-$Hg_{0.8}Cd_{0.2}Te$ in the electric quantum limit, the usual and infallible means of determining the density of free carriers $N_s$ in a surface layer, it did not have the value expected from measured capacitance of the MOS structure /3/. Looking in more detail at the capacitance-voltage relation /4/ it has become clear that the small density of states and the consequently rapidly rising $E_F$ vs. $N_s$ is responsible for the discrepancies. In the quantum limit the $E_F$ for $4 \times 10^{11}$ cm$^{-2}$ carriers approximately equals the $E_g$ of 60 meV. It follows that the depletion charge which is derived from the total band bending will increase significantly. Examining Fig. 2 closely we see that the depletion layer length $z_{dep}$ grows as $\sqrt{E_g - E_A + E_F + aE_o}$. Here a is a numerical factor between 0 and 1 and depends on the increased band bending in the inversion layer region. Both $z_{dep}$ and $N_{dep}$ are functions of $N_s$. Alternatively, it is clear that the density of states for a narrow-gap semiconductor subband in the quantum limit is easily determined from the capacitance. This point is explored further in Ref. /4/.

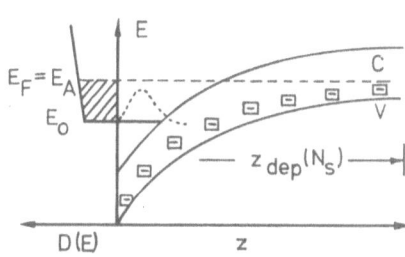

Fig. 2:
Depletion layer depth $z_{dep}$ and total band bending including the $E_o$ and $E_F$ contributions. The depletion layer charge grows with $E_o$ and $E_F$.

## IV. THE SURFACE LEVEL AS A RESONANCE

The subband $E_o$, degenerate with and strongly coupled to the filled valence band continuum, represents something qualitatively new and different. It is no longer a sharply defined quantum level with a discrete energy. Depending on the strength of the tunneling interaction the surface band is broadened and shifted. As the wave function is not confined

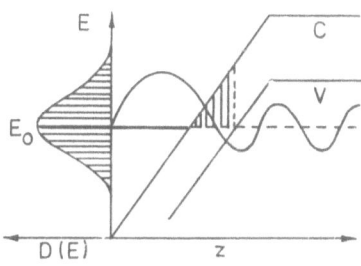

Fig. 3:
The surface level $E_o$ in the presence of Zener-tunneling is a shifted and broadened resonance. The tunneling barrier is triangular with height $E_g$.

to the surface layer anymore, but extends over the whole crystal, there are problems regarding the normalization of the wave function and the definition of the surface charge density $e|\psi_0|^2$ to be used in the Poisson equation.

This point is discussed at length in Ref. /5/. As an approximate way of dealing with the difficulties that arise in the computation, an auxiliary boundary condition is introduced. The wave function $\psi_0$ is required to vanish in the mid-gap position as shown schematically in Fig. 3 above. This yields a sharp state that can be normalized to represent a surface charge density in the usual way. The density of states distribution typical of the resonant state is replaced by a sharp peak at $E_0$. The approximation itself amounts to selecting the one state with maximum amplitude in the surface layer to represent the entire distribution of possible states of the resonance. The question of how well this serves to describe measured energies is explored in Refs. /6,7/, where also numerical estimates of the amount of broadening and the shift can be found.

An alternative approach, which explicitly recognizes the resonant character of the surface band, is given in Ref. /8/. This calculation treats the tunneling interaction as a perturbation on the self-consistent potential evaluation in Ref. /1/. Shift and broadening appear as the real and imaginary parts of a Green's function.

In the sense of the schematic drawing in Fig. 3 the tunneling barrier is triangular with height $E_g$ and a width that decreases as the inverse surface electric field. A simple estimate of the width of the resonance is provided by calculating the tunneling probability and thus the lifetime of the $E_0$ state. The broadening is Planck's constant h divided by the lifetime.

## V. SUBBAND CALCULATIONS IN THE SIX-BAND SCHEME

Having defined the boundary condition for the resonant surface levels, it is now possible to proceed with the calculation in terms of the multi-band formalism of Refs. /5,9/. For $Hg_{0.8}Cd_{0.2}Te$ it is appropriate to consider a 6 x 6 Kane Hamiltonian $\mathcal{H}$ that couples the 2-fold, spin-degenerate $\Gamma_6^C$ bands with the 4-fold degenerate valence states $\Gamma_8^V$. The electrostatic surface potential appears on the diagonal so that the Schrödinger equation has the form $(\mathcal{H} + V(z)I_{6\times6})\Psi = E\Psi$. The function $\Psi$ has 6 components with envelopes $f_i(z)$. The charge density that is employed in the Poisson equation is

$$\rho(z) = e \sum_{occ.states} \sum_{i=1}^{6} \left| f_i(z) \right|^2$$

where the sum extends over all occupied states. The calculation is repeated for each value of the parallel momentum $k_{\parallel}$. Parameters in the Kane-Hamiltonian are those appropriate for bulk $Hg_{0.8}Cd_{0.2}Te$. Details of the calculation are to be found in Refs. /6,7/. The emphasis in that work is on evaluating as precisely as possible the case of a single filled band in the limit of strong overlap. The central approximation is that of replacing the resonance by a sharp state defined by the boundary condition discussed above. The calculation is done for parameters realized in the experiments. Moreover it takes an $E_g$ which remains constant throughout the surface region. Wave function $f_1(z)$, which describes a $\Gamma_6^C$ state of the conduction band, is required to vanish at z = 0. It is a choice of boundary condition which in the limit of large $E_g$ recovers the one-band calculation familiar for Si(100). Because other components $f_i(z=0)$ do not vanish, there is a net charge at the interface to the anodic oxide.

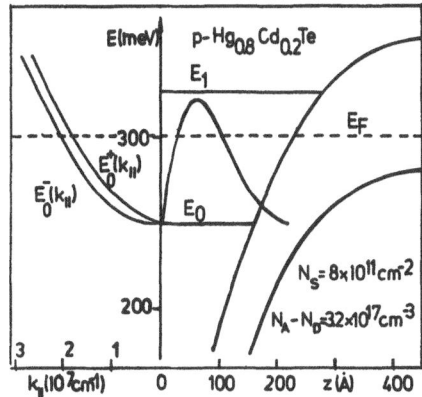

<u>Fig. 4:</u> Surface band for the electric quantum limit (x = 0.208, p-type). Note the large spin-splitting into ± states. Energies are measured from mid-gap at the surface (z = 0).

Fig. 4 gives an example for the potential, the wave function $f_1(z)$ for the $k_\parallel = 0$ state, the Fermi energy, the spin-split bands $E_0^\pm(k_\parallel)$, the next higher $E_1$ energy for $k_\parallel = 0$, all as calculated by Nachev in Refs. /6,7/. Comparing with the result in Fig. 1, the tighter binding of the electrons to the surface is evident. There is increased overlap of $E_0$ with filled states of the valence band, and by design only the ground-state subband is occupied. The substantial splitting of the bands into + and - components is the result of including the spin-orbit term in the Hamiltonian. For a wide range of $k_\parallel$ the splitting is nearly constant. In the limit $k_\parallel \to 0$ the bands coalesce. The doping is chosen as $N_A - N_D = 3.2 \times 10^{17}$ cm$^{-3}$. The surface charge is $N_s = 8 \times 10^{11}$ cm$^{-2}$.

## VI. DETERMINING SUBBAND ENERGIES

Strong band-band coupling requires a strong surface depletion field. The subband splitting, such as that between the $E_0$ groundstate and $E_1$, is correspondingly large. The typical value $E_{01}$ is equal to and larger than the band-gap. This makes it difficult to measure the resonance spectroscopically, in particular using the tunable infrared spectrometer that we have employed in previous experimental work. The reason is that illumination with band-gap radiation will alter the effective depletion charge. Quasi-Fermi levels are established with the optical excitation of electron-hole pairs. Thus the process of measuring itself affects the level spacing. While it may yet prove possible to maintain the depletion charge by a suitable short-circuit of the illuminated inversion layer, the present data shows a significant decrease of the depletion charge under band-gap excitation conditions.

The work of R. Sizmann in Ref. /3/ and a recent paper /7/ take an alternative approach. Conductivity data, in particular the derivative $d\sigma/dV_g$, shows distinct and easily identifiable structures that signify the onset of occupancy of the next surface band. Thus in Fig. 5 the arrows

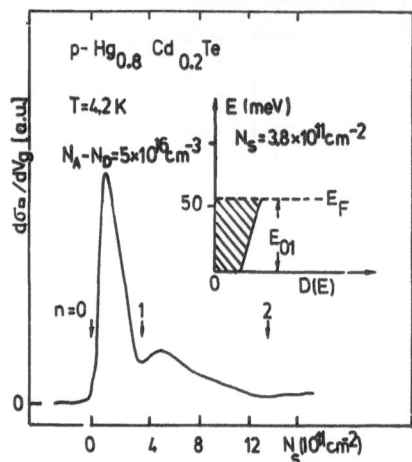

Fig. 5: Conductivity derivatives
vs. $N_S$ showing structures for
occupancy onset of subbands
n = 0, 1, and 2 (Refs. /3,7/).

marked n = 0, 1, and 2 give the $N_S$ values where $E_0$, $E_1$, and $E_2$ respectively
just touch $E_F$. A position such as n = 1 at 3.8 x $10^{11}$ cm$^{-2}$, measures $E_{01}$
when the density of states is known. The insert makes clear that the
integral of D(E) over the range $E_0$ to $E_1$ must give the measured $N_S$. Using
cyclotron masses as measured in other experiments we evaluate the splitting
as 54 meV. Alternatively one can compare directly the calculated and
experimental $N_S$ values at onset of occupation. In Ref. /7/ we show that
good agreement with the calculation exists when the depletion field is
not too high. There is a significant difference for the case of very strong
binding such as that in the previous Fig. 4. The filling of n = 1 begins
somewhat below the $N_S$ value that is predicted from the calculations.
A more complete discussion is contained in Ref. /7/

VII. MAGNETOTRANSPORT EXPERIMENTS AND SPIN-SPLITTING

Counting carefully Shubnikov-de Haas peaks and quantum-Hall-effect
plateaus R. Sizmann in Ref. /3/ reached an interesting conclusion. For
the case where only two subbands were known to be occupied as in Fig. 6,
it is possible to observe in low magnetic fields a coincidence of three
levels. Two of these must necessarily originate from the same subband,
i.e. a pair of Landau levels with opposite spin are degenerate. R. Sizmann
observed more generally that in low magnetic field various other overlaps
occur even in the limits of a single occupied band.

Such data suggested to us that we search more diligently in samples
with even larger depletion field for the level crossings. In a current
paper /10/ we explore the limit of purely electric splitting between the
bands by examining data in low magnetic fields, where the separation of

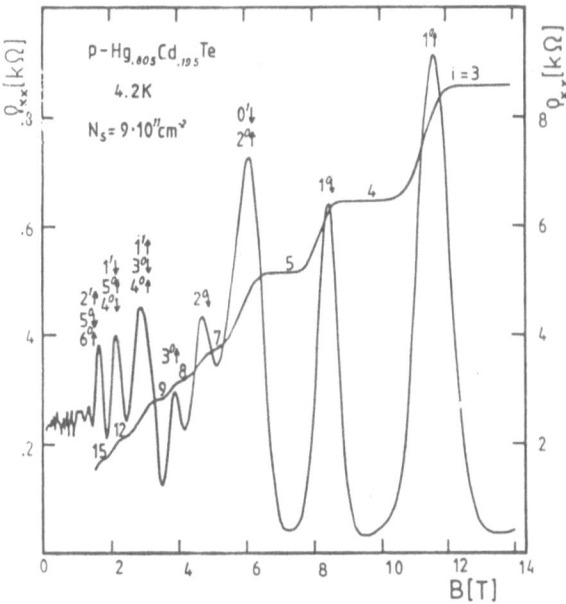

**Fig. 6:** Quantum Hall effect and Shubnikov-de Haas oscillations for two occupied surface bands

Landau levels is essentially given by the $E^+$, $E^-$ splitting such as that calculated in Fig. 4. We argue that for the strong depletion field of a $3.2 \times 10^{17}$ cm$^{-3}$ doped sample and B-fields $\lesssim$ 2 T, that the electric splitting dominates. In this limit it is legitimate to construct spin-split quasi-classical cyclotron orbits simply by marking the relevant, quantized $k_n$ values on the dispersion curves $E_0^\pm$ ($k_{\parallel}$). The $k_n$ are calculated from the orbit radii in real space that just enclose the quantized flux $(\ell + 1/2)h/e$ for $\ell = 0, 1, 2,...$ etc. Scaling back to k-space according to the known factor $eB/h$ one arrives at the quantized $k_n$.

We proceed with the construction of Landau levels working with the dispersion curves previously shown in Fig. 4. Our interest is focussed on possibly overlapping Landau levels such as $0^+$ and $1^-$, $1^+$ and $2^-$, and so on. This requires that $k_0$ intersects the + curve in Fig. 7 at the same energy as $k_1$ cuts the $E_0^-$ ($k_{\parallel}$) relation. The construction requires a field B = 1.2 T for the $0^+/1^-$ overlap, and B = 2.2 T for the $1^-/2^+$ coincidence. As calculated, the dispersion curve is that for $8 \times 10^{11}$ cm$^{-2}$ carriers in the surface. The Fermi energy actually lies above the predicted overlap energies and they would not be observable in a Shubnikov-de Haas experiment with $8 \times 10^{11}$ cm$^{-2}$. However, the dispersion curve and the splitting change only little with $N_s$ in the quantum limit because the depletion field is large. It follows that for a simple first order look at level crossing in Shubnikov-de Haas data the construction in Fig. 7 is sufficient.

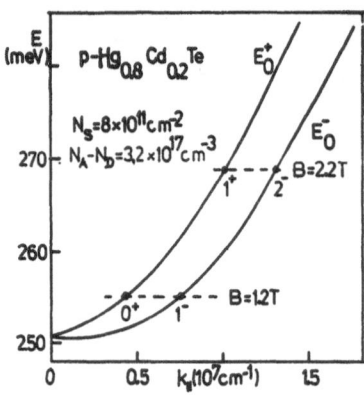

Fig. 7:
For the calculated surface bands $E_0^+$ and $E_0^-$ the Landau levels $0^+$ and $1^-$ are degenerate for fields 1.2 T, Landau levels $1^+$ and $2^-$ for 2.2 T. The dots mark the calculated Landau orbit radii in k-space for the given magnetic field.

The predicted level-crossings have actually been found in Ref. /10/ to be in good agreement with the construction in Fig. 7. We conclude to have thus determined the electric splitting of the subbands. It agrees with the calculated value, which in turn speaks for the 6-band model calculation that Nachev has made in Ref. /6/.

VIII.  CONCLUDING REMARKS

My lecture has sought to highlight current work on subbands for that classical narrow-gap semiconductor $Hg_{0.8}Cd_{0.2}Te$. The central theme is that the electric quantum limit of p-type material is the case that can best be analyzed and is most sensitive to real surface conditions.

I have cited liberally from the work of coworkers and students, from publications that are in preparation and will appear elsewhere. Particular thanks go to students R. Sizmann, R. Wollrab, and I. Nachev. I thank U. Rössler (Univ. Regensburg) for many discussions. Samples have been provided by J. Ziegler and H. Maier of Telefunken Electronic in Heilbronn.

REFERENCES:

1/ Y. Takada, K. Arai, Y. Uemura, Physics of Narrow Gap Semiconductors, Lecture Notes in Physics Vol. 152, ed. E. Gornik et al., Berlin: Springer, p. 101, 1982.
2/ F. Koch, Proc. of the Int. Winter-School Mauterndorf, Austria, Springer Series in Solid State Science Vol. 53, ed. G. Bauer et al., p. 20, 1984.
3/ R. Sizmann, Diploma thesis, TU Munich (1985), unpublished
4/ V. Mosser, R. Sizmann, F. Koch, to be published
5/ A.E. Marques, L.J. Sham, Surface Sci. 113, 131 (1982); and A.E. Marques, Ph.D. Thesis, University of California, San Diego, 1982, unpublished
6/ I. Nachev, Doctorial Dissertation, TU Munich, in preparation
7/ I. Nachev, R. Wollrab, R. Sizmann, F. Koch, A. Ziegler, U. Rössler, H. Maier, J. Ziegler, to be published
8/ W. Brenig, H. Kasai, Z. Physik B 54, 191 (1984)
9/ W. Zawadski, J. Phys. C 16, 229 (1983)
10/ R. Wollrab, R. Sizmann, F. Koch, H. Maier, J. Ziegler, to be published

# THE FAR INFRA-RED MAGNETOTRANSMISSION OF ACCUMULATION

## LAYERS ON n-(Hg,Cd)Te

John Singleton, Firoz Nasir and Robin Nicholas

The Clarendon Laboratory
Parks Road, Oxford
OX1 3PU, England

A description is given of the use of far-infrared magnetotransmission to observe the subband electrons in accumulation layers at anodic oxide films on $n$-$Hg_{0.8}Cd_{0.2}Te$. The measurements show that the accumulation layer electrons exhibit many novel features, such as enhanced resonant 2D magneto-polarons and skipping orbits, and as such are of great interest in the study of the two dimensional electron gas.

## INTRODUCTION

The narrow gap semiconductor Mercury cadmium telluride ($Hg_{1-x}Cd_xTe$) has found a wide range of applications as an infrared detector material.[1] In many (Hg,Cd)Te devices, the surface area to volume ratio is very large, and the application of some form of surface passivant is necessary, in order to reduce the surface recombination velocity and 1/F noise. This is usually achieved by the growth of $\sim 700$ Å of anodic oxide film in aqueous solution[1,2,3]: ionised impurities in the anodic oxide cause the conduction and valence bands to be bent down to form a one-dimensional potential well close to the surface. The potential well can contain a degenerate two-dimensional electron gas (2DEG) with the electrons bound in the direction perpendicular to the surface. The eigenstates for this direction are a set of discrete levels known as electric subbands, with energies $E_0$, $E_1$ . . . $E_i$, whereas the motion in the plane of the surface will be essentially unconfined. In surface space charge layers on (Hg,Cd)Te, the low effective mass and small 2D density of states[4] can lead to the population of several subbands[5,6] for modest values of the total surface carrier concentration, $N_s$, In this paper, we shall report some of the magneto-optical measurements which have been useful in characterising the subbands in accumulation layers at anodic oxide films, and describe some of the novel modes of behaviour exhibited by the surface electrons.

## SAMPLE DETAILS AND TRANSPORT RESULTS

The samples used in the measurements described here are $\sim 2.5$ mm $\times$ 1 mm $\times$ 10 $\mu$m bars of high-grade n-(Hg,Cd)Te, with x from 0.194 to 0.217, bulk 77K mobilities $\sim 2 \times 10^5$ $cm^2/V_s$ and bulk carrier densities in the range $4 \times 10^{14}$ $cm^{-3}$ to $3 \times 10^{15}$ $cm^{-3}$. Both front and back surfaces are anodically oxidised to form two accumulation layers both of which contribute to the conductivity[5]. Although the structures are not gated, $N_s$ (typically $0.6 \times 10^{12}$ $cm^{-2}$ < $N_s$ < $2.5 \times 10^{12}$ $cm^{-2}$) can be varied at low temperatures, using UV illumination, which repopulates the ionised oxide impurities[7,8].

Shubnikov-de Haas and parallel-field magnetoresistance measurements performed on these samples have shown that five or six subbands are populated in the accumulation layers, and that these subbands have very well-defined occupancies in spite of the ± 0·01 variation in x and the large variation in bulk carrier density. The virtual x-independence of the subband occupancies can be easily explained using a simple model of the accumulation layer due to Ando[9], in which Poisson's equation is solved self consistently with the WKB condition: the final solution depends only on Eg measured in effective Rydbergs. As the effective Rydberg $\propto 1/m^*$, Eg is roughly constant for all x, and the only slight difference in the occupancies is due to non-parabolicity (figure 1)[9,10]. Indeed, Shubnikov-de Haas (SdH) measurements made on accumulation layers on n-$Hg_{0.7}Cd_{0.3}Te$ indicate that the i=0 subband occupancy is only ~5% different from that measured in the x=0.2 samples[7].

## CYCLOTRON RESONANCE IN PERPENDICULAR MAGNETIC FIELD

The far-infra-red (FIR) magnetotransmission of each sample was examined in the Faraday geometry, with the magnetic field perpendicular to the accumulation layer plane, at a number of wavelengths in the range 251 μm to 38 μm. Typical results are shown in figure 2: the cyclotron resonances (CR) appear as minima in the transmission. The strong absorption at low fields is due to the bulk electrons, and can be used to deduce an accurate x-value for the sample[11], whilst the varying number of resonances at higher fields are due to the surface electrons, which have larger effective masses[12,6]. The number of subband CR resolved varies as a function of energy, and at one point (96 μm) six are observed: this variation is due to the strong polaron effect in the higher (i=3,4) subbands[13], which will be discussed below.

If $N_s$ is varied using the UV illumination technique mentioned above[7], simultaneous SdH and CR measurements can be made, so that the subband effective masses in the low-energy limit can be deduced as a function of $N_s$. Typical data are shown in figure 3, plotted alongside the theoretical values of Takada et al.[14] and the experimental values are seen to be in reasonable agreement with the theory. However, some samples exhibit effective masses which appear to be much too low when compared with the theory[10], and this is thought to be due to the anodisation process leaving the surface Hg-rich in certain cases: the reason for this variation between samples prepared in the "same" way is not known.

## RESONANT 2D MAGNETOPOLARONS

There is a great deal of interest in the roles that screening and confinement play in the polaron effect in a quasi two-dimensional electron gas (Q2DEG). In an infinitely thin, unscreened Q2DEG, theoretical models predict an enhancement of the electron-optic phonon coupling[16]: the inclusion of screening and finite potential width quickly destroys this enhancement[17]. Available experimental data seem system and sample dependent, with enhanced 2D polaron effects reported in some cases (InSb[18]) but not in others (GaAs[19,20], (Hg,Cd)Te[12]).

(Hg,Cd)Te exhibits two distinct reststrahlen bands due to " HgTe-like" phonons and higher energy "CdTe-like" phonons[21]. The latter modes are weak, and Faraday geometry CR of bulk electrons shows resonant polaron coupling only at the "HgTe-like" LO phonon energy[22,23]. Returning to the accumulation layer electrons in figure 2, strong polaron effects in the highest (i=4,5) subbands are visible in the raw data: as the energy approaches the "HgTe-like" reststrahlen region(73 μm <λ<83 μm),the large polaron contribution to the

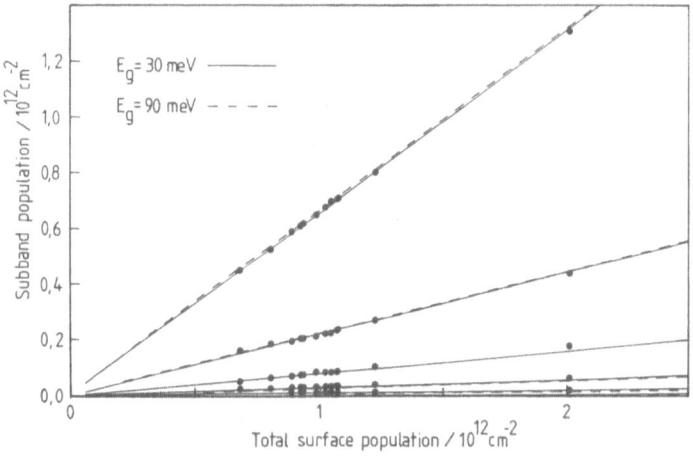

Fig.1. Subband populations plotted against total surface carrier density for accumulation layers on n-(Hg,Cd)Te (x=0.2: data=points, theoretical model results=lines).

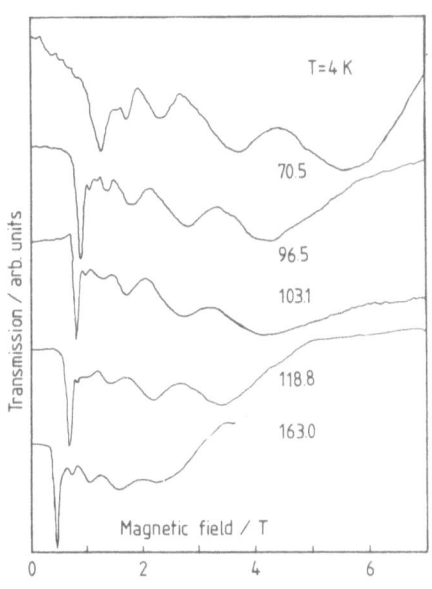

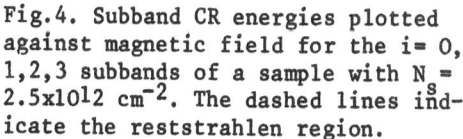

Fig.2. FIR magnetotransmission of n-(Hg,Cd)Te sample at various wavelengths (shown in microns). (T=4 K)

Fig.4. Subband CR energies plotted against magnetic field for the i= 0, 1,2,3 subbands of a sample with $N_s$ = $2.5 \times 10^{12}$ cm$^{-2}$. The dashed lines indicate the reststrahlen region.

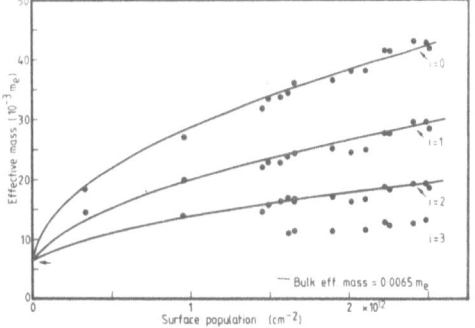

Fig.3. Effective masses of surface electrons plotted as function of $N_s$ (points=data: curves=theory[14]).

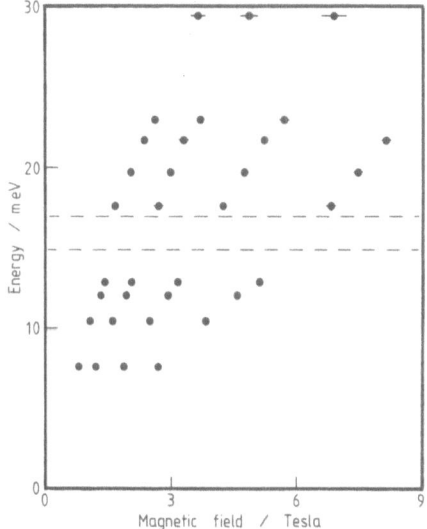

i=4,5 subband effective masses progressively separates their CR from the bulk CR. This effect disappears above the reststrahlen band (ie the resonant condition).

If the subband CR energies are plotted as a function of magnetic field (figure 4), resonant polaron coupling is observed as a displacement of the CR positions at energies above the "HgTe-like" phonons to lower magnetic fields than expected[18]. From inspection, the coupling strength appears to increase with subband index.

In order to compare the relative strengths of the bulk and surface polaronic effects, the subband effective masses from a sample with $N_s$ =2.0x10$^{12}$ cm$^{-2}$ are normalised and plotted alongside normalised bulk data[22] in figure 5. The measured effective mass of each subband should be roughly energy independent until the ultraquantum limit for that subband is reached: subsequently the effective mass will increase with increasing energy. The effect of any resonant polaron coupling will be superimposed on the above effects, and be observed as as discontinuity in the effective mass[18]. For the i=0 and i=1 subbands the polaronic effects are weaker than those in the bulk, being very small for i=0. In the case of the i=2 subband, the coupling is around the same strength as that in the bulk. Finally, in the case of the i=3 subband, its ultraquantum limit is reached at a field corresponding to $\hbar\omega_c$ = 12 meV, so that the effective mass starts to increase due to band non- parabolicity at around this energy. Superimposed on this is the resonant polaron effect, which is enhanced over that in the bulk.

The behaviour of the i=2 and i=3 subbands is similar in all the samples studied: for example figure 6 shows the subband effective masses from a sample with $N_s$ = 0.9x10$^{12}$ cm$^{-2}$. The populations of the i=2 and i=3 subbands are such that they reach their ultraquantum limits at $\hbar\omega_c \approx$10 meV and $\hbar\omega_c \approx$ 7 meV respectively and so the effective masses increase above these energies due to non-parabolicity: again, discontinuities in the i=2 and i=3 subband effective masses close to the "HgTe-like" reststrahlen band indicate resonant polaron coupling. In all of the samples studied, the phonon to which the i=2 and i=3 subbands are coupling appears to be closer in energy to the "HgTe-like" LO phonon than to the "HgTe-like" TO phonon: thus it is tentatively suggested that the i=2 and i=3 subbands couple to the LO phonon as do the bulk electrons[22-13].

The behaviour of the lower (i=0,1) subbands is somewhat different. At high values of $N_s$ (figures 4 and 5), very weak resonant coupling appears to occur at the "HgTe-like" LO phonon energy, and the interaction is stronger for the i=1 subband than for the i=0. In contrast, if the effective masses of the i=0 and i=1 subbands in the low $N_s$ samples are examined as a function of energy (eg figure 6) a sharp increase is seen just below the "HgTe-like" TO phonon energy, behaviour more consistent with a resonant interaction at the TO phonon frequency. This increase is not due to band non-parabolicity, as, for example, the ultraquantum limit of the i=1 subband in figure 6 is reached at $\hbar\omega_c \approx$30 meV.

Recent experiments on a variety of (Ga,In)As-InP and (Ga,In)As-(Al,In)As heterostructures showed that weak resonant polaron effects occurred at the TO frequency in high carrier density heterojunctions, but that the dominant interaction was at the LO frequency in lower carrier density quantum wells[24-25]. These results were interpreted as due to the difference in screening in the two cases. The subbands on n-(Hg,Cd)Te described above are perhaps behaving in the same way: the higher subbands are analogous to the weakly screened electrons in the quantum wells, so that the resonant coupling occurs at the LO frequency. Likewise, the i=0,1 subbands are strongly screened as in the heterojunctions, so that at low $N_s$ weak coupling at the TO frequency occurs, and further increases in $N_s$ destroy this interaction. Attempts

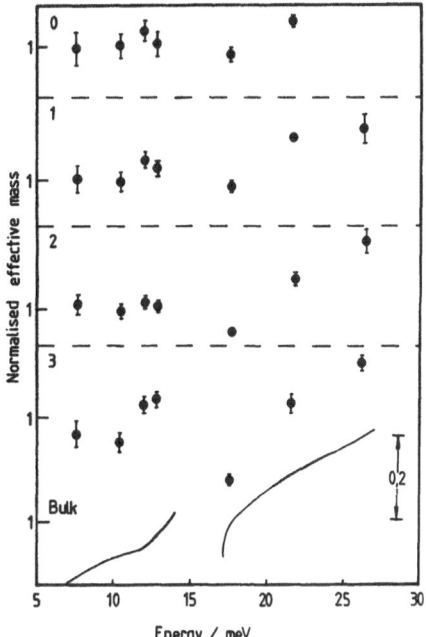

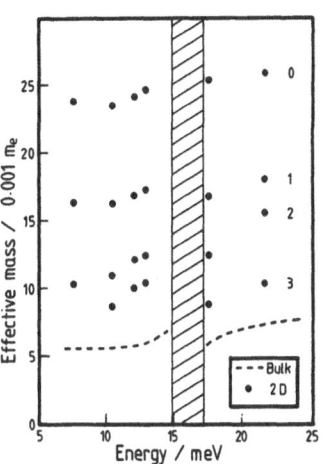

Fig.5. Normalised subband effective masses plotted alongside normalised bulk data[22] (see text).

Fig.6. Subband effective masses in a sample with $N_s=0.9 \times 10^{12} cm^{-2}$ plotted as function of energy: the shaded area is the reststrahlen band.

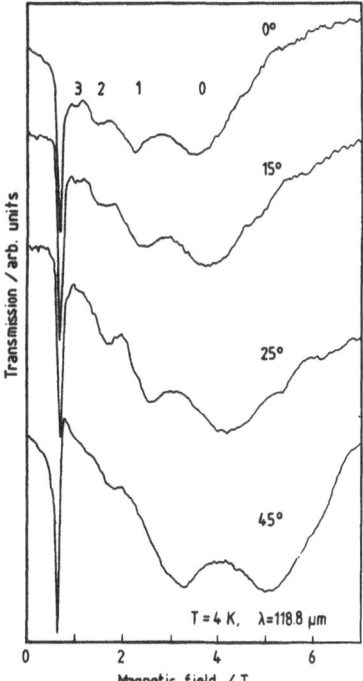

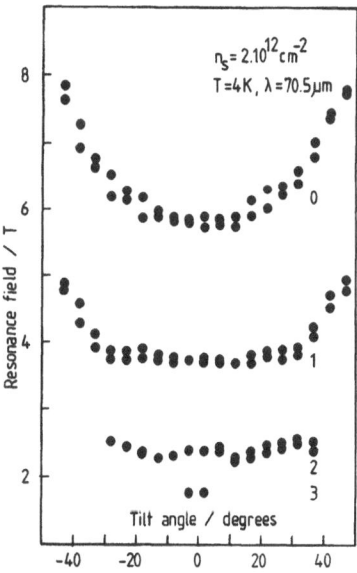

Fig.7. The effect of tilting the magnetic field on the FIR magneto-transmission of a sample with $N_s = 2.01 \times 10^{12} cm^{-2}$. The photon energy is 10.4 meV.

Fig.8. CR magnetic fields as a function of tilt angle, for a sample with $N_s = 2.01 \times 10^{12} cm^{-2}$. ($\hbar\omega_c = 17.58$ meV)

to explain this behaviour using static screening have been unsuccessful[13], and it is believed that a model incorporating full dynamical screening of the electron-phonon interaction[17] and the effects of a high magnetic field will be necessary in order to account for the observed results.

## TILTED FIELD CYCLOTRON RESONANCE

When a magnetic field is applied perpendicular to a two dimensional electron gas, highly degenerate Landau levels are created on each of the electric subbands, and the energy spectrum becomes completely discrete[26]. However, if the magnetic field is tilted away from the surface normal, the electron motions parallel and perpendicular to the surface will couple, and hybrid mixed levels will be formed[27]. These effects have been observed in Silicon[28], PbTe[29], InAs[30] and InSb[31] MIS structures and in InAs-GaSb quantum wells[32]. In the limit where the electric subband energies exceed the Landau level separations by a large factor, the component of field parallel to the layer will essentially act as a perturbation, shifting the subband energies by a small amount[33]. The largest effect in this case will be a movement of phenomena observed with B perpendicular to the layer to higher total fields, approximately as $1/\cos\theta$, as the tilt angle $\theta$ is increased.

The effect of tilting the magnetic field on the FIR magnetotransmission of a sample with $N_s = 2.0 \times 10^{12}$ cm$^{-2}$ is shown in figure 7 for $\hbar\omega_c = 10.4$ meV. As the sample is tilted out of the magnetic field, the i=3 subband CR ceases to be resolved, and that due to the i=2 subband decreases in intensity. This is due to the diamagnetic shift of the subbands, caused by the component of the magnetic field in the plane of the accumulation layer, which causes the higher subbands to depopulate[4,5]. Meanwhile, the lower (i=1,0) subband CR move to higher total field as the tilt angle increases.

On plotting the resonance magnetic fields as a function of angle, for $\hbar\omega_c = 17.6$ meV (figure 8), however, it is plain that the CR are not moving to higher fields in the simple $1/\cos\theta$ manner described above. Only the most deeply bound i=0 subband appears to be 2D: the other subbands, higher in the potential well, show an increasing degree of three-dimensionality. At lower fields, the classical cyclotron orbits will be larger, and so one would expect that the subbands would look more two-dimensional. In figure 9a, the cyclotron energy is decreased to 10.4 meV, increasing the cyclotron radius: the i=1 subband is now 2D in character whilst the i=2 and i=3 subbands still appear to be 3D. On decreasing $\hbar\omega_c$ further (figure9b) to 7.6 meV, all the subbands except the i=3 appear to be 2D in character. Measurements of this type can be used to estimate the subband widths[34], by assuming that the behaviour of the electrons in a subband will be 3D-like until the classical cyclotron orbit "touches" one or both sides of the confining potential. Using the results in figures 8 and 9, and the accompanying SdH data, we obtain subband widths of 300±100 Å, 500±100 Å and 700±100 Å for the i=1,2 and 3 subbands respectively[34]. In spite of the crudeness of this estimate, these values are in good agreement with the self-consistently determined confinement lengths deduced using the model of Takada et al.[14]

In tilted field and partially circularly polarised light, the "2D-like" CR are found to shift when the magnetic field is reversed (figure 10). With the field in the direction which should produce electron cyclotron motion which will couple to the predominant circular polarisation of the light, the subband CR occur at a higher magnetic field and exhibit narrower linewidths than with the field reversed. This lack of symmetry with respect to field reversal is thought to be due to the electrons having two distinct modes of behaviour. Classically, the parallel field component will cause the cyclotron orbits to be displaced so that the electrons encounter one or the other of confining potentials of the accumulation layer. One set, confined by the depressed conduction band, will perform complete orbits

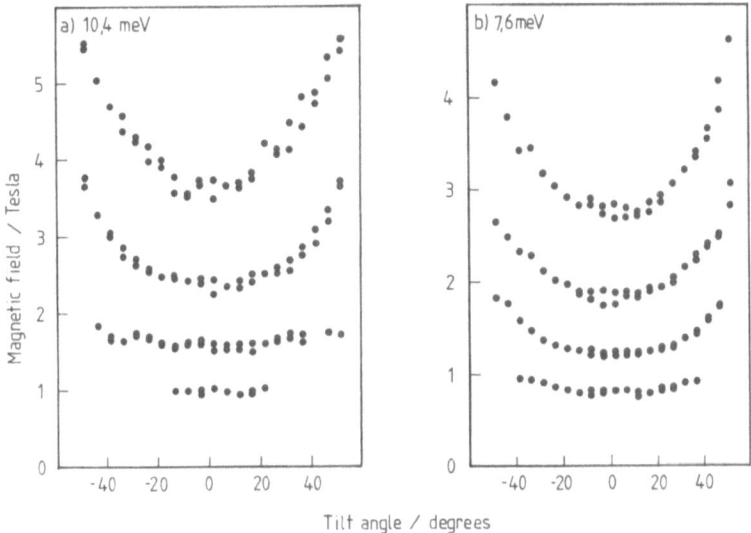

Fig.9. CR magnetic fields for the i=0,1,2,3 subbands as a function of tilt angle for a: $\hbar\omega_c$ = 10.4 meV and b: $\hbar\omega_c$ = 7.6 meV. $N_s$ =2.01x10$^{12}$cm$^{-2}$

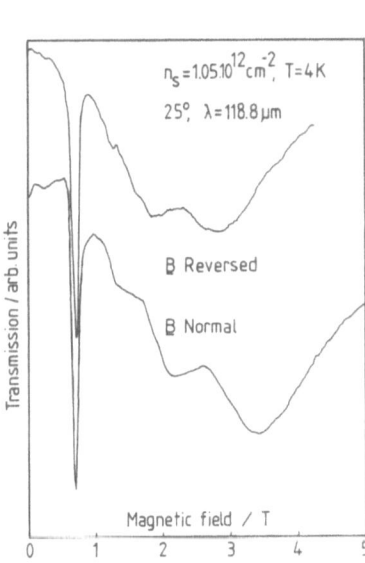

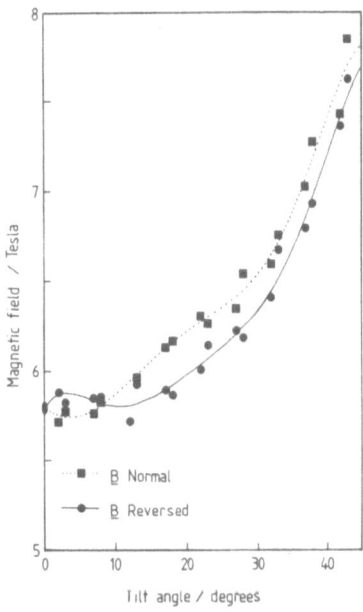

Fig.10. The effect of reversing the field direction in tilted field and partially circularly polarised light: the CR broaden and move to lower B.

Fig.11. CR magnetic fields for the i= 0 subband of a sample with $N_s$=2.01x 10$^{12}$cm$^{-2}$ shown as a function of tilt angle. points for both field directions are shown: the lines are polynomial fits to act as a guide to the eye. (CR energy = 17.58 meV)

whereas the other set will be periodically specularly reflected from the semiconductor-oxide interface, and perform a skipping orbit, first observed in metals[35]. The skipping orbits will occur at higher frequencies than the complete orbits and the associated CR will show a larger linewidth due to the closer proximity of the surface. In addition, the skipping orbits will couple very differently to radiation[35-36].This hypothesis explains qualitatively the shift of the CR to lower fields and larger linewidths on field reversal (figure 10),and it is believed that this is the first direct observation of skipping orbits in a semiconductor system.

The detailed behaviour of the two types of resonance is shown as a function of tilt angle for the i=0 subband of a sample with $N_s \approx 2.01 \times 10^{12} cm^{-2}$ in figure 11, with $\hbar\omega_c = 17.58$ meV. The assymetry between the two field directions starts at very small tilt angles (the error in the mechanism is $\pm 1^o$) and the resonance identified as due to "complete" orbits initially stays at lower fields than that due to skipping orbits: at $\sim 10^o$ the two swap over to give the situation described above. The former effect may be due to the initial displacements of the two types of electron due to the small parallel-field component: the set of electrons which are displaced into the bulk of the crystal will have a larger overlap with the valence band states than before, which will tend to lower the observed effective mass[14-37]. All of the samples examined in tilted field behaved in a similar manner, with the exception that the percentage shift in resonance position is larger ($\sim 10\%$: see figure 10) for $N_s \approx 1 \times 10^{12} cm^{-2}$ than for $N_s \approx 2 \times 10^{12} cm^{-2}$ ($\sim 4\%$:see figure 11). For fixed $N_s$, the percentage shift appears to be approximately energy independent over the range of energy studied ($\hbar\omega_c = 7.6$ meV, 10.4 meV, 17.6 meV), as is the "swap-over" of the two types of resonance at low angles.

The behaviour of the accumulation layer electrons on the n-$Hg_{0.8}Cd_{0.2}Te$ in tilted field is thus qualitatively rather different than that observed in InSb inversion electrons[38-39]. In the case of (Hg,Cd)Te, the mixing of the electric and magnetic quantization is manifested as the small assymmetry in CR position between the forward and reverse field directions, essentially a perturbation[34], whereas in InSb, the mixing is strong, and a "bulk-like" CR evolves as the tilt angle increases[38]. At present, existing simple models do not account well for this difference[36].

CONCLUDING REMARKS

It is hoped that this brief review has given some idea of the interesting physics which can be performed with the accumulation layer electrons on n-(Hg,Cd)Te. Not only are the material parameters of (Hg,Cd)Te such that high-field limits, reststrahlen bands etc can be reached with reasonable magnetic fields: in the accumulation layers we have the added bonus of around six subbands, ranging from the deeply bound i=0, to the i=3,4,5 subbands, which are bound by a few meV and penetrate several hundred Ångstroms into the bulk material. A complete range of experimental conditions therefore exist on one sample, and the influence of confinement and screening on the 2DEG can be probed.

ACKNOWLEDGEMENT

We are grateful to Drs G.T. Jenkin, P.Knowles and L.K.Nicholson of the GEC Hirst Research Centre, Wembley, UK, for the provision of the (Hg,Cd)Te samples.

REFERENCES

1. P.C.Catagnus and C.T.Baker: US patent 3,977,018 (1976)
2. Y.Nemirovsky and E.Finkman: J.Electrochem.Soc. 126 768 '(1979)

3. J.F.Wager and D.R.Rhiger: J.Vac.Sci.Technol. A3 212 (1985)
4. W.Q.Zhao, F.Koch, J.Ziegler and H.Maier: Phys.Rev. B 31 2416 (1985)
5. J.Singleton, R.J.Nicholas, F.Nasir and C.K.Sarkar: J.Phys.C 19 35 (1986)
6. J.Singleton, F.Nasir and R.J.Nicholas: Surface Science 170 409 (1986)
7. F.Nasir, J.Singleton and R.J.Nicholas: to be published.
8. R.B.Schoolar, B.K.Janousek, R.L.Alt, R.C.Carscallen, M.J.Daugherty and A.A.Fote: J.Vac.Sci.Technol. 21 164 (1982)
9. T.Ando: J.Phys.Soc.Jpn. 54 2676 (1985)
10. J.Singleton, F.Nasir and R.J.Nicholas: Proc.SPIE 659 23 (1986)
11. M.Weiler: Semiconductors and Semimetals 16 119 (1981)
12. J.Scholz, F.Koch, J.Ziegler and H.Maier: Solid State Commun. 46 665 (1983)
13. J.Singleton, R.J.Nicholas and F.Nasir: Solid State Commun. 58 833 (1986)
14. Y.Takada, K.Arai and Y.Uemura: Springer Lecture Notes in Physics 152 101 (1982)
15. F.Koch: Springer Series in Solid State Sciences 53 20 (1984)
16. D.M.Larsen: Phys.Rev. B 30 4595 (1984)
17. W.Xiaoguang, F.M.Peeters and J.T.Devreese: Phys.Stat.Sol. b 133 229 (1986)
18. M.Horst, U.Merkt and J.P.Kotthaus: Phys.Rev.Lett.50 754 (1983)
19. W.Seidenbusch, G.Lindemann, R.Lassnig, J.Edlinger and E.Gornik: Surf. Sci. 142 375 (1984)
20. M.Horst, U.Merkt, W.Zawadzki, J.C.Maan and K.Ploog: Solid State Commun. 53 403 (1985)
21. J.Baars and F.Sorger: Solid State Commun. 10 875 (1972)
22. M.A.Kinch and D.D.Buss: J.Phys.Chem.Solids 32 Supplement 1 461 (1971)
23. L.Swierkowski, W.Zawadzki, Y.Guldner and C.Rigaux: Solid State Commun 27 1245 (1978)
24. R.J.Nicholas, L.C.Brunel, S.Huant, K.Karrai, J.C.Portal,M.A.Brummell, M.Razeghi, K.Y.Cheng and A.Y.Cho: Phys.Rev.Lett. 55 883 (1985)
25. L.C.Brunel, S.Huant, R.J.Nicholas, M.A.Hopkins, M.A.Brummell, K.Karrai, J.C.Portal, M.Razeghi, K.Y.Cheng and A.Y.Cho: Surface Science 170 542 (1986)
26. T.Ando,A.B.Fowler and F.Stern: Rev.Mod.Phys. 54 437 (1982)
27. T.Ando: Phys.Rev. B 19 2106 (1978)
28. W.Beinvogl and J.F.Koch: Phys.Rev.Lett. 40 1736 (1978)
29. H.Schaber and R.E.Doezema: Phys.Rev. B 20 5257 (1979)
30. R.E.Doezema, M.Nealon and S.Whitmore: Phys.Rev.Lett. 45 1593 (1980)
31. J.H.Crasemanm and U.Merkt: Solid State Commun. 47 917 (1983)
32. J.C.Maan, Ch.Uihlein, L.L.Chang and L.Esaki: Solid State Commun. 44 653 (1982)
33. F.Stern and W.E.Howard: Phys.Rev. 163 816 (1967)
34. J.Singleton, F.Nasir and R.J.Nicholas: Solid State Commun. 59 879 (1986)
35. R.E.Prange and T.W.Nee: Phys.Rev. 168 779 (1968)
36. U.Merkt: Phys.Rev. B 32 6699 (1985)
37. W.Brenig and H.Kasai: Z.Phys. B 54 191 (1984)
38. J.H.Crasemann, U.Merkt and J.P.Kotthaus: Phys.Rev B 28 2271 (1983)
39. M.Horst and U.Merkt: Solid State Commun. 54 559 (1985)

# CYCLOTRON RESONANCE OF INVERSION ELECTRONS ON InSb

U. Merkt

Institut für Angewandte Physik, Universität Hamburg
Jungiusstr. 11, D-2000 Hamburg 36, F.R.G.

## 1. INTRODUCTION

Quasi two-dimensional electron systems as realized in space-charge layers on semiconductors have extensively been studied with spectroscopic methods on various materials in recent years.[1] The most interesting feature on narrow-gap semiconductors like InSb or $Hg_{1-x}Cd_xTe$ is the coupling of valence and conduction band which results from their small gap energy. As a consequence the conduction band is strongly nonparabolic, i.e., the apparent mass strongly increases with energy. This behavior is well-known for three-dimensional narrow-gap semiconductors and also has been observed in two-dimensional systems of inversion or accumulation layers. In such systems the motion of electrons is restricted in one spatial dimension perpendicular to an interface by the action of the surface electric field. Due to the presence of this field new effects[2] result from the interaction of electron states in the conduction band and hole states in the valence band: electrons in inversion layers can tunnel into the valence band. This effect leads to a resonant state, i.e., the electron in the inversion layer and the hole in the bulk are coupled.[3,4] The subband energies are decreased as compared to parabolic semiconductors with large gaps. This effect has been studied experimentally with intersubband spectroscopy in accumulation layers on $Hg_{1-x}Cd_xTe$.[5] Excitation of intersubband resonances with light polarized parallel to the interface becomes possible as a result of a coupling between the free motion parallel to the interface and the quantized motion perpendicular to it. This has been demonstrated experimentally in inversion layers on InSb and InAs.[6]

Here we review our work on cyclotron resonance in inversion layers on InSb in magnetic fields which have been applied perpendicular[7] and parallel[8,9] to the layers. We describe the experimental results by a simple model, namely a three-level k·p-model in the triangular-well potential.[10] In the model we assume a constant surface electric field to achieve simple analytical expressions for the eigenenergies. This model cannot compensate for a self-consistent description but it provides a rather vivid picture which allows direct physical insight into the intricate problems of nonparabolicity in spatially confined systems. Unlike in previous theoretical work[10] we do not solve the effective Schrödinger equation of the k·p-model semiclassically, but we give quantummechanical solutions in terms of parabolic cylinder (Weber) functions. In magnetic fields perpendicular to the interface we study the nonparabolic increase of subband masses with two-dimensional electron density and with magnetic field strength in various electric subbands.[7] In magnetic fields parallel to the layers we observe cyclotron resonance in crossed electric and magnetic fields as the electric field is directed perpendicular to the plane.[8,9] In particular, we could verify in this system the predicted destruction of the Landau quantization itself[11] when the ratio of field strengths E/B exceeds a critical value $u=(\mathcal{E}_g/2m_0^*)^{1/2}$ which is the maximum velocity possible in the conduction band according to the two-band model.

## 2. THEORY: NONPARABOLIC SUBBANDS IN THE TRIANGULAR-WELL POTENTIAL.

### 2.1 Purely electric subbands

In the approximation of the triangular-well potential a constant surface electric field E inside the semiconductor and an infinitely high potential barrier at the semiconductor-oxide interface are assumed as shown in Fig.1(a). In the one-band effective-mass approximation (EMA) the resulting subband energies are[1]

$$\mathcal{E} = \frac{\hbar^2 k_\parallel^2}{2m_0^*} + \left(\frac{9\pi^2}{8m_0^*}\right)^{1/3} (e\hbar E)^{2/3} (i+\tfrac{3}{4})^{2/3} \qquad . \tag{1}$$

The first term is the kinetic energy $\mathcal{E}_\parallel$ of the free motion parallel to the interface and the second term is the discrete subband energy $\mathcal{E}_i^{EMA}$ of the quantized motion perpendicular to it. As long as only the ground subband i=0 is considered, a fairly realistic value for the electric field strength E is obtained by putting equal the average distances of electrons away from the interface $\langle z_0 \rangle$ as calculated in the Fang-Howard variational approach[1] and in the triangular-well potential,[1] respectively. This gives an effective field

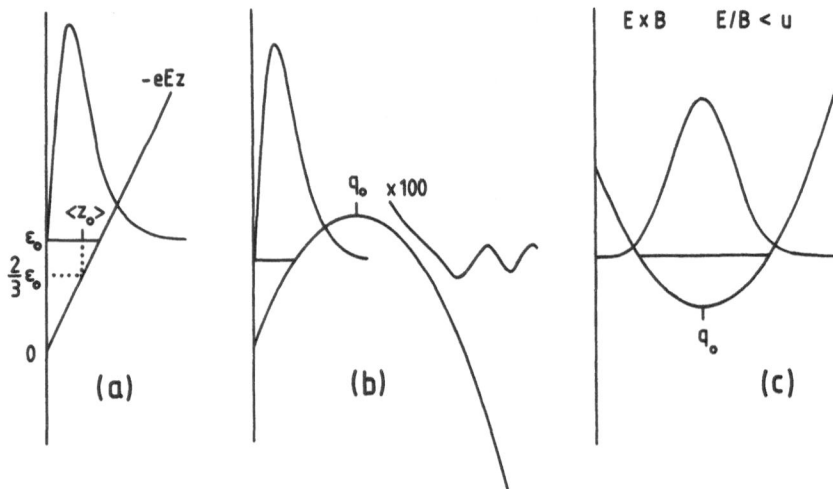

Fig.1. (a) Triangular-well potential and wave function for the ground subband in the effective mass approximation (EMA). The average distance of electrons away from the interface in any subband i is given by $\langle z_i \rangle = 2\mathcal{E}_i/3eE$. (b) Resulting effective potential and envelope function in the k·p-model. The oscillatory holelike part of the wave function denotes tunneling through the energy gap. (c) Effective potential and envelope function in the crossed field configuration (magnetic case).

$$E = \frac{\pi^2}{12}\left(n_{depl} + \frac{11}{32}n_s\right)\frac{e}{\varepsilon_0\kappa} \qquad (2)$$

with depletion charge $n_{depl}$, two-dimensional electron density $n_s$, and dielectric constants $\varepsilon_0$ and $\kappa$ of vacuum and semiconductor, respectively. The triangular-well approximation has also been used to describe nonparabolic subbands on narrow-gap semiconductors with InSb-type bandstructure.[10] Three levels at the $\Gamma$ point are taken into account in a k·p-model: a $\Gamma_6$ conduction level separated by the gap energy $\mathcal{E}_g$ from a $\Gamma_8$ valence level, this in turn separated from a $\Gamma_7$ valence level by the spin-orbit interaction energy $\Delta$. The choice of the electric field direction $E \parallel z$ allows one to write the wave functions in the form (index l runs over the bands)

$$\Psi = \sum \Phi_l(z)\ e^{i(k_x x + k_y y)}\ u_l(\vec{r}) \qquad (3)$$

with envelope functions $\Phi_l$ which slowly vary over a unit cell and with Luttinger-Kohn functions $u_l$ which are periodic. The free electron term is neglected and the final effective Schrödinger equation for the z-dependent envelope function $\Phi_s(z)$ which describes the bound motion perpendicular to the interface and which is related to the S-like conduction band reads

$$\left\{ -\frac{\hbar^2}{2m_0^*}\frac{d^2}{dz^2} - \alpha z - \frac{m_0^*}{2}\left(\frac{2e^2 E^2}{\mathcal{E}_g m_0^*}\right)z^2 - \frac{\hbar^2}{2m_0^*}\frac{eE}{\mathcal{E}+\mathcal{E}_g/2 - eEz}\frac{d}{dz} \right\}\Phi_s = \lambda\Phi_s \qquad (4)$$

with abbreviations $\alpha = -2eE\mathcal{E}/\mathcal{E}_g$ and $\lambda = \mathcal{E}^2/\mathcal{E}_g - \mathcal{E}_g/4 - \mathcal{E}_\parallel$. The zero of the energy scale is chosen at the center of the energy gap. Equation (4) cannot be solved analytically, but below we will derive a power series for the subband energies $\mathcal{E}$ which is valid for not too large energies ($\mathcal{E} \ll \mathcal{E}_g$). For InSb the subband energies are small compared to the gap energy up to relatively high electron densities $n_s$ (see Ref.6). The presumption of small energies also allows one to apply the simple boundary condition $\Phi_s(z=0)=0$ for the envelope function related to the S-like conduction band.[10] The intricate problems of proper boundary conditions in the k·p-approach are discussed in some detail in Refs. 2 and 3.

Previously, the semiclassical WKB quantization scheme has been employed in Ref.10 to solve Eq.(4). Here we treat the eigenvalue problem in terms of parabolic cylinder (Weber) functions of second kind. For this, we introduce a dimensionless space coordinate $q=z/l_E$ via an electric oscillator length $l_E=(\hbar/2m_0^*\omega_E)^{1/2}$ related to an electric frequency $\omega_E=(2e^2 E^2/m_0^*\mathcal{E}_g)^{1/2}$. However, in spite of these definitions, the eigenvalue problem is not the one of the usual harmonic oscillator since the quadratic term in Eq.(4) is negative. Defining a positive parameter

$$a = \frac{1}{\mathcal{E}_g \hbar\omega_E}\left[\left(\frac{\mathcal{E}_g}{2}\right)^2 + \mathcal{E}_g \mathcal{E}_\parallel\right] \qquad (5)$$

we obtain the differential equation

$$\left[ \frac{d^2}{dq^2} + \frac{1}{4}(q-q_0)^2 - a + \left(\frac{\hbar\omega_E}{\mathcal{E}_g}\right)^{1/2} \frac{1}{1+2\frac{\mathcal{E}}{\mathcal{E}_g} - \left(\frac{\hbar\omega_E}{\mathcal{E}_g}\right)^{1/2}q} \frac{d}{dq} \right] \Phi_s(z) = 0 \qquad (6)$$

with a center coordinate $q_0 = \mathcal{E}/eEl_E$ and the so-called Zener term[10] which is proportional to the first derivative. Provided the electric field is not too strong the Zener term can be ignored at first instance, since it is of order $(\hbar\omega_E/\mathcal{E}_g)^{+1/2}$ whereas the parameter $a$ is of order $(\hbar\omega_E/\mathcal{E}_g)^{-1}$. Therefore, we first treat the eigenvalue problem without the Zener term and then calculate the corresponding correction in first order perturbation theory. Thus we first get the differential equation

$$\left[ \frac{d^2}{dq^2} + \frac{1}{4}(q-q_0)^2 - a \right] W(a,q) = 0 \qquad (7)$$

for the Weber functions $W(a,q)$ which are discussed in Ref.12. From the two linearly independent solutions $W(a,\pm q)$ we only take into account functions $W(a,+q)$ as they have an oscillatory part of small amplitude inside the semiconductor $(q\to\infty)$ which is depicted in Fig.1(b). Qualitatively spoken, admixtures of functions $W(a,-q)$ are responsible for a resonant broadening of subband states, i.e., for a broadening even in the absence of scattering. This is a very interesting feature of surface quantization in space-charge layers on narrow-gap semiconductors.[3] However, this broadening of eigenstates is small on InSb as compared to linewidths of real samples and, therefore, we do not treat this effect here.

As a consequence of our boundary condition, the negative center coordinates $-q_0$ become zeros of the Weber functions $W(a,+q)$. For these zeros a power series has been derived which immediately gives $(\mathcal{E}=q_0 el_E E)$ a series for the energies[13]

$$\mathcal{E} = \sqrt{\left(\frac{\mathcal{E}_g}{2}\right)^2 + \mathcal{E}_g \mathcal{E}_\parallel} \left[ 1 + 2\left(\frac{\mathcal{E}_i}{\mathcal{E}_g}\right)^{EMA} \frac{1}{\left(1+\frac{4\mathcal{E}_\parallel}{\mathcal{E}_g}\right)^{2/3}} - \frac{2}{5}\left(\frac{\mathcal{E}_i}{\mathcal{E}_g}\right)^{EMA^2} \frac{1}{\left(1+\frac{4\mathcal{E}_\parallel}{\mathcal{E}_g}\right)^{4/3}} \pm \ldots \right]. \qquad (8)$$

The kinetic energy $\mathcal{E}_\parallel = \hbar^2 k_\parallel^2 / 2m_0^*$ is calculated with the bulk effective mass $m_0^*$ at the conduction band edge. The subband energies $\mathcal{E}_i^{EMA}$ are the ones of the EMA calculated from Eq.(1) for momentum $\hbar k_\parallel = 0$. Note, that the zero of the energy scale in the EMA has been chosen at the conduction band edge as shown in Fig.1(a). This figure depicts the wave function of the ground electric subband in the EMA. In Fig.1(b) the corresponding wave function and the effective potential in the $k \cdot p$-approach are visualized. The effective potential $-m_0^* \omega_E^2 (z-z_0)^2/2$ results from the triangular-well potential and appears in the effective Schrödinger equation Eq.(4). The wave function in the $k \cdot p$-model has a "holelike" oscillatory part of small amplitude. This corresponds to tunneling from the conduction to the valence band and demonstrates that such band mixing effects are obtained in the absence of the Zener term.

Equation (8) has some very attractive features: in the three-dimensional limit $(E=0)$ it reduces immediately to the well-known Kane formula

$$\mathcal{E} = \sqrt{\left(\frac{\mathcal{E}_g}{2}\right)^2 + \mathcal{E}_g \frac{\hbar^2 k_\parallel^2}{2m_0^*}} \qquad (9)$$

and it describes the most important feature of nonparabolic subbands ($E \neq 0$) in a rather transparent way: the free motion parallel and the quantized motion perpendicular to the interface are coupled as the coefficient in front of the subband energy $\varepsilon_i^{EMA}$ depends on the kinetic energy $\varepsilon_\parallel$. At high kinetic energies ($\varepsilon_\parallel \gg \varepsilon_g$) this leads to subband energies $\varepsilon = \hbar k_\parallel u$ that no longer depend on the subband index which is in clear contrast to the EMA (see Fig.2). The velocity $u = (\varepsilon_g/2m_0^*)^{1/2}$ is the maximum velocity possible in the conduction band according to the two-band model. At very small kinetic energies ($\varepsilon_\parallel \ll \varepsilon_g$) and subband energies ($\varepsilon_i^{EMA} \ll \varepsilon_g$) we recover the EMA result given in Eq.(1).

For not too high energies $\varepsilon_\parallel$ and $\varepsilon_i^{EMA}$ we can calculate the effective subband masses

$$m_i^*(\varepsilon) \approx m_0^* \left( 1 + 2 \frac{\frac{1}{3} \varepsilon_i^{EMA} + \varepsilon_\parallel}{\varepsilon_g} \right) \quad . \tag{10}$$

This means, that the nonparabolic mass increase due to the subband quantization is less effective than the increase due to the kinetic energy. Intuitively, this can be explained by the fact that in any electric subband i the electron on an average is $\frac{1}{3}\varepsilon_i^{EMA}$ away from the conduction band edge as depicted in Fig.1(a). This just gives the proper factor $\frac{1}{3}$ in Eq.(10). Whereas at the subband edges ($k_\parallel = 0$) the effective masses increase with subband index, at any given Fermi energy $\varepsilon_F$ the effective masses are less in higher subbands: $m^*_{i+1}(\varepsilon_F) < m^*_i(\varepsilon_F)$. This has in fact been observed in space-charge potentials of triangular shape.[7,14]

Equation (8) also accounts qualitatively for the decrease of the subband energies with decreasing gap energy $\varepsilon_g$ as has been found experimentally in accumulation layers on $Hg_{1-x}Cd_xTe$.[5] For the subband edges ($k_\parallel = 0$) we obtain

$$\varepsilon_i \approx \frac{\varepsilon_g}{2} + \varepsilon_i^{EMA}\left(1 - \frac{1}{5} \frac{\varepsilon_i^{EMA}}{\varepsilon_g} \right) \quad . \tag{11}$$

Intuitively, this may be explained by the form of the wave function with the periodic "holelike" part inside the semiconductor. This part becomes more and more important as the gap energy is decreased, the particle becomes less confined, and the eigenenergy is correspondingly lowered. This lowering of the subband energy $\varepsilon_i$ is more pronounced in higher subbands.

We now discuss the effect of the Zener term on the subband energies. The leading correction[15] of the eigenvalue, namely $\Delta a = \hbar \omega_E/2\varepsilon_g$ is obtained if the Zener term is developed into a geometrical series. The corresponding correction of the eigenenergies $\Delta \varepsilon$ again is found via the zeros of the Weber functions:

$$\Delta \varepsilon \approx \frac{128}{81\pi^2} \left( \frac{\varepsilon_0^{EMA}}{\varepsilon_g} \right)^3 \varepsilon_g \quad . \tag{12}$$

This correction is small and independent of the subband index i. Only in fourth order the Zener term leads to observable shifts of the subband spacings. Therefore, the eigenenergies which are given in Eq.(8) up to second order are not affected by the Zener term.

## 2.2 Landau levels in perpendicular magnetic fields

Landau levels of two-dimensional subbands in magnetic fields applied perpendicular to the interface can also be described with the present model when one makes the substitution

$$\varepsilon_\| = \frac{\hbar^2 k_\|^2}{2m_0^*} \rightarrow \hbar\omega_C\left(n+\frac{1}{2}\right) \pm \frac{1}{2}g_0^*\mu_B B \qquad (13)$$

in Eq.(8). This is justified provided the spin–orbit interaction energy is much larger than the gap energy ($\Delta \gg \varepsilon_g$) which is approximately the case for InSb.[10] The cyclotron frequency $\omega_C = eB/m_0^*$ and the effective Lande' factor $g_0^*$ are both taken at the conduction band edge, n is the Landau index.

The Landau energies $\varepsilon_{i,n}$ obtained from Eqs.(8) and (13) depend on subband index i, Landau index n, and spin orientation $\pm$ as shown in Fig.3 for three electric subbands. Effects of nonparabolicity are obvious: the energy of a particular Landau level increases less than linearly with magnetic field and the spacings between adjacent Landau levels decrease when the Landau index $n^\pm$ is increased at a fixed magnetic field. Both effects are different in different electric subbands and the cyclotron energy $\varepsilon_{i,n+1}-\varepsilon_{i,n}$ depends on magnetic field, Landau index $n^\pm$, and electric subband index i. From the cyclotron energy a cyclotron mass

$$m^*_{i,n} = \frac{e\hbar B}{\varepsilon_{i,n+1} - \varepsilon_{i,n}} \qquad (14)$$

can be calculated. This mass is a very sensitive measure for nonparabolicity, since only in the EMA this mass is a constant independent of quantum numbers and magnetic field strength.

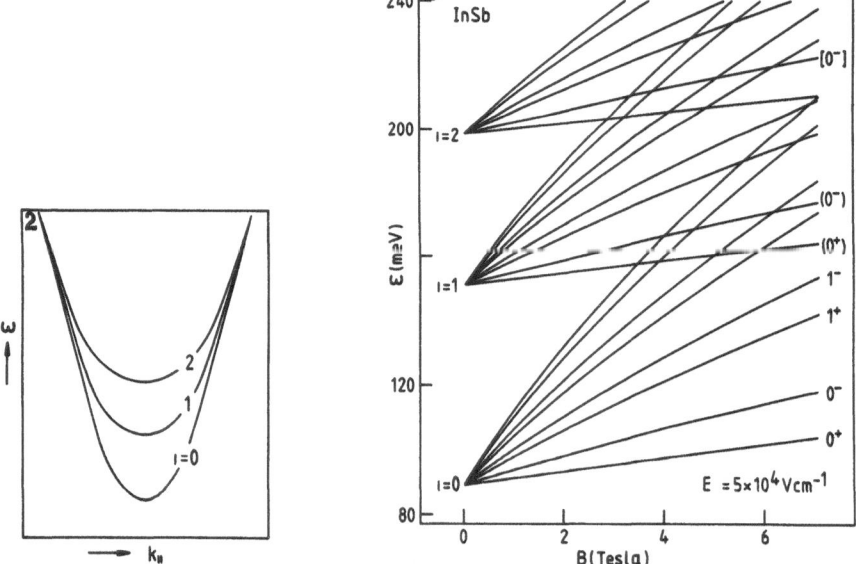

Fig.2. Electric subbands on a narrow-gap semiconductor.

Fig.3. Spin-split Landau levels of electric subbands calculated with InSb band parameters. From Ref.7.

## 2.3 Landau levels in parallel magnetic fields

In a strictly two-dimensional electron system cyclotron resonance is not possible in magnetic fields that are applied parallel to the plane of free motion. However, it is in fact possible in inversion layers on InSb provided the magnetic field (B∥x) is strong and the electron density is low, i.e., the surface electric field (E∥z) is weak.[8,9] Then the cyclotron orbit fits into the inversion channel and one observes cyclotron resonance in crossed electric and magnetic fields since the electric field direction is perpendicular to the interface. In the following we only briefly outline the theoretical description that has been developed in Ref.8.

Again we start from the effective Schrödinger equation but already without Zener term

$$\left[ -\frac{\hbar^2}{2m_0^*} \frac{d^2}{dz^2} - \alpha' z + \frac{m_0^*}{2} \left[ \left(\frac{eB}{m_0^*}\right)^2 - \left(\frac{2e^2E^2}{\varepsilon_g m_0^*}\right) \right] z^2 \right\} \psi_s = \lambda' \psi_s \tag{15}$$

and with abbreviations $\alpha' = \alpha + \hbar\omega_C k_y$ and $\lambda' = \lambda \pm \frac{1}{2} g_0^* \mu_B B$. We introduce the dimensionless space variable $q = z/l_{eff}$ via an effective oscillator length $l_{eff} = (\hbar/2m_0^* \omega_{eff})^{1/2}$ with an effective frequency

$$\omega_{eff} = \sqrt{\omega_C^2 - \omega_E^2} \; = \; \omega_C \sqrt{1 - \delta^2} \; . \tag{16}$$

The ratio $\delta$ is the one of the drift velocity $v_d = E/B = \hbar k_d / m_0^*$ in crossed fields and the maximum velocity $u = (\varepsilon_g / 2m_0^*)^{1/2}$ of the semiconductor.[11] We here restrict the discussion to the magnetic case ($\delta < 1$) since the electric case ($\delta > 1$) is very similar to the purely electric one (B=0) that has been discussed in the previous section. If a negative parameter

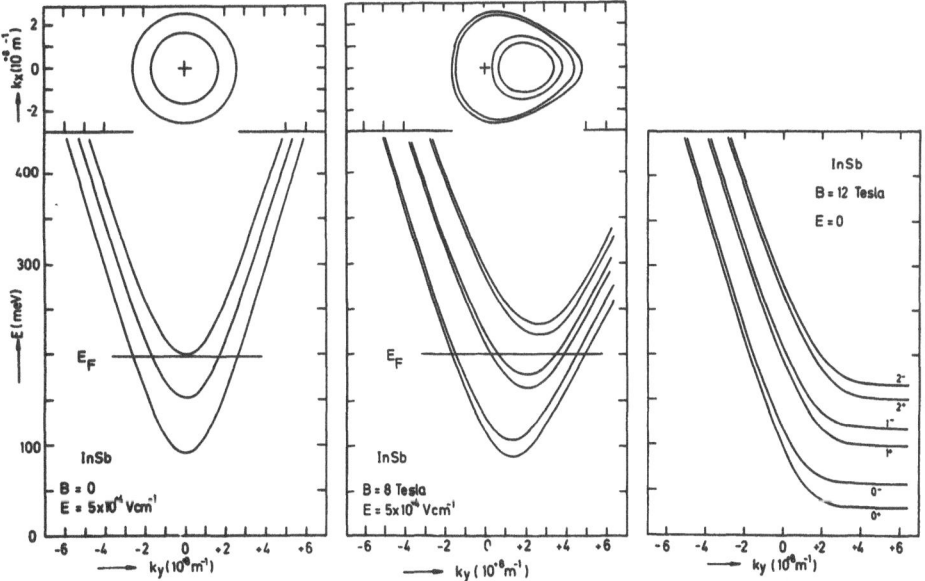

Fig.4. Subband structure in magnetic fields parallel to the interface (crossed field configuration). The transition from a purely electric to a purely magnetic band structure is shown. The Fermi lines are determined for an electron density $n_s = 1.5 \times 10^{12}$ cm$^{-2}$. From Ref.8.

$$a = -\frac{1}{\hbar\omega_{eff}} \left( \lambda' + \frac{\alpha'^2}{2m_0^*\omega_{eff}^2} \right) \tag{17}$$

is defined and the center coordinate

$$q_0 = \frac{\sqrt{2}}{(1-\delta^2)^{3/4}} (k_y - 2k_d\mathcal{E}/\mathcal{E}_g)l \tag{18}$$

is introduced with cyclotron radius $l=(\hbar/eB)^{1/2}$, we obtain the differential equation

$$\left[ \frac{d^2}{dq^2} - \frac{1}{4}(q-q_0)^2 - a \right] U(a,q) = 0 \tag{19}$$

of the parabolic cylinder (Weber) functions of first kind.[12] The linearly independent solution $V(a,q)$ does not satisfy the boundary condition inside the semiconductor, since it is divergent there.[12] There is no oscillatory part of the wave function inside the semiconductor [see Fig.1(c)] in contrast to the electric case ($\delta>1$) and in contrast to the case of purely electric subbands [see Fig.1(b)]. In the magnetic case ($\delta<1$) the magnetic field is strong enough to localize the electron.

The eigenenergies in the magnetic case ($\delta<1$) can be calculated analytically

$$\mathcal{E} = \hbar k_y v_d + \sqrt{1-\delta^2} \sqrt{(\frac{\mathcal{E}_g}{2})^2 + \mathcal{E}_g D_{n,\pm}} \tag{20}$$

$$D_{n,\pm} = -a(q_{0i})\hbar\omega_c(1-\delta^2)^{1/2} + \frac{\hbar^2 k_x^2}{2m_0^*} \pm \frac{1}{2}g_0^*\mu_B B$$

in terms of the parameter $a$. For the relations $a(q_{0i})$ good analytical approximations exist.[8,9] Calculated surface band structures are shown in Fig.4 for the purely electric case [B=0, Eq.(8)], the crossed field case [$\delta<1$, Eq.(20)], and the purely magnetic case [E=0, Eq.(20)]. In the magnetic cases ($\delta<1$) cyclotron resonance can be observed at large positive wave vectors $k_y$. Here the subbands run parallel to each other and we have the cyclotron transition energy ($k_x=0$)

$$\hbar\omega = \sqrt{1-\delta^2} \left[ (\frac{\mathcal{E}_g}{2})^2 + \mathcal{E}_g\left( \hbar\omega_c\sqrt{1-\delta^2} \pm \frac{1}{2}g_0^*\mu_B B \right) \right]^{1/2} . \tag{21}$$

Note that the momentum $\hbar k_y$ is conserved in an optical transition. We like to emphasize that at large positive momentum $\hbar k_y$ one has bulk cyclotron resonance in crossed electric and magnetic fields between states that are practically not affected by the presence of the interface. From Eq.(21) it follows immediately that cyclotron resonance in crossed fields only exists as long as the ratio $\delta<1$ as predicted in Ref.11.

# 3. EXPERIMENTS: CYCLOTRON RESONANCE

The experiments were performed on p-type InSb ($N_A \sim 3 \times 10^{14} cm^{-3}$) samples with $SiO_2$ gate oxides and semitransparent NiCr gate contacts.[16] Cyclotron resonance was measured with far-infrared lasers and a Fourier transform spectrometer at liquid-helium temperatures ($T \approx 4K$) and the magnetic field was applied with its direction perpendicular and parallel to the inversion layers.

## 3.1 Perpendicular magnetic fields

Cyclotron masses that have been extracted from laser spectra (see Fig.5) at a relatively low laser energy ($\hbar\omega = 17.6 meV$) and correspondingly low magnetic fields ($B \sim 2-4T$) are displayed in Fig.6. Masses are shown for three electric subbands i=0,1, and 2. Such a definition of cyclotron masses independent of Landau index is only meaningful in weak magnetic fields, i.e., in the limit $B \to 0$. Then we observe subband masses $m^* = \hbar^2 k_\parallel (\partial \mathcal{E}/\partial k_\parallel)^{-1}$ that can be compared with Eq.(10). On the other hand, in the experiments we have to take into account the condition $\omega_c \tau > 1$ which prevents the observation of sharp cyclotron resonance at lower magnetic fields (B<1T).

The masses in all subbands increase with electron density $n_s$ as a result of subband nonparabolicity. The experimental points show some deviations from monotonic increase by the influence of quantum oscillations as has been discussed previously.[17] The increase of the masses is different in different subbands and is strongest in the ground subband i=0. Qualitatively, these features can be understood with Eq.(10). Since, both, the

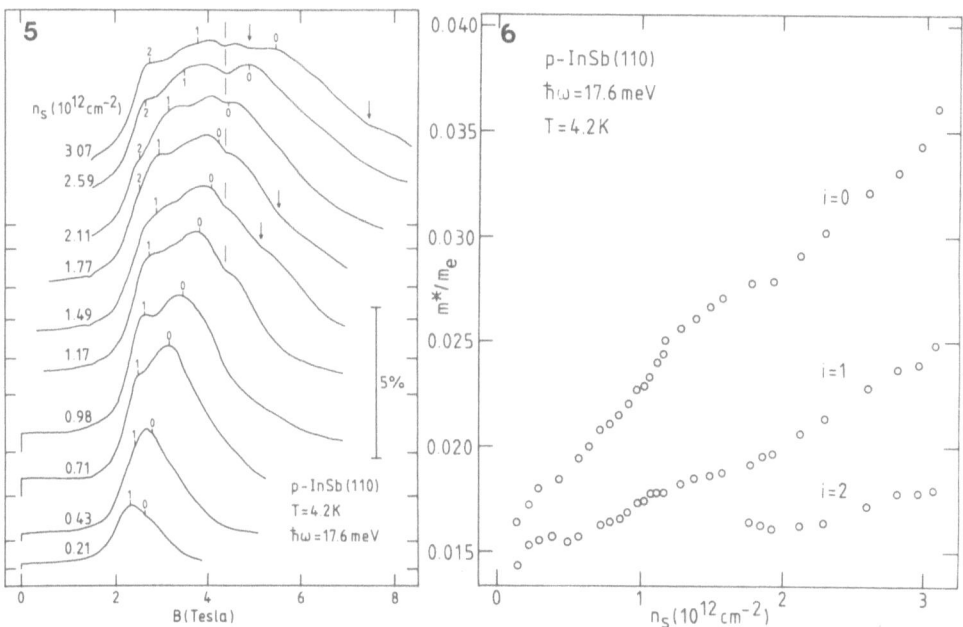

Fig.5. Experimental cyclotron spectra at a fixed laser energy $\hbar\omega$ and various electron densities $n_s$. The resonance positions of subband cyclotron resonances as obtained from theoretical fits are indicated (i=0,1,2). The arrows mark more pronounced quantum oscillations, the dashes cyclotron resonances of bound holes in the p-type substrate. From Ref.7.

Fig.6. Subband cyclotron masses in perpendicular magnetic fields vs electron density for three electric subbands (i=0,1,2) measured at a fixed laser energy $\hbar\omega$. From Ref.7.

subband energy $\mathcal{E}_i{}^{EMA}$ and the kinetic energy $\mathcal{E}_{\parallel}$ in a particular subband increase with electron density $n_s$ we cannot directly separate the nonparabolic mass increase due to the two energies. In spite of this, we find qualitative agreement between experiments and our simple theory: The masses in all subbands increase with electron density since subband energy and occupation, i.e., kinetic energy at the Fermi energy increase. The masses are less in higher subbands since the subband energies are higher. This immediately follows from Eq.(10) if the nominator $\frac{1}{2}\mathcal{E}_i{}^{EMA}+\mathcal{E}_{\parallel}$ is replaced by $\mathcal{E}_F-\frac{3}{2}\mathcal{E}_i{}^{EMA}$. In the limit of zero electron density ($n_s=0$) the subband energy $\mathcal{E}_i{}^{EMA}$ and the kinetic energy $\mathcal{E}_{\parallel}$ vanish and we observe the bulk mass of InSb at the conduction band edge $m_0^*=0.014m_e$.

We have also studied cyclotron masses in higher magnetic fields with special emphasis on the magnetic quantum limit when only tranitions $0^+\rightarrow1^+$ in the ground electric subband i=0 are allowed.[13] At the low electron density $n_s=2\times10^{11}\,cm^{-2}$ this is the case at magnetic fields $B\gtrsim8.3T$. However, even at lower magnetic fields the spectra are dominated by the $0^+\rightarrow1^+$ transition and we could extract the corresponding cyclotron masses over a wider range of magnetic fields (see Fig.7). At low fields $B\lesssim2.5T$ many Landau levels $n^{\pm}$ in the ground and first excited electric subband are occupied, however, the peak of the experimentally observed resonance is caused by cyclotron resonance in

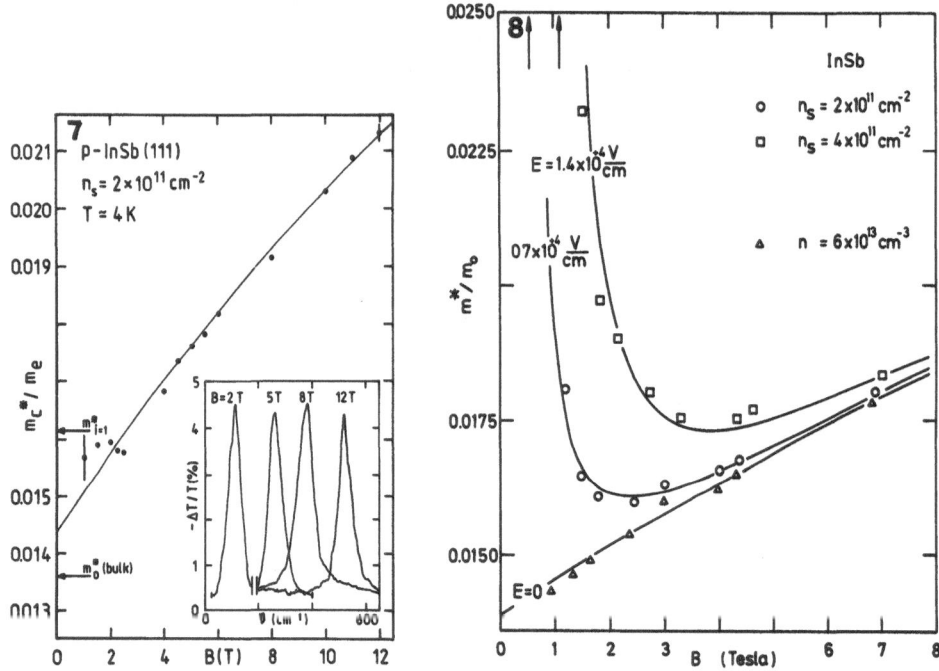

Fig.7. Experimental cyclotron masses in perpendicular magnetic fields. The solid line is calculated with InSb band parameters for $0^+\rightarrow1^+$ transitions which are observed at magnetic fields B>4T. The arrows indicate the bulk band edge mass $m_0^*$ and the calculated electric (B=0) subband mass $m_{i=1}^*$ of the first excited subband, respectively. Typical spectra measured with a Fourier spectrometer at fixed magnetic fields B=2, 5, 8 and 12T are shown in the inset.

Fig.8. Cyclotron masses in crossed electric and magnetic fields (circles and squares) and in the absence of an electric field (triangles). The solid lines are calculated for $0^+\rightarrow1^+$ transitions, the arrows indicate the critical magnetic field strengths where cyclotron resonance vanishes and the mass diverges in the absence of scattering for the two field strengths E≠0. From Ref.8.

the excited subband i=1 (see also Fig.5) and the mass $m_1^*$ calculated with Eq.(10) agrees rather well with the experimental values. In the range B≈2.5–4T we could not observe cyclotron resonance because of strong rest-strahlen absorption. At magnetic fields B≳4T the experimental data agree well with the solid line calculated for $0^+ \to 1^+$ transitions in the ground subband. We emphasize that no adjustable parameter is involved in this theoretical description, except for a slight modification of the gap energy that has been discussed previously.[8] In particular, this means that the surface electric field of Eq.(2) is in fact a good approximation for the ground subband at low electron densities.

## 3.2 Parallel magnetic fields

Cyclotron resonance of inversion electrons in magnetic fields applied parallel to the layers (Voigt-configuration) has been studied in Refs.8 and 9. In the experiments the light is incident perpendicular to the sample and is polarized perpendicular to the magnetic field. It is interesting to note that the resonances in parallel magnetic fields are sharper than in perpendicular fields (Faraday-configuration) since the electrons complete their Landau cycloides inside the semiconductor and do not feel interface or oxide charges as severe as in the Faraday-configuration.[18]

Apparent cyclotron masses measured at two electron densities $n_s$, i.e., electric field strengths E≠0 and in a n-type bulk InSb sample of low doping (n=6x10^{13}cm^{-3}, E=0) are depicted in Fig.8. The solid lines have been calculated for $0^+ \to 1^+$ transitions from Eqs.(14) and (21). The cyclotron masses in crossed fields approach the E=0 values at high magnetic fields but strongly differ at lower ones: Whereas the E=0 masses extrapolate to the band edge mass $m_0^*=0.014m_e$, the masses in crossed fields show a steep increase. Theoretically, they diverge as described by the relation $m^* \approx m_0^*(1-\delta^2)^{-1}$ which is obtained from Eq.(21) in the limit $\delta \to 1$. The divergence is indicated by the arrows for the two electric field strengths in Fig.8. Experimentally, the disappearance of the cyclotron maximum $0 \to 1$ corresponds to the condition $\omega\tau < 1$ since we have a finite electron relaxation time $\tau$ (see Ref.8).

The divergence of the cyclotron mass and the disappearance of cyclotron resonance are spectacular manifestations of the relativistic analogy that has been discussed in Ref.8. The factor $(1-\delta^2)^{-1}$ in the denominator of the apparent mass $(\delta \to 1)$ is a consequence of two relativistic effects: the magnetic field has to be Lorentz transformed to the system moving with the magnetic drift velocity E/B which gives one factor $(1-\delta^2)^{-1/2}$. Since the energy has to be transformed back to the laboratory system we have another factor $(1-\delta^2)^{-1/2}$ that corresponds to the relativistic transverse Doppler shift.

## 4. CONCLUSIONS

Electrons in space-charge layers allow to study the motion of semiconductor electrons in external electric and magnetic fields under equilibrium conditions. For this, the presence of an interface is essential. Otherwise hot electron effects, such as the emission of longitudinal optical phonons, would govern the behavior in strong electric fields.[19] Also, the presence of a barrier enables to define a density of states in the absence of scattering processes.

Without a magnetic field there is quantization into electric subbands. Nonparabolic effects that arise from the coupling of valence and conduction band and that cannot be accounted for in the EMA are clearly present: dispersion relations $\varepsilon_i(k_{\parallel})$ are different in different subbands and merge into each other at high wave vectors $k_{\parallel}$. Physically, this means that the electrons no longer feel their relationship to a particular subband at high energies ($\varepsilon \gg \varepsilon_g$). In the EMA the dispersion relations of all subbands are identical and they are always separated by a constant intersubband energy.

The electrons are not really confined to the interface as they have oscillatory holelike parts of their wave functions inside the semiconductor. Related to this tunneling between conduction and valence band is the fact that, strictly spoken, we do not have discrete subband quantization ($k_{\parallel}=0$) in narrow-gap semiconductors: In the presence of an electric field there are states at all energies, i.e., there is no energy gap in the system. However, on InSb the coupling of bands is not so strong and the resonant character of states is not so severe to prevent subband quantization. However, we like to speculate that this could in fact occur in strong electric fields on semi-conductors of very narrow gap energies. If an effective Compton wavelength $\lambda_c = \hbar/m_0^* u$ is defined, electric fields $E \gg \varepsilon_g/e\lambda_c$ are required. This estimate follows from Eq.(5) and the assumption that the parameter $a \ll 1$. The observation of subband spacings that strongly decrease with gap energy[5] is a first experimental step in this direction.

In small magnetic fields ($\hbar\omega_c \ll \varepsilon_F$) perpendicular to the surface subband masses can be measured by cyclotron resonance. These masses show features that are characteristic for nonparabolic subbands in triangular-shaped potentials. The masses increase with electron density $n_s$, i.e., with electric field E and with kinetic energy $\varepsilon_{\parallel}$ of the motion parallel to the interface. The masses are highest in the lowest subband. In the magnetic quantum limit the apparent mass increases with magnetic field similar to the increase in the bulk of the semiconductor. However, the mass is always higher due to the influence of the electric field.

In the crossed-field configuration that is established when the magnetic field is applied parallel to the inversion layers, cyclotron resonance only can be observed as long as the electric field is not too strong ($0 < \delta < 1$). Above the critical electric field $E \gg uB$ the Landau quantization itself is destroyed as has been predicted a long time ago.[11] This is one of the most spectacular manifestations of the coupling of bands in narrow-gap semiconductors and it could be described in close analogy to the behavior of relativistic electrons in free space. Above the critical electric field strength diamagnetically shifted subbands are observed.[20] In principle, such subbands are similar to purely electric ones with which we started our discussion.

## 5. ACKNOWLEDGMENTS

This review is based on work which I did together in course of time with M. Horst, S. Klahn, J. P. Kotthaus, S. Oelting and W. Zawadzki. I thank F. Koch, L. J. Sham, and U. Rössler for valuable discussions and the Deutsche For-schungsgemeinschaft for financial support.

# REFERENCES

1. For a comprehensive review see T. Ando, A. B. Fowler, and F. Stern, Rev. Mod. Phys. 54:437 (1982).

2. Y. Takada, K. Arai, N. Uchimura, and Y. Uemura, J. Phys. Soc. Japan 49:1851 (1980); Y. Takada, J. Phys. Soc. Japan 50:1998 (1981).

3. G. E. Marques and L. J. Sham, Surf. Sci 113:131 (1982); G. E. Marques, Dissertation, University of California (San Diego), 1982.

4. W. Brenig and H. Kasai, Z. Phys. B 54:191 (1984).

5. J. Scholz, F. Koch, J. Ziegler, and H. Maier, Surf. Sci. 142:447 (1984).

6. K. Wiesinger, H. Reisinger, and F. Koch, Surf. Sci. 113:102 (1982) and references therein.

7. A. Daerr, J. P. Kotthaus, and J. F. Koch, Solid State Commun. 17:455 (1975); U. Merkt, M. Horst, T. Evelbauer, and J. P. Kotthaus, Phys. Rev. B 34 (1986), in press.

8. W. Zawadzki, S. Klahn, and U. Merkt, Phys. Rev. Lett. 55:983 (1985); Phys. Rev. B 33:6916 (1986).

9. U. Merkt, Phys. Rev. B 32:6699 (1985).

10. W. Zawadzki, J. Phys. C 16:229 (1983); Surf. Sci. 37:218 (1973).

11. W. Zawadzki and B. Lax, Phys. Rev. Lett. 16:1001 (1966).

12. J. C. P. Miller, in: "Handbook of Mathematical Functions", M. Abramowitz and I. A. Stegun, ed., Dover, New York(1965), Chap. 19, pp. 685-720; F. W. J. Olver, J. Res. Nat. Bur. Stand. Sect. B, 63:131 (1959).

13. U. Merkt and S. Oelting, to be published. The series for the $i^{th}$ zero is $q_{0i}=2a^{1/2}(1-t_i a^{-2/3}/2 - t_i^2 a^{-4/3}/40 \pm \cdots )$ with zeros $t_i$ of the Airy function.

14. F. Koch, in: "Two-Dimensional Systems, Heterostructures, and Superlattices", G. Bauer, F. Kuchar, and H. Heinrich, ed., Springer, Berlin (1984) pp.20-31.

15. In the derivation of Eq.(12) we make use of the identities $\langle W|dW/dq\rangle=0$ and $\langle W|q|dW/dq\rangle=-1/2\langle W|W\rangle$ which are valid for any wave function.

16. U. Mackens and U. Merkt, Thin Solid Films 97:53 (1982).

17. M. Horst, U. Merkt, and K. G. Germanova, J. Phys. C 18:1025 (1985).

18. J. H. Crasemann, U. Merkt, and J. P. Kotthaus, Phys. Rev. B 28:2271 (1983).

19. A. A. Andronov, V. A. Kozlov, L. S. Mazov, and V. A. Valov, J. Phys. C 13:6287 (1980); S. Komiyama, Adv. Phys. 31:255 (1982).

20. S. Oelting, U. Merkt, and J. P. Kotthaus, Surf. Sci. 170:402 (1986).

# OPTICAL, MAGNETO-OPTICAL AND TRANSPORT INVESTIGATIONS

# OF THE NARROW-GAP SYSTEM $InAs_xSb_{1-x}$ *

F. Kuchar[a], Z. Wasilewski[b], R.A. Stradling[c], and R.J. Wagner[d]

[a] Inst.f.Festkörperphysik, University and L.Boltzmann-Institut
Vienna, Austria
[b] High Pressure Research Center, Warsaw, Poland
[c] Physics Department, Imperial College, London, England
[d] Naval Research Laboratory, Washington, D.C., USA

## 1. INTRODUCTION

Mixed crystals of III-V semiconducting compounds are of considerable interest as regards their fundamental properties as wells as applications in electronic and optoelectronic devices. One of the fundamental properties of a semiconductor - the minimum optical bandgap - is usually smaller in the mixed crystals than the concentration weighted average of the binary constituents ("bandgap bowing"). In the mixed crystal system $InAs_xSb_{1-x}$ with $0 < x < 0.7$ the bandgap exhibits values which are smaller than at x=0 being the smallest values appearing in III-V semiconductors.[1] At x=0.4 the energy gap is 0.1 eV at 300K, increasing to 0.15 eV at 0K. This property makes the mixed crystals with low x values extremely interesting as detectors for the 8-12μm spectral range (atmospheric window). The compound with x=0.91 ($\varepsilon_G$=0.33eV at T=77K) has potential applications for another atmospheric window between 3 and 5 μm and for fiber-optics communications at relatively long wavelengths. It can be grown lattice matched on GaSb. In such narrow-gap semiconductors, the conduction-electron parameter most directly related to the band gap is the effective mass which can be deduced from far-infrared magneto-optical spectra.

In this paper we present results of far-infrared optical and magneto-optical experiments as well as some transport experiments on thin films of mixed crystals with x=0.07 and 0.145. They concern cyclotron resonance, Reststrahl absorption, and transport effects like Shubnikov-de Haas effect and magnetic freeze-out. Also, the influence of hydrostatic pressure on the magneto-optical and transport properties is studied. From the cyclotron resonance data the first direct determination of effective mass values in $InAs_xSb_{1-x}$ in the composition range of low x values was possible.

Section 2 will deal with the preparation of $InAs_xSb_{1-x}$ mixed crystals and the experimental techniques used for the transport and far-infrared measurements. In Chapter 3 we will review some properties which are relevant for the present investigation. There, also a few general remarks on the bandgap bowing will be made. In Chapter 4 results of the transport, optical and

---

*The main part of the work on $InAs_{0.145}Sb_{0.855}$ was performed at the Physics Department of the University of St.Andrews, Scotland.

magneto-optical experiments will be presented, as well as the effect of hydrostatic pressure. Chapter 5 describes recent developements in the crystal growth and gives an outlook on possible two-dimensional electronic systems.

## 2. EXPERIMENTAL

The $InAs_xSb_{1-x}$ films used in this investigation were grown by A.R. Clawson[2,3] on glass substrates using a hot-wire zone-recrystallization method. The films consist of large crystallites about 20 μm in width and up to 1 cm in length. Some of their properties are superior to those of films or bulk crystals grown by other methods. Those included gradient-freeze,[4] horizontal Bridgman,[4] zone recrystallization[4] and annealing of quenched samples[5] for bulk crystals. Growth of thin films by liquid phase epitaxy (LPE),[6,7] by organometallic-chemical vapor deposition (OM-CVD),[8,9] and by molecular beam epitaxy (MBE)[10,11] has been reported more recently. The MBE work will be discussed in more detail in Chapter 7. OM-CVD and MBE yielded films with high crystalline quality; no electrical properties have been reported. For performing magneto-optical experiments pure samples with low free carrier concentration and high mobility at low temperature are neccessary. This was widely achieved by the hot-wire zone-recrystallization method. The glass substrates were lapped off down to a thickness of about 10 μm.

Magneto-optical transmission and photoconductivity experiments in Faraday geometry were performed using optically pumped far-infrared laser systems (wavelengths between 570 and 33.1 μm). For transmission experiments in the Reststrahl range a Fourier transform spectrometer (Beckman IR720) was used. High hydrostatic pressures were applied in an optical pressure cell (sapphire window, liquid pressure medium) capable of producing about 13.5 kbar at room temperature and 10.5 kbar at liquid helium temperature. Transport experiments were also performed in a cell without the window up to 17 kbar at 4.2K. Due to the brittleness of the samples, for transport and photoconductivity experiments just two In contacts were soldered to the sample (x=0.145).

Table I. Some properties of the $InAs_xSb_{1-x}$ films. Electrical data at T=77K.[12]

| Composition x | Film thickness (μm) | $N_D-N_A$ ($cm^{-3}$) | $\mu_H$ ($cm^2$/Vs) |
|---|---|---|---|
| 0.07 | 3.5 | $4.3 \times 10^{15}$ | $1.1 \times 10^5$ |
| 0.145 | 3.3 | $1.1 \times 10^{16}$ * | $3.5 \times 10^4$ |

* The Shubnikov-de Haas experiments at 4.2K yielded a value of $6.6 \times 10^{15}$.

## 3. PROPERTIES OF THE $InAs_xSb_{1-x}$ SYSTEM

In Table II some properties of the binary constituents of the mixed crystals are listed.

Most of the early work on optical and transport properties of the mixed crystals was done by Woolley and coworkers. The data concerning the minimum bandgap $E_g$ shown in Fig.1 were reported by Coderre and Woolley[1]. The values were deduced from optical absorption[15] as well as electrical experiments (Hall effect, conductivity, thermoelectric power)[4]. For optoelectronic applications it is also interesting to note that the intrinsic carrier concentration can be as high as $1 \times 10^{17} cm^{-3}$ at room temperature (x=0.4). The band-

Table II. Some properties of InSb and InAs. Phonon frequencies from reflectance measurements at 4.2K.[13] Other data from Ref.14.

| Material | Lattice constant (Å) | Melting point (°C) | Density (g/cm$^3$) | Phonon-frequencies (cm$^{-1}$) | |
|---|---|---|---|---|---|
| | | | | TO | LO |
| InAs | 6.036 | 943 | 5.68 | 219 | 243 |
| InSb | 6.478 | 530 | 5.78 | 185 | 197 |

| Material | Bandgap (eV) | | m*/m$_o$ (4.2K) electrons | Carrier mobilities at 300K (cm$^2$/Vs) | |
|---|---|---|---|---|---|
| | 300K | 0K | | electrons | holes |
| InAs | 0.32 | 0.4 | 0.024 | $3.0 \times 10^4$ | $4.0 \times 10^2$ |
| InSb | 0.18 | 0.235 | 0.0139 | $7.8 \times 10^4$ | $7.5 \times 10^2$ |

gap bowing is most pronounced of all III-V mixed crystal systems (Fig.1a)[1]. The minimum value of $E_g$ is reached at x=0.4 being close to 0.1eV at room temperature. For all compositions there is a temperature $T_C$ below the solidus temperature at which the minimum bandgap passes through zero (Fig.1b). This is interpreted as an interchange of the $\Gamma_6$ conduction band with the $\Gamma_8$ light-hole valence band, all three bands being degenerate at $T_C$. This variation of the conduction and valence band structure with temperature is similar to the dependence on composition x in $Hg_{1-x}Cd_xTe$. The main difference is that $InAs_xSb_{1-x}$ remains an open-gap semiconductor at all x values at technically interesting temperatures. Therefore, extremely small bandgaps of bulk crystals below 0.14eV (T=77K) as occur with $Hg_{1-x}Cd_xTe$ cannot be reached there.

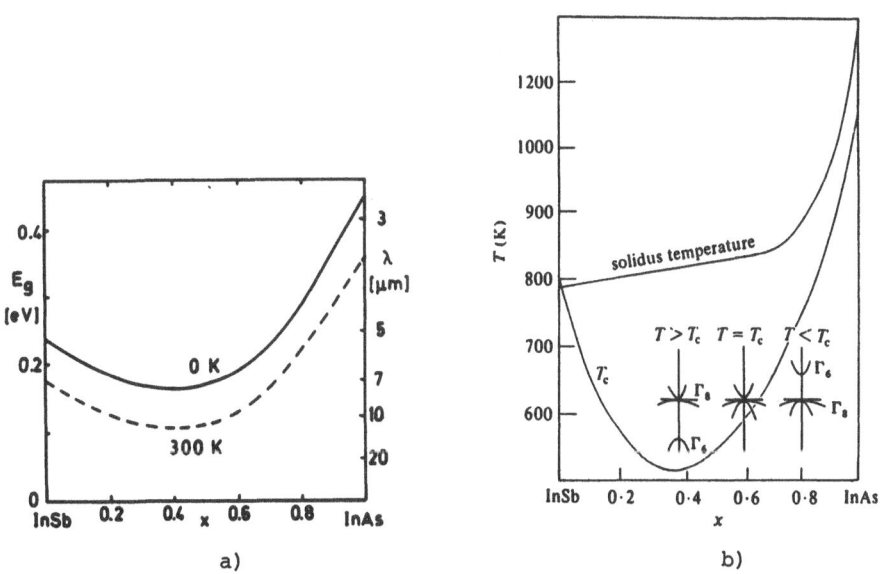

Fig.1 (a) Variation of the energy gap $E_g$ of $InAs_xSb_{1-x}$ at T=0K and 300K with composition x; (b) variation of the transition temperature $T_c$ and solidus temperature with x (after Ref.1).

Data of the spin-orbit splitting $\Delta$ can be found in Refs.16 and 17. A bowing as a function of composition was also observed with $\Delta$.

For comparison with the magneto-optical results of the present work it is interesting to list effective mass values obtained from previous, more indirect experiments. Those were plasma reflectance,[18] Faraday rotation,[19] and magneto-thermoelectic power.[20] The data are compiled in Ref.18 and show a minimum of $m^*/m_o \approx 0.01$ at $x \approx 0.4$, the composition of the minimum of $E_g$.

### 3.1  Some Remarks on the Bandgap Bowing in III-V Mixed Crystals

Various theoretical treatments of the problem assumed chemical[21,22] or positional disorder[23] as being responsible for the observed bowing of the gap $E_g$ as a function of composition x. Van Vechten et al.[21,22] attributed the difference between $E_g^{th}$ as calculated using the virtual crystal approximation (VCA) and $E_g^{exp}$ as observed experimentally to the effect of an aperiodic contribution to the crystal potential $(V^a)$ as a consequence of the chemical (atomic) disorder. Since the aperiodic term breaks the crystal symmetry, mixing of conduction and valence band states at the $\Gamma$ point occurs. It is suggested that the effective masses are changed to the same extent as the band mixing occurs. The main numerical uncertainty arises from a band width parameter A. This was found by comparison with experimental $E_g$ values to be 1eV for all alloy systems studied.

Positional disorder and the resulting strain was considered by Siggia.[23] As a consequence of the strain conduction and valence band states are mixed yielding a reduction of the interband momentum matrix element P. Effective masses were obtained by using the Kane formula with the modified $P^2$ and the experimentally determined $E_g$. Because of the uncertainty in the relevant deformation potential the results can only be crude extimates.

Van Vechten et al. as well as Siggia could produce good fits to room-temperature effective mass values in various mixed crystal systems including $InAs_xSb_{1-x}$.[22] Hermann and Weisbuch,[24] however, reported agreement with the predictions of the k.p. theory when including interaction with higher bands without considering disorder effects. For their procedure they used experimental values of effective masses as well as the g factors. This discrepancy with the disorder theories might be due to using low temperture $E_g$ values for fitting room temperature $m^*$ values. The three treatments are discussed by Nicholas et al.[25] including a determination of the deformation potential appearing in Siggia's work.

Recently, Zunger and Jaffe[26] pointed out that the assumption of a large disorder contribution neccessary to explain the bowing of the energy gap is not justified according to coherent-potential-approximation calculations by Chen and Sher.[27] Furthermore Zunger and Jaffe[25] suggested a new structure model based on EXAFS studies on GaAs-InAs by Mikkelsen and Boyce.[28] The EXAFS results demonstrated that although the lattice constant closely follows Vegard's rule the anion-cation bond lengths of the binary constituents do not average to a single bond length but remain close to their respective values. This bond alteration corresponds to a structural distortion to a local chalcopyrite coordination around the common ion of the two binary constituents. This finding excludes the applicability of the structural model underlying the virtual crystal approximation used by van Vechten et al.[21,22] Zunger and Jaffe,[25] therefore propose (i) that the structurally induced contribution to the bowing of $E_g$ is controlled by bond alteration (ii) that the disorder contribution results from compositional disorder around the common ion and that there exists a distribution of the displacement of the mixed ions around the ideal zinc blende site.

# 4. EXPERIMENTAL RESULTS AND DISCUSSION

## 4.1 Reststrahl Range

Transmission spectra measured on the $InAs_xSb_{1-x}$ samples with x=0.07 and 0.145 at T=4.2K in the range 100-300cm$^{-1}$ are shown in Fig.2a. Indicated are the transverse (TO) and longitudinal optical (LO) phonon frequencies of InSb and InAs (Table I). A comparison with reflection data of the binary constituents[13] shows that the TO frequencies are on the low-frequency side of the transmission minima and the LO frequencies are similarly placed with respect to the maxima. The phonon frequencies are plotted in Fig. 2b as a function of composition x. Although the experimental uncertainty particularly for the lowest TO phonons is at least $\pm 2$ cm$^{-1}$ two important features are observed: (i) InSb-type as well as InAs-type phonons exist in $InAs_xSb_{1-x}$. (ii) A linear variation of the phonon frequencies with x can be fitted. The frequency of the InAs-type phonons increases, whereas that for the InSb-type phonons decreases with x. A qualitatively similar behavior was found in $InAs_{1-x}P_x$ by Nicholas et al.[29] from magnetophonon experiments.

## 4.2 Shubnikov-deHaas Effect and FIR Photoconductivity

Up to three Shubnikov-deHaas (SdH) oscillations could be observed in the magnetoresistance $\rho_B$ as well as in the far-infrared photoconductivity (P.C.) signal (optical SdH) of the sample x=0.145. In $\rho_B$ the oscillations are very weak, in the photoconductivity a peak at about 1.1T dominates which corresponds to the Fermi energy coinciding with the maximum of the density of states of lowest Landau level. Spin splitting is not resolved. The lower edges of the Landau levels are certainly not sharp at the doping level of our sample.

The positions of the maxima of the $\rho_B$ and P.C. oscillations are plotted versus the reciprocal magnetic field in Fig.4. The field position of the prominent peak at about 1.1T shows only very little scatter for the measuring wavelengths of 53.5, 96.5, 118.8, 163, and 570 $\mu$m. From the periodicity $\Delta(B^{-1})=0.9T^{-1}$ of the oscillations on the $B^{-1}$ scale an electron concentration of $6.6 \times 10^{15}$cm$^{-3}$ was calculated. This analysis and the agreement with the $\rho_B$ oscillation proves that the oscillations of the P.C. signal are an optical SdH effect.

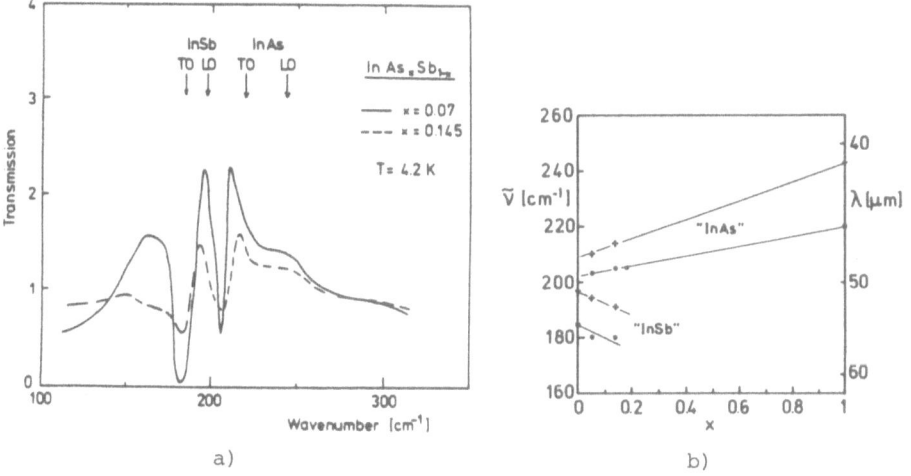

a)                                          b)

Fig.2 (a) Transmission spectra of $InAs_xSb_{1-x}$ in the Reststrahl region.
(b) Phonon frequencies (+...LO, •...TO) obtained from (a). The straight lines connect the data with the corresponding frequencies of the binary constituents.

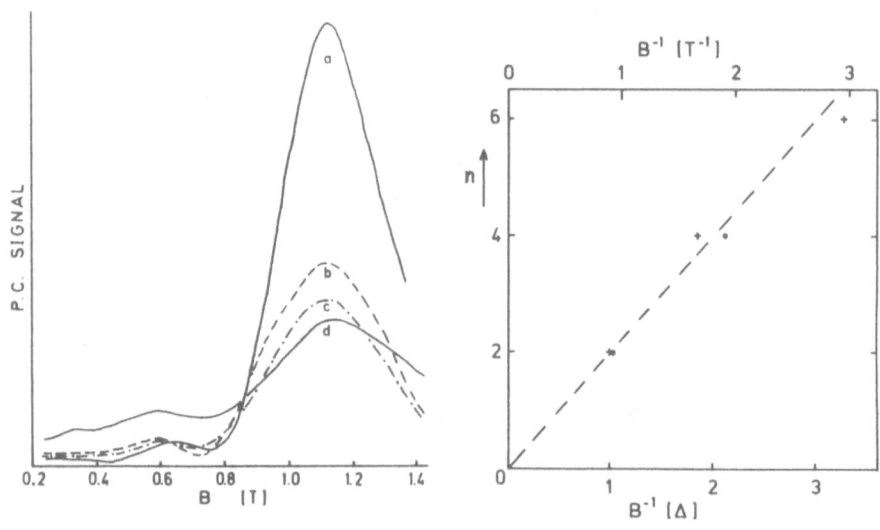

Fig.3

Fig.4

Fig.3   Photoconductivity signal as a function of magnetic induction for
different laser wavelengths: (a) 118.8 μm, (b) 96.5 μm, (c) 163 μm,
and (d) 570 μm. T=4.2K. The lowest-field peak (≈0.33T) is observed
at the longest wavelength only.

Fig.4   Integer numbers versus inverse field positions of the SdH maxima of
the magnetoresistance (dots) and of the photoconductivity signal
(crosses). Δ is the period obtained from the i=2 peak of the SdH
oscillations. The cross at n=6 represents the 570 μm results. The
crosses at n=2 and n=4 are average values of  measurements with
λ=70.51, 96.5, 118.8, 163, 570 μm. The scatter of the data is
±0.02Δ(n=2), ± 0.045Δ(n=4).

## 4.3  Magneto-Optical Experiments

The magneto-optical experiments concern the spectral region of the cyclotron resonance. Since the photoconductivity spectra are dominated by the
optical SdH effect, cyclotron resonance has to be measured in transmission.
In Fig.5 spectra are shown obtained for $InAs_{0,145}Sb_{0,855}$ with various laser
wavelengths. Fig.6 shows a plot of the corresponding phonon energies versus
the magnetic field of the transmission minima. Indicated in this figure are
the LO phonon frequencies for the sample with x=0.07 as obtained from the
Fourier transform spectra (Ch.4.1). There, coupled cyclotron-phonon modes
can occur[30] leading to a deviation from the curve extrapolated from the
lower-frequency data. This seems to happen in the case of the 46.8 μm data,
possibly also in the 36.6 μm data which is in the region of multiphonon
absorption.

The discussion of these data regarding a determination of effective mass
values follows in the next chapter together with the results obtained under
hydrostatic pressure.

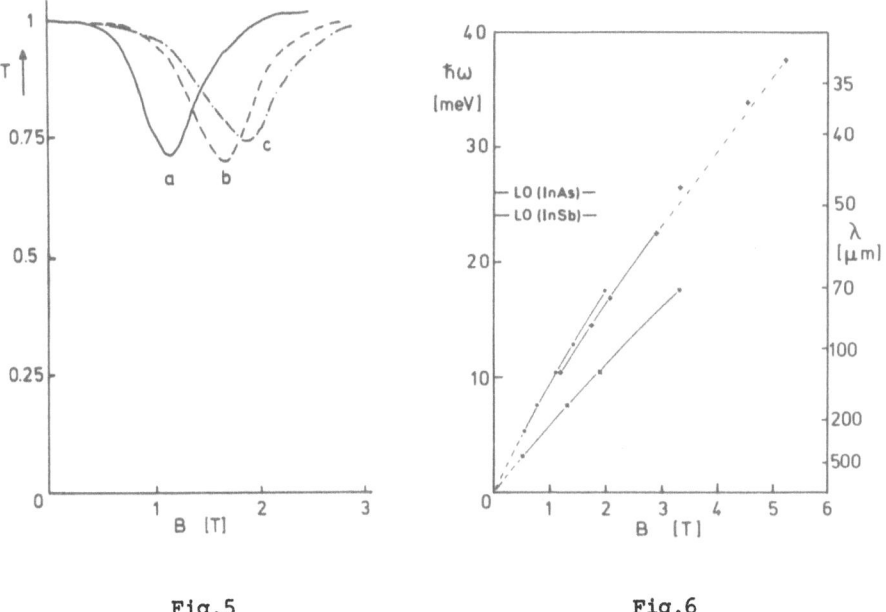

Fig.5                              Fig.6

Fig.5 Magneto-optical spectra (unpolarized radiation) of InAs$_{0.145}$Sb$_{0.855}$
λ=118.8 μm, T=4.2K. (a) 0 kbar  (b)  6.7 kbar, and (c) 10.3 kbar
hydrostatic pressure. The pressure data are discussed in Chapter 4.4.

Fig.6 Laser frequency versus magnetic field position of the transmission
minima of the magneto-optical spectra. +...x=0.07; ● and x...x=0.145,
0 and 10.3 kbar, resp. The LO phonon frequencies for x=0.07 are
indicated.


## 4.4   The Effect of Hydrostatic Pressure

It is well known from experiments on InSb that the application of hydrostatic pressure can have a tremendous effect on the electrical[31] and magneto-
optical properties.[32] This is caused by the relative lowering of the energy
of a deep impurity level associated with a satellite conduction-band valley.
The consequences are a reduction of the free carrier concentration (metal-
insulator transition) and a drastic sharpening of line-widths in magneto-
optical spectra (cyclotron resonance, shallow donor transitions). Thus, it
is interesting to investigate whether a similar behaviour can be also
induced in InAs$_x$Sb$_{1-x}$ by hydrostatic pressure.


4.4.1 Magnetoresistance and FIR Photoconductivity. Fig.7 shows the
effect of hydrostatic pressure on the zero-field resistance $R_o$ = R(B=0) and
on the magnetoresistance $\Delta\rho/\rho_o$=[R(B)-R(B=0)]/R(B=0). R(B=0) increases by
more than 3 orders of magnitude at pressures up to 14.5 kbar and at 17 kbar
becomes immeasurably high. Despite this strong increase there is no change
observed in the positions of the oscillation maxima of the photoconductivity
signal  up to the pressure of  10.3 kbar achievable in the optical cell. This
means, that up to this pressure there is no reduction of the carrier concentration at least at magnetic fields of about 1.5T. The change of the effective mass m* with pressure (see next section) does not change the period of
the SdH oscillations. The strong increase of the resistance can only be to
a small measure due to the change of m*. In the present stage of the experimental information we can just argue that the interaction with the deep level

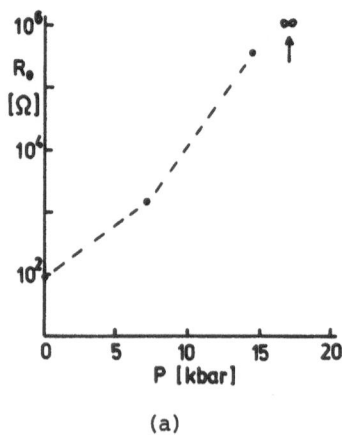 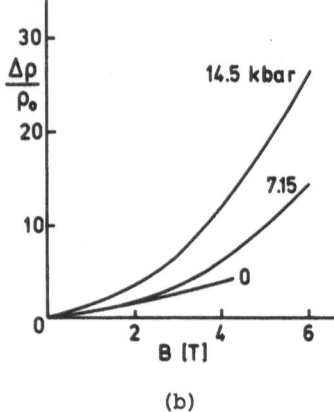

(a)                          (b)

Fig.7 (a) Variation of the zero-field resistance R(B=0) of the x=0.145
sample with hydrostatic pressure P. (b) Magnetoresistance $\Delta\rho/\rho_0$ for
different pressure values. T=4.2K.

mentioned above leads to a strong increase of the scattering of the electrons,
as long as it is energetically degenerate with the $\Gamma$ valley. Another possibility is an effect of the polycrystalline nature of the film (Chapter 2).

The change to an insulating state between 14.5 and 17 kbar probably
corresponds to the metal-insulator transition observed in InSb. There, the
electrons from the central valley are transferred to the deep level which
lies within the minimum energy gap at these high pressures.

The increase of $\Delta\rho/\rho_0$ with magnetic field (Fig.7b) can be due to a
magnetic freeze-out in the upper field range only (the Fermi energy equals
the energy of the lowest Landau level at about 1.1T). In this field range
also the photoconductivity singal increases monotonically, reflecting the
increase of $\Delta\rho/\rho_0$.

4.4.2 Magneto-Optical Spectra. Magneto-optical experiments under hydrostatic pressure were performed for the $InAs_{0.145}Sb_{0.855}$ sample only. In order
to work out the variation with pressure in a clear way, "effective mass values" are plotted in Fig.8. For zero-pressure also the x=0.07 data are included. The m* values were obtained from the magnetic field positions $(B_c)$ of
the transmission minima by using $\omega=(e/m^*)B_c$. It is important to notice that
they not necessarily have the meaning of an effective mass as will be shown
below.

At low magnetic fields the Fermi level is well within the conduction
band. An extrapolation (dotted lines) to zero field yields the effective
mass values at the Fermi level. The Fermi energies are obtained from the
field positions of the n=2 peak of the SdH oscillations and the corresponding cyclotron energies $(E_F= \hbar\omega_c/2)$. The use of the two-band equation

$$m^*(E) = m_0^* \ (1+ \frac{2E}{E_g})$$

(1)

yields approximate values of the band-edge effective mass $m_0^*$ as shown in
Table III. The pressure dependence of $E_g$ was estimated from room-temperature resistance measurements to be 6% $(kbar)^{-1}$. $E_g=0.19eV$ at zero-pressure
was taken from Ref.1. For x=0.07 we obtain $m_0^*=1.13 \times 10^{-2}$ at zero pressure
by using the carrier concentration given in Table I and $E_g=0.22eV$.

226

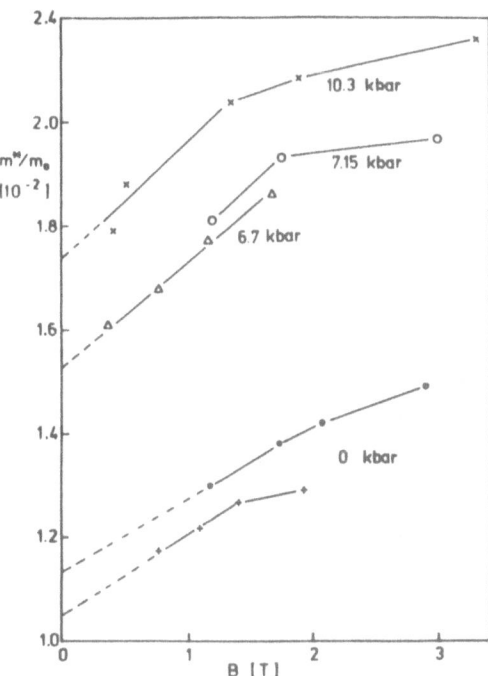

Fig.8 "Effective mass" values as a function of magnetic induction. The curves are drawn through the data points. Dots: x=0.07, other symbols: x=0.145.

A better estimate of the band edge masses is possible by extrapolating the low-field slopes of the $m*/m_0$ curves of Fig.8 to zero-energy. This is possible since both cyclotron resonance and SdH data are available; as mentioned above the Fermi energy $E_F$ equals $\hbar\omega_c/2$ at B=1.1T. The average energy where the cyclotron resonance transition between the Landau levels N=0 and N=1 occurs can be set equal to $\hbar\omega_c$ ($=E_F+\hbar\omega_c/2$ at 1.1T). So, an "extrapolated" band-edge mass $m*_{00}$ is obtained (Table III, last column). The P=0 value is significantly smaller than that of Ref.18. The pressure dependence of $m*_{00}$ is 7% (kbar)$^{-1}$ which compares quite well with that one of $E_g$ at 300K given above.

Table III. Fermi energies and effective mass values for x=o.145. Details see text.

| P (kbar) | $E_F$ (meV) | $10^2 m*/m_0$ (B→0) | $10^2 m*_0/m_0$ | $10^2 m*_{00}/m_0$ |
|---|---|---|---|---|
| 0 | 5.3 | 1.05 | 0.99 | 0.88 |
| 6.7 | 3.7 | 1.53 | 1.49 | 1.31 |
| 10.3 | 3.2 | 1.73 | 1.69 | 1.49 |

At magnetic fields above those corresponding to the last peak of the SdH oscillation ($\approx$ 1.1T), a distinct change of the slope of the magnetic field dependence of m* is observed. The slope at low field can be attributed to the non-parabolicity of the conduction band. We assume that magnetic freeze-out occurs at high fields (compare section 4.4.1). There, also shallow-donor transitions which are not resolved in the spectra can contribute. In this particular case, it is the so-called impurity shifted cyclotron resonance (the $1s \to 2p_{+1}$) transition in low-field notation). Since the energy difference between the $2p_{+1}$ level and the N=1 Landau level is smaller than between the 1s and the N=0 Landau level, the values of the parameter "m*" at high fields are below those extrapolated from the low-field region.

The zero-pressure effective-mass values for $InAs_{0.145}Sb_{0.855}$ at low fields were compared with a calculation based on an eight-band model[33]. In this calculations experimental values for $E_g$[1] and the spin-orbit splitting[17] were used. The value of the momentum matrix element was linearly interpolated between those of the binary constituents. For the interaction with higher conduction bands - treated in perturbation theory - the parameters of InSb were used. The calculations gave $m^*(B \to 0)$ values by about 20% higher than experimentally observed, the variation with field was weaker than in the experiment. This shows that the compositional disorder in the mixed crystal (see Chapter 3.1) causes an additional mixing of states of different bands. It leads to a lower effective mass value at zero-field and a stronger non-parabolicity than obtained from the InSb-type eight-band calculation. This is opposite to the behaviour in $InAs_{1-x}P_x$[25].

## 5.   OUTLOOK FOR 2D STRUCTURES

As mentioned in Chapter 2, recently some success was made in growing epitaxial layers of $InAs_xSb_{1-x}$ with high crystalline quality. This was achieved by OM-CVD[8,9] and MBE[10,11]. Particularly, the MBE results are most promising for the growth of herostructures and superlattices. As regards metal-insulator-semiconductor structures we refer to the article by Dr. Wieder in this volume.

Here, we would like to mention two types of layer structures as candidates for 2D structures with $InAs_xSb_{1-x}$ in the range of low x values.

(i) CdTe is closely lattice matched to InSb ($\Delta a/a \approx 0.05\%$ at 25°C). Theoretical studies[34,35] predict conduction and valence band offsets of 0.3eV and 0.9eV, respectively, at T=77K. This is based on electron affinity values of 4.28eV for CdTe[36] and 4.57eV for InSb.[37] Since the use of the electron affinity for exactly calculating band offsets has been criticized recently[38,39] the values cited above should be considered preliminary . MBE growth of either CdTe or InSb has been reported.[40] A problem of the growth of CdTe/InSb heterostructures and superlattices is the preferential Cd loss or CdTe/InSb interdiffusion. This could be avoided using a two-step growth technique for CdTe on (100) InSb at growth temperature up to 310°C.[41] The results from SIMS investigations of these layers do not have, however, the high resolution neccessary to exclude the existence of interface dipoles. Similar problems as with CdTe/InSb are to be expected when suggesting CdTe for heterostructures and superlattices with $InAs_xSb_{1-x}$. Additionally, the excellent lattice matching is lost and only strained layer superlattices (SLS) could be grown. The great advantage of having a window like layer ($E_g$=1.44eV) on top of the narrow-gap semiconductor remains.

(ii) Recently $InAs_xSb_{1-x}$/InSb superlattices with good crystalline perfection could be grown.[11] This shows the route to $InAs_{0.39}Sb_{0.61}/InAs_xSb_{1-x}$ strained layer superlattices (SLS) which could extend the spectral response of the bulk material (x=0.39) from 9 µm to about 12 µm at 77K. Superlattices of this type were treated theoretically by Osbourn.[42] The essential points of his work are the following. The material with the smallest gap, x=0.39, is sandwiched between larger-gap alloys with x < 0.39. The slight lattice mis-

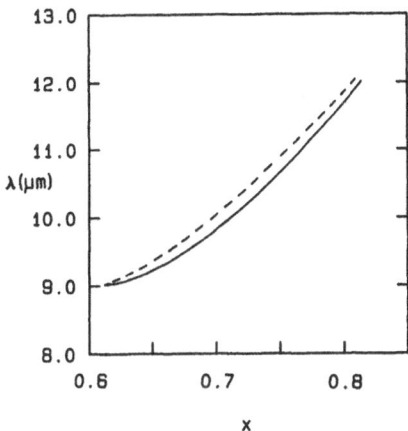

Fig.9 Calculated wavelength values at 77K associated with conduction band to light hole band transition in $InAs_{0.39}Sb_{0.61}$/ $InAs_xSb_{1-x}$ SLS's with 500 Å layers which are equally thick as a function of x. The two sets of results correspond to different band offset: $\Delta E_c$=0.04 (0.39-x) (solid line), $\Delta E_c$=0.21 (0.39-x) (dashed line).

match is accomodated by strains in the x=0.39 layer if it is thin enough. The resulting SLS structure contains expansive hydrostatic and compressive (100) uniaxial strain components in the (100) oriented x=0.39 layers. The net effect of the conduction and valence band shifts (essentially due to the hydrostatic component) is a reduction of the minimum band gap. Osbourn's results are shown in Fig.9 for two values of the conduction band offsets $\Delta E_c = E_c(InAs_xSb_{1-x})-E_c(InAs_{0.39}Sb_{0.61})$. For a thickness of the x=0.39 layer of 500Å the quantum size corrections to the values of Fig.9 are very small. Because of the shallow electron and light hole quantum wells ($\leq$ 40meV) expected in this SLS's, the discrete ground levels in the wells have energies only few meV above the band edges. Advantages of this mixed crystal SLS's over bulk $Hg_{0.82}Cd_{0.2}Te$ for the same wavelength range are the metallurgical properties, the generally easier processing of III-V semiconductors, weaker dependence of the band gap on composition (inhomogeneities), and reduced band-to-band tunnelling.

Acknowledgement   The stay of one of the authors (F,K,) at the University of St. Andrews was supported by the SERC,

References

1. W.M.Coderre and J.C.Woolley, Can.J.Phys.46:1207 (1968); J.Phys.Chem. Solids 32 (supplement 1): 535 (1971).
2. A.R.Clawson, Thin Solid Films 12:291 (1972).
3. H.H.Wieder and A.R.Clawson, Thin Solid Films 15:217 (1973).
4. W.M.Coderre and J.C.Woolley, Can.J.Phys. 46:1207 (1968).
5. N.N.Sirota and E.I.Bolvanovich, Doklady Akd.Nauk B.SS.R. 11:593 (1967).
6. G.B.Stringfellow and P.E.Greene, J.Electrochem.Soc. 118:805 (1971).
7. J.R.Skelton and J.R.Knight, Solid Sate Electr. 28:1166 (1985).
8. P.K.Chiang and S.M.Bedair, J.Electrochem.Soc. 131:2422 ((1984).
9. T.Fukui and Y.Horikoshi, Jpn.J.Appl.Phys. 19:L53 (1980).
10. W.T.Tsang, T.H.Chiu, D.W.Kisker, and J.A.Ditzenberger, Appl.Phys.Letters 46:283 (1985).
11. G.S.Lee, Y.Lo, Y.F.Lin, S.M.Bedair, and W.D.Laidig, Appl.Phys.Letters 47:1219 (1985).

12. A.R.Clawson (private communication).
13. M.Hass and B.W.Henvis, J.Phys.Chem.Solids 23:1099 (1962).
14. D.R.Lovett, "Semimetals and Narrow-BAndgap Semiconductors", Pion Ltd., London (1977).
15. J.C.Woolley and J.Warner, CanJ.Phys.42:1879 (1964).
16. S.S.Vishnubhatla, B.Eyglunent, and J.C.Woolley, Can.J.Phys.47:1661 (1969).
17. O.Berolo and J.C.Woolley, Proc.11$^{th}$ Int.Conf.Phys.Semicond. Warsaw (Polish Scientific Publishers, 1972), p.1420.
18. M.B.Thomas and J.C. Woolley, Can.J.Phys.49:2052 (1971).
19. E.H.van Tongerloo and J.C.Woolley, Can.J.Phys. 46:1199 (1968).
20. M.J.Aubin and J.C.Woolley, Can.J.Phys. 46:1191 (1968).
21. J.A.Van Vechten and T.K.Bergstresser, Phys.Rev.B:3351 (1970).
22. O.Berolo, J.C. Woolley, and J.A.Van Vechten, Phys.Rev.B8:3794 (1973).
23. E.D.Siggia, Phys.Rev.B10:5147 (1974).
24. C.Hermann and C.Weisbuch, Phys.Rev.B15:816, 823 (1977).
25. R.J.Nicholas, R.A.Stradling, and J.C.Ramage, J.Phys.C12:1641 (1979).
26. A.Zunger and J.E.Jaffe, Phys.Rev.Letters 51:662 (1983).
27. A.B.Chen and A.Sher, Phys.Rev.Letters 40:900 (1978) and Phys.Rev.B23: 5360 (1981).
28. J.C.Mikkelsen and J.B.Boyce, Phys.Rev.Letters 49:1412 (1983).
29. R.J.Nicholas, R.A.Stradling, J.C.Portal, and S.Askenazy, J.Phys.C12: 1653 (1979).
30. See the review by B.D.McCombe and R.J.Wagner, Adv.in Electronics and Electron Physics 37:1, 38:1 (1975).
31. S.Porowski, L.Konczewicz, A.Raymond, R.L.Aulombard, J.L.Robert, and M.Baj, Springer Lecture Notes in Physics 177:357 (1983).
32. Z.Wasilewski, A.M.Davidson, R.A.Stradling, and S.Porowski, Ref.31, p.233.
33. M.Kriechbaum, Physica 117B&118B:444 (1983) and private communication.
34. B.Rabin, C.Scharager, M.Hage-Ali, O.Siffert, F.V.Wald, and R.O.Bell, Phys.Stat.Sol.(a)62:237 (1980).
35. R.G.van Welzenis and B.K.Ridley , Solid State Electron.27:113 (1984).
36. T.Swank, Phys.Rev. 153:844 (1967).
37. S.Haneman, J.Phys.Chem.Solids 11:205 (1959).
38. W.A.Harrison, in Springer Series in Solid State Sciences (ed.G.Bauer, F.Kuchar, and H.Heinrich) 67:62 (1986).
39. H.Heinrich and J.M.Langer, Ref.38, p.83.
40. See papers cited in Ref.41.
41. G.M.Williams, C.R.Whitehouse, N.G.Chew, G.W.Blackmore, and A.G.Cullis, J.Vac.Sci.Technol. B3:704 (1985).
42. G.C.Osbourn, J.Vac.Sci.Technol. B2:176 (1984).

# NARROW BANDGAP SEMICONDUCTOR DEVICES

H. H. Wieder

Electrical Engineering and Computer Sciences Department
University of California, San Diego
La Jolla, California  92093

## ABSTRACT

A review of past and current research on electronic devices based on the modulation of the surface potential of depletion, accumulation or inversion layers of bulk or thin film elemental or compound semiconductors whose fundamental bandgaps, $E_g < 1$ eV, reveals that the characteristics of the semiconductor-gate insulator interfaces determine, to a large extent, their charge carrier transport properties; in the case of metal-insulator-semiconductor (MIS) structures the energy levels, density and capture cross-sections of interface states as well as the type, density and spatial distribution of traps within the insulator affect their DC drain current stability, their transconductance and their gain-bandwidth products. At this time, semiconductor-quasi-insulator-semiconductor heterostructures have superior properties compared to MIS structures and most of the MIS-related problems are also absent in modulation-doped two-dimensional electron gas heterojunction structures.

## INTRODUCTION

A field-effect transistor (FET) is a three-terminal device which depends on the electrostatic modulation of the current, $I_{DS}$, which flows between its source and drain electrodes. Control of the source-drain conductance is implemented by the gate voltage, $V_g$, applied to a control gate situated above and in between the source and drain electrodes of the FET shown, in Fig. 1. Such transistors are usually made by means of photolithographic, etching and liftoff procedures. They employ semiconducting layers deposited or grown on insulating or semi-insulating (SI) substrates by vacuum deposition, chemical vapor phase deposition (CVD), organometallic vapor phase epitaxy (OMVPE), molecular beam epitaxy (MBE) or by direct ion implantation into available SI substrates. The source and drain contacts might be alloyed, diffused or ion implanted ohmic contacts or junctions. The low surface barrier height, $\psi_B$, of narrow bandgap semiconductors prevents the use of metal Schottky barrier gate electrodes. A great deal of effort has been expended in attempting to circumvent this problem by the use of p-n junction or heterojunction gates or by the use of dielectrically insulated gate structures. Figure

2 shows typical low frequency characteristics of FET and represents the dependence of $I_{DS}$ on the applied source-drain voltage, $V_{DS}$, with $V_g$ as a fixed independent parameter. It shows that $I_{DS}$ is essentially linear in $V_{DS}$ for low values of $V_{DS}$ and reaches a saturated value, $I_{DSS}$, in large $V_{DS}$. An FET which has a quiescent $I_{DS}(V_g = 0) \sim 0$ is an enhancement mode transistor in contrast with a normally conducting FET which can be

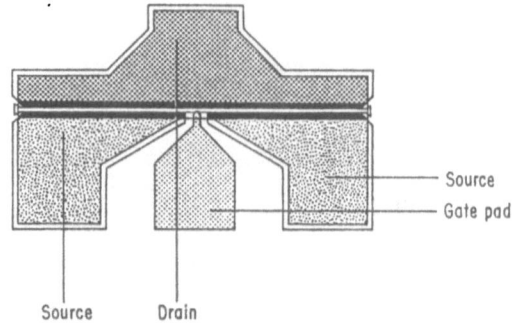

Figure 1.  Configuration of a split source FET.

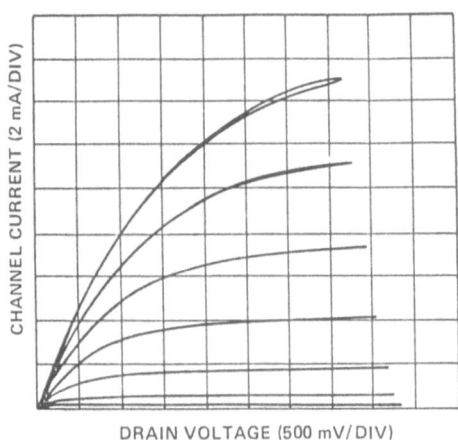

Figure 2.  Typical low frequency FET characteristics, horizontal axis is applied source-drain voltage, vertical axis is drain current with gate voltage as a parameter.

modulated in depletion. From data such as that shown in Fig. 2 the transconductance, $g_m = (\partial I_{DS}/\partial V_g)$ can be obtained from both the saturated and linear $I_{DS}(V_{DS})$ regimes. A figure-of-merit of such transistors is their current gain cutoff frequency, $f_T$, where the output current is equal to the input current, and is also defined as the gain-bandwidth product,

$$f_T = g_m \cdot (2\pi C_{gs})^{-1} \tag{1}$$

where $C_{gs}$ is the gate to source capacitance. It is related to the transit time, $\tau_t$, by $f_T = (2\pi\tau_t)^{-1}$. The electron transit time is nearly independent of $I_{DS}$ or $V_g$ over a substantial portion of the range in

232

which $g_m(I_{DS})$ and $C_{gs}(I_{DS})$ have essentially the same slopes. An additional figure-of-merit is the frequency at which the transistor output power is equal to the input power; this is the maximum frequency of oscillation, where $g_i$ and $g_D$ are, respectively, the input and output

$$f_{max} = (f_T/2) \cdot (g_i/g_D)^{\frac{1}{2}} \tag{2}$$

conductance of the FET. In order to reduce $\tau_t$ the gate length, $l_g$, of the FET, its low electric field mobility, $\mu_0$ and the intervalley gap, $\Delta\Gamma L$, between the conduction band minimum and the next higher conduction band valley are primary considerations. For a large $f_T$, $\mu_0$ and $\Delta\Gamma L$ should be as large as possible and $l_g$ as small as possible. Two-dimensional computer simulations of conventional FET reveal that $g_m$ increases only slowly with decreasing $l_g$ while $g_D = (\partial I_{DS}/\partial V_{DS})$ also increases. The limit for the useful reduction of $l_g$ is when it is approximately equal to the channel thickness, d. To keep $(l_g/d) > 1.5$, a value chosen as the lower limit, the channel thickness must be reduced as well as the gate length. However, to keep $I_{DS}$ within reasonable bounds the electron density must also be increased. Such an increase, produced by increasing the donor density implies an increase in impurity scattering with a corresponding reduction in $\mu_0$ as well as the possibility of interband tunneling or barrier breakdown. Additional constraints are imposed on the source-drain channel length. To first order, the portion of the channel not covered by the gate represents a series resistance, $R_s$, which reduces $g_m$ so that

$$g_m = g_m(1 + g_m R_s)^{-1} \tag{3}$$

The source and drain resistances also depend on their contact and spreading resistances; elaborate metallurgical methods and self-aligned techniques are used to minimize these and to make the fraction of the channel not covered by the gate negligible. A less useful but often quoted FET figure-of-merit is the field-effect mobility, $\mu_{fe}$, derived by fitting $I_{DS}(V_g)$ vs $V_{DS}$ low frequency measurements to the gradual channel approximation model.

## Thin Film Transistors

Among the earliest applications considered for InSb thin films was their use in thin film transistors (TFT). Such a device consists of an InSb polycrystalline film vacuum-deposited on a glass substrate and shaped in the form of a narrow channel between ohmic source and drain electrodes. The channel conductivity is modulated by a potential applied to a metal gate which is insulated from the channel by an intermediate dielectric layer. Frantz[1] made such a TFT using flash-evaporated InSb with an electron density, $n = 3.7 \times 10^{17}/cm^3$ and mobility $\mu = 560 cm^2/V\text{-}s$. He obtained conductivity modulation at room temperature in both depletion and enhancement but did not obtain saturation of the drain current nor did he obtain channel pinchoff. Subsequently Luo and Epstein[2] used a similar procedure to construct coplanar TFT with InSb layers 0.03 to 0.05 μm in thickness, electron density $n = 5 \times 10^{17} cm^3$ and mobility $\mu = 250 cm^2/V\text{-}s$. Gate insulators were 0.04 μm thick vacuum-deposited $SiO_x$ layers. Vacuum-deposited In or Sb was used for source and drain electrodes and Al as the gate electrode. At room temperature the $I_{DS}$ vs $V_{DS}$ characteristics of these TFT were essentially the same as those obtained by Frantz; at 77°K they found the expected saturation of $I_{DS}$ and channel pinchoff as a function of $V_g$. For $l_g = 25$ μm they obtained a maximum $g_m = 6$ mS.

Lile and Anderson[3] have investigated the properties of structurally inverted InSb TFT. The surface of an aluminum gate vacuum-deposited on its glass substrate was anodized to a thickness of ~ 0.015 μm thus providing the gate insulating layer upon which 0.2 to 0.3 μm thick InSb was vacuum-deposited through an aperture mask. They demonstrated that source and drain series resistance reduce $g_m$ and, although they obtained a well defined $I_{DSS}$, their $g_m < 1$ mS was attributed primarily to the low $\mu_0 \sim 300$ $cm^2$/V-s. Van Calster[4] investigated the properties of dual gate InSb TFT with 0.15 μm thick $SiO_x$ gate insulators and tried various thermal annealing procedures to improve the mobility of vacuum-deposited InSb layers. The latter is a function of thickness and is, typically, only $10^3$ $cm^2$/V-s for a thickness, d = 0.1 μm, decreasing sharply with d. Due to the strong degeneracy of the electron gas he observed transistor action down to liquid helium temperatures.

In contrast with the InSb TFT are the results obtained by Brody and Kunig[5] on InAs TFT. By controlling the As/In vapor flux ratio they deposited, in vacuum, InAs films on glass and sapphire substrates with Hall mobilities of $3 \times 10^3$ $cm^2$/V-s for d < 0.1 μm and $8 \times 10^3$ $cm^2$/V-s for d > 0.3 μm while the electron densities were in the range between $10^{17}$ and $2 \times 10^{18}$ $cm^3$. A coplanar TFT with a channel length of 100 μm and width of 1400 μm employing a 0.1 μm InAs layer and a 0.15 μm thick $SiO_x$ gate insulator was found to have a $g_m = 10$ mS and a $f_T = 8$ MHz. In view of the degeneracy of the electron distribution such a TFT is essentially temperature independent. However, they found the DC characteristics to be unstable and attributed this to charge redistribution in the insulator.

Vacuum-deposited PbS layers have been used for TFT.[3,6] Evaluation of their properties is complicated by inter- and intragrain variations in stoichiometry and by the presence of oxides of both Pb and Te. Near intrinsic conductivity was required in order to observe conductivity modulation thus introducing a strong temperature dependence of the TFT characteristics. Effective electron mobilities were found to be low; of the order of 300 $cm^2$/V-s and $g_m < 1$ mS.

## Metal-Insulator-Semiconductor Field-Effect Transistors

Metal-insulator-semiconductor field-effect transistors (MISFET) represent, in most respects, a more advanced stage of device evolution than TFT. MISFET employ single crystal semiconductors usually for inversion mode transistors. Their parameters are, therefore, not dependent on the size and distribution of grains or the electrical properties of intergrain barriers of polycrystalline layers nor are they subject to charge carrier scattering at the channel-substrate hetero-interface as are TFTs because the substrate is isolated from the inversion layer by a depletion region. The source and drain contacts have a conductivity opposite to that of the substrate; they are, therefore, isolated from each other unless the gate voltage exceeds a threshold, $V_{th}$, sufficient to establish a conducting inversion layer channel between them. Figure 3 illustrates, schematically, the structure of such a device.

InSb MISFET

In order to investigate the oscillatory magnetoconductivity and negative photo-conductivity of quantized electrons in the surface inversion layer of InSb Katayama et al[7,8] made n-channel MISFETs of $p = 10^{14}/cm^3$ single crystal InSb. Gate insulating layers were made by the deposition of $SiO_2$ using chemical vapor phase disproportionation of $(C_2H_2O)_4Si$. Vacuum-deposited Al was used as the gate; source and drain electrodes were made by Rh plating and In-Sn alloy was used to attach leads to them. The MISFET were found to have electron mobilities in excess of $10^4$ $cm^2/V$-s at 4°K. They were used to evaluate the Shubnikov-de Haas oscillations obtained as a function of the applied transverse magnetic induction and their dependence on $V_g$. They observed a negative photoconductivity attributed to resonant absorption between surface quantum levels in the spectral range between 13 and 28 μm and suggested that optically-induced electronic transitions between sub-bands might provide the basis for a gate voltage-tunable photosensor. Shappir et al[9] have demonstrated the feasibility of p-channel inversion-mode MISFET operating at 77°K. For this purpose they used Te-doped InSb, $N_D = 8.5 \times 10^{14}/cm^3$, in which Cd was diffused to form the source and drain contacts and 0.1 μm thick $SiO_x$ was deposited, at 215°C, to form the gate

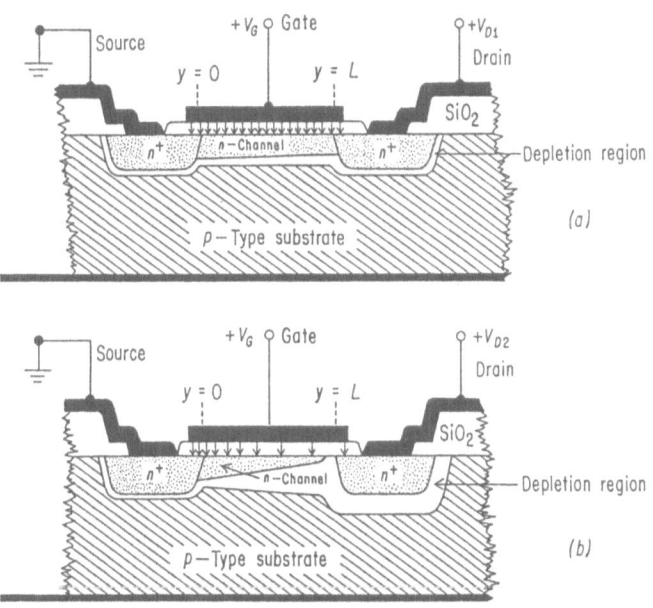

Figure 3. Cross-section of inversion-mode MISFET; a) applied gate voltage is above threshold for channel conduction; b) with gate voltage greater than channel pinchoff value.

insulating layer. The MISFET were made by means of photolithographic techniques with a width to length ratio, w/l = 13.7; the source and drain contacts were made by vacuum-deposition of Cr-Au alloy layers.

From the MISFET characteristics they calculated an effective hole mobility, $\mu_p = 330$ cm$^2$/V-s. Experimentally observed hysteresis in $I_{DS}$ vs $V_{DS}$ was attributed to modulation of the positive charge in the oxide; with the application of a negative $V_g$ tunneling of electrons was considered to take place from the oxide, where they are trapped, into the semiconductor; when $V_g$ is removed the electrons tunnel back into the traps driven by the electric field associated with the positive oxide charges. Below $V_{th}$ they found an early breakdown of the drain junction attributed to $V_g$-dependent tunneling of electrons from the valence band of the drain p-region into the n-type inversion layer formed under that portion of the gate which overlaps the reverse biased drain region.

Fujisada[10] described a p-channel inversion-mode MISFET made by selective Be ion implantation into Te-doped n = $6 \times 10^{14}$/cm$^3$ InSb with low temperature activation of the implanted species for the source and drain contacts. He used a composite gate insulator made of a ~ 0.03 µm thick anodic oxide grown on the InSb surface by wet anodization procedures on which a 0.36 µm thick $A\ell_2O_3$ is vacuum-deposited. Such a composite insulator allows the surface potential to be modulated from accumulation through flatband and depletion into inversion with an interface state density, at midgap, of ~ $3$-$4 \times 10^{11}$/cm$^2$-eV and a slight flatband shift of ~ 0.03 V. However it cannot be operated effectively in saturation because of the large increase in drain-substrate current with $V_{ds}$; furthermore, the DC value of its $I_{DS}$ is unstable and it has a $V_{ds}$ and $V_g$ dependence of its $V_{th}$. Wei et al[11] have obtained good results with a direct deposition of $SiO_x$ on InSb for linear and two-dimensional charge injection devices as well as MISFET. The gate insulator, ~ 0.1 µm thick, was produced by the pyrolitic reaction of oxygen and silane in ratio of 1:10$^3$ in $N_2$ carrier gas at ~ 200°C. Photolithographic techniques were used to form p-channel, planar, circular gate MISFET, with the source and drain electrodes made by Be$^+$ ion implantation, through the insulator, with a fluence of ~ $5 \times 10^{14}$/cm$^2$ and with post-implantation anneal performed in Argon at ~ 200°C. Transient capacitance measurements performed at 78°K on such reverse biased p-n junctions indicated two deep levels: $E_c - 0.05$ eV and $E_c - 0.11$ eV. The interface state density was found to have a minimum of $5 \times 10^{10}$/cm$^2$-eV in the upper half of the bandgap increasing to $5 \times 10^{11}$/cm$^2$-eV in the lower half of the bandgap. The flatband voltage shift < 0.2 eV. The channel hole mobility, derived from the MISFET characteristics, was ~ 310 cm$^2$/V-s. Ohashi et al[12] have made n-channel inversion-mode MISFET of molecular beam epitaxially-grown p-type 2 to $5 \times 10^{17}$/cm$^3$ InSb on GaAs substrates with 0.08 µm thick $SiO_2$ for gate insulators. The gate length was varied from 3 to 100 µm and the properties of such MISFET, at room temperature, were evaluated and compared to theoretical expectations. The calculated $\mu_{fe}$ as a function of the epilayer thickness, is between $1.5 \times 10^3$ and $4 \times 10^3$ cm$^2$/V-s, considerably smaller than the bulk Hall mobility and $I_{DS}$ did not saturate as a function of $V_{DS}$; $g_m$ = 6 mS/mm and did not increase with decreasing $l_g$ as expected from elementary theory. The interface state density measured at 77°K was 2 to $6 \times 10^{12}$/cm$^2$-eV at midgap; however, no account was taken of its effect on $g_m$. The low $\mu_{fe}$ of the inversion layer was attributed to interfacial scattering between the gate insulator and the epilayer surface.

Tunneling in a gate-controlled junction diode made of InSb was investigated by Margalit et al[13] at 77°K. The n-inversion layer at the surface of the p-diffused region is controlled by the surface potential which is a function of the gate voltage applied to the metal gate on the ~ 0.2 μm thick $SiO_2$ gate insulating layer. For tunneling to occur in the reverse biased junction the bottom edge of the conduction band at the surface must overlap the bulk valence band edge and the p acceptor density in the bulk must be large enough so that the depletion layer width is sufficiently thin to provide an appreciable tunneling probability. For tunneling in the forward biased junction an additional requirement is imposed: for $V_g$ = 0 the Fermi level, $E_F$, must cross below the valence band edge in the bulk and above the inverted surface conduction band edge. In either case in order to obtain modulation of the surface potential and hence of the tunneling current, the surface state and interface state density at the dielectric-semiconductor interface must be small.

Fujisada and Sasase[14] and subsequently Fujisada and Kawada[15] have also investigated the properties of InSb gate-controlled p-n junction diodes. Gate insulators were either ~ 34 nm thick anodized oxide (MOS) or composite ~ 50 nm thick anodic oxide with a superposed ~ 0.18 μm thick $Al_2O_3$ layer (MAOS structures). Reverse biased current vs voltage measurements made at 77°K on both MOS and MAOS diodes indicated that this current is essentially independent of $V_g$ and that it increased gradually with voltage up to ~ the value of $V_g$. Thereafter it depends strongly on $V_g$. An exponential increase in current is obtained when the junction voltage exceeds $V_g$; the current increases one order of magnitude for every 0.3 V for the MAOS device and it increases by the same amount for 0.07 V applied to the MOS device.

InAs MISFET

InAs has properties advantageous for MISFET applications because of its low effective electron mass and relatively high energies of its satellite conduction band minima. However, its fundamental bandgap which is only 0.38 eV at 77°K restricts the maximum $V_{DS}$ because of ionization-induced breakdown of the channel. The surface of n-type InAs is normally accumulated and that of p-type InAs is normally inverted. Baglee et al[16] have investigated inversion layer charge transport in InAs at 77°K using a gated Van der Pauw clover-leaf-type structure. Acceptor doped, $p = 2.5 \times 10^{17}/cm^3$, (111B)-oriented InAs was used as the substrate and gate insulators, nominally 0.1 μm thick, were made either by wet anodization in various electrolytes or by the sputter deposition of $SiO_x$ in vacuum. Al was used as the gate electrode and In was used for the contacts. Gated Hall measurements were made by pulsing the source-drain current in order to avoid Joule heating. Figure 4a shows the surface electron density, $n_s$, as a function of $V_g$. Evidently the specimen anodized in KOH has a lower $n_s(V_g)$ than the others employing different gate insulators. However its peak mobility, shown in Fig. 4b, is higher although its dielectric breakdown strength is lower and its leakage current is higher than those of the other insulators. The decrease in the Hall mobility with increasing $n_s$ in Fig. 4b has been interpreted by Moore and Ferry[17] as scattering from Coulombic centers localized at the semiconductor-oxide interface while at higher $n_s$ surface roughness is considered to be the dominant mobility limiting mechanism. For the data in Fig. 4b a Coulombic scattering density of $1.3 \times 10^{11}/cm^2$ and an rms surface roughness of 1.5 nm was used to match

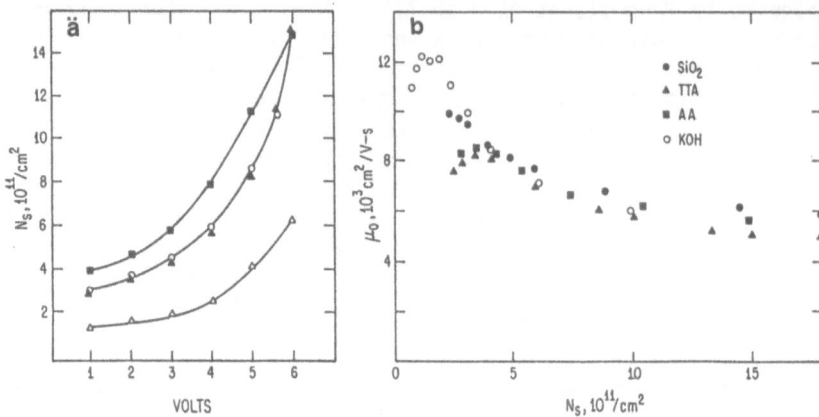

Figure 4.   a)  Gated surface galvanomagnetic properties of InAs at 77°K;
surface electron density as a function of gate voltage and various gate
insulators (after Baglee et al, ref. 16);   b)  Mobility of surface
charge carriers of InAs at 77°K as a function of surface electron
density and type of gate insulator; •, sputtered $SiO_2$; ▲, anodized in
tartaric acid solution; ■, anodized in arsenic acid; o, anodized in KOH
(after Baglee et al, ref. 16).

the KOH oxide data; it is in good agreement with fixed charge in the
oxide calculated from C-V measurements in terms of the measured shift of
the onset of inversion which yielded $N_{fix} = 9 \times 10^{10}/cm^2$.  Similar results
were also obtained for the other oxides.  However, there is a discrepancy between the theoretically calculated density of scattering centers
for the $SiO_x$ gate insulator and that determined from C-V data; the
latter, $N_{fix} = 2 \times 10^{12}/cm^2$ ought to produce an effective mobility of 5600
$cm^2$/V-s instead of the value measured experimentally, 9800 $cm^2$/V-s.  A
possible reason for the discrepancy might be some form of, as yet unde-
termined, screening of the Coulomb potential in the oxide accompanied
perhaps by an extended spatial distribution of the fixed charge.  Figure
5 shows the good fit between the theoretically calculated and experimentally measured inversion channel mobilities determined by Moore and
Ferry[17] and includes the data obtained by Kawaguchi[18] using a mylar film
gate insulator.  Reich and Ferry[19] have made a two-dimensional computer
simulation of a narrow, $l_g$ = 0.25 μm InAs Schottky barrier gate FET
operating at 77°K.  Using a finite difference two-dimensional numerical
analysis to solve the $I(V_g, V_{DS})$ characteristics in the linear and
velocity saturation regimes, they came to the conclusion that such
devices might provide performance competitive with superconductive
Josephson junction devices.

Borrello et al[20] have investigated the interaction between the InAs
depletion regions formed by surface states and impurity diffusion.  Mead
and Spitzer[21] found from C-V measurements made on p-InAs at 77°K that
the hole barrier is 0.47 eV while $E_g$ = 0.44 eV.  This implies that the
Fermi level is ~ 0.03 eV above the conduction band edge indicating
degenerate inversion.  The surface depletion region is dependent on the
carrier concentration through the Debye screening length and the surface

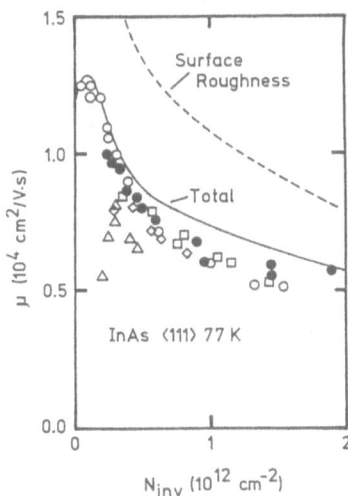

Figure 5. Dependence of measured and theoretically calculated inversion channel mobility of InAs as a function of its electron density and type of gate insulator: □ , anodized in arsenic acid; o, anodized in KOH; ◊ , anodized in tartaric acid; ● , sputter deposited $SiO_2$; Δ , mylar sheet (after Moore and Ferry, ref. 17).

depleted layer can be made to merge with a bulk depleted layer formed by impurity diffusion. This was done by Cd diffusion into n-type InAs forming a p-n junction in which the Cd concentration at the surface is between $10^{18}$ and $10^{19}/cm^3$ dropping to ~ $10^{17}cm^3$ at a depth of 2 μm and declining thereafter with a complementary error function profile. A metal contact is deposited on the etched junction surface. In the band diagram of Fig. 6, w is the diffusion depth less the barrier width. If a sufficient amount of doped material is removed by etching, then the position of zero electric field is altered from w to a plane which permits an externally applied potential to influence the entire structure. Eventually, the valence band edge is several kT from the Fermi level and the hole concentration in the potential well is no longer determined by the impurity concentration. The hole lifetime can influence directly, via the barrier height, the electron current produced by an external potential applied to such a junction.

## Mercury Cadmium Telluride Gate-controlled Diodes

Kolodny and Kidron[22] have investigated the properties of gate-controlled ion-implanted p-n junctions of mercury cadmium telluride. For this purpose they used p-type $Hg_{0.71}Cd_{0.29}Te$ wafers with hole densities of 1 to $5x10^{16}/cm^3$ and $\mu_0$ ~ 250 $cm^2/V$-s at 77°K. Ion implantation of B, Aℓ, P, and Ar was performed at room temperature with fluences of $10^{13}$ to $10^{15}/cm^2$ and energies of $10^2$ to $3x10^2$ keV. After post-implantation annealing in vacuum up to 140°C the junction depth was, typically, < 1 μm . Gate-controlled diodes were made with vacuum-deposited indium gate electrodes overlapping the edge of the mesa diodes and insulated by ~ 0.5 μm thick ZnS or by an anodic oxide of HgCdTe, as

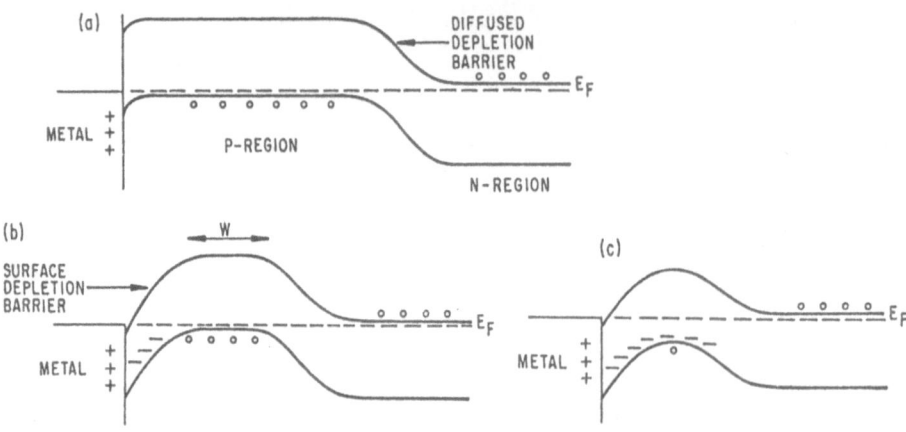

Figure 6. Surface and diffusion potential barrier interactions in
InAs; a) energy band diagram following diffusion of Cd into n-type
InAs; b) reducing thickness of p-type surface layer by etching, w, is
diffusion depth less surface barrier width; c) further reduction of
surface layer with valence band edge several kT from Fermi level (after
Borrello et al, ref. 20).

shown schematically in Fig. 7. If $V_g$ is more negative than the flatband
voltage, $V_{fb}$, then the surface on the p-side of the junction is accumulated and that on the n-side is inverted. The interface charge was

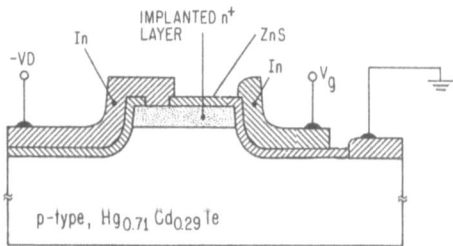

Figure 7. Schematic cross-section of HgCdTe gate-controlled junction
diode (after Kolodny and Kidron, ref. 22).

determined to be negative ~ $3 \times 10^{11}/cm^2$. Figure 8 shows the characteristic properties of such a device attributed to reverse and forward
tunneling currents with the forward injection currents large enough to
mask the differential negative resistance. For $V_G > V_{fb}$ an n-type
surface inversion layer forms on the p-side of the junctions. In
reverse bias the large leakage currents saturate. This saturation
current, through the inversion layer, is attributed to reverse and
forward tunneling currents with the forward injection currents large
enough to mask the differential negative resistance. For $V_g > V_{fb}$ an n-
type surface inversion layer forms on the p-side of the junction. In

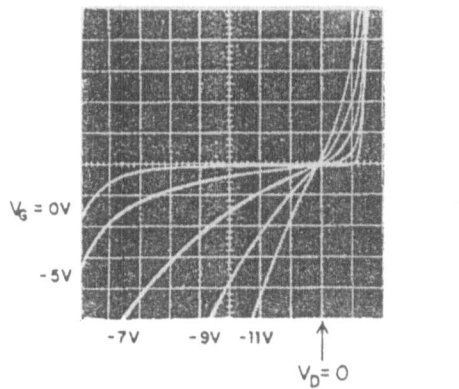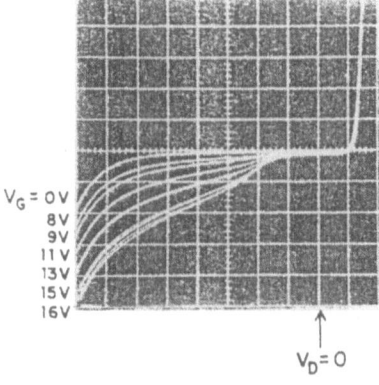

Figure 8.  Current vs voltage of gate-controlled junction diode at 77 K.  Left:  $V_g < V_{fb}$ (p-side accumulated; Right:  $V_g > V_{fb}$ (p-side inverted) [after Kolodny and Kidron, ref. 22].

reverse bias the large leakage currents saturate.  This saturation current, through the inversion layer, is attributed to channel pinchoff in a manner similar to that of silicon gate-controlled diodes.  The dielectric properties of the oxides of HgCdTe grown by wet anodization were investigated by Bertagnolli[23] using KOH-ethylene solutions.  Such oxides grown at room temperature have a fixed positive charge ~ $10^{12}/cm^2$, a 0.1 V hysteresis attributed to slow surface states ~ $10^{11}/cm^2$, and a low frequency dispersion of its dielectric properties.  No such dispersion is found in the higher quality oxides grown at 50°C.  These exhibit a fixed charge ~ $1.4x10^{12}/cm^2$, a smaller hysteresis, ~ $5x10^{10}/cm^2$, a fast surface state density ~ $10^{12}/cm^2$-eV and a high surface recombination velocity.  Oxides grown at 75°C, on the other hand, were found to have a pinned surface Fermi level.

## Germanium MISFET

Germanium has a room temperature electron mobility $\mu_n$ = 3900 $cm^2$/V-s and a hole mobility $\mu_p$ = 1800 $cm^2$/V-s compared to silicon with the same impurity concentration, $\mu_n$ = 1400 $cm^2$/V-s and $\mu_p$ = 450 $cm^2$/V-s  and is, therefore, of particular interest for use in integrated circuits employing complementary n-channel and p-channel MISFET.  However, the native oxides of Ge are volatile at the oxidation temperature, they are usually composed of suboxides rather than $GeO_2$ and their interface charge densities are considerably greater than $10^{12}/cm^2$.  Rzhanov and Neizvestny[24] found that Ge MISFET with composite $SiO_2$-$Si_3N_4$ gate insulators have more favorable electrical properties.  The room temperature $\mu_{fe}$ dependence on $n_s$ of such MISFET is qualitatively similar to that of Si MOSFET; it rises from ~ 750 $cm^2$/V-s at 300°K to ~ $10^3$ $cm^2$/V-s at 140°K decreasing thereafter with a further decrease in temperature.  Further investigations on Ge n-channel inversion-mode MISFET have been made by Rosenberg[25].  He used (100)-oriented p-doped Ge with gate insulators made by the thermal reaction at ~ 700°C of Ge with nitrogen to form ~ 0.01 μm thick $Ge_3N_4$ layers.  These were found to have a breakdown strength in excess of $5x10^6$ V/cm and an interface density of less than $2 \times 10^{11}/cm^2$. Source and drain electrodes were made by the ion implantation of As and vacuum deposited Aℓ was used for metallization of the gate, source and drain. Such a prototype MISFET had a $\mu_{fe} = 1.9 \times 10^3$ $cm^2/V$-s although its drain resistance was high and it had large junction leakage currents.

## Ternary III-V Alloy MISFET

The ternary alloy $In_{0.53}Ga_{0.47}As$ with a fundamental room temperature bandgap of 0.75 eV, $\mu_0 = 8 \times 10^3$ $cm^2/V$-s for $n = 10^{17}/cm^3$ is well suited for a variety of FET applications and it can be grown in the form of epitaxial layers on SI InP substrates. Wieder et al[26] have shown that such layers with a hole density, $p = 2 \times 10^{17}/cm^3$ can be used to make inversion mode MISFET with source and drain electrodes of alloyed 20:80 Sn-Au and gate insulating layers made of 0.1 μm thick $SiO_2$ grown by the low temperature plasma assisted pyrolisis of silane in the presence of $NO_3$. Subsequently, Liao et al[27] described such an inversion-mode MISFET with a $Si_3N_4$ gate insulator with a $g_m = 3$ mS/mm. Considerably better results were obtained by Ishii et al[28] with enhancement-mode MISFET employing a composite gate insulator of ~ 0.01 μm thick anodic oxides of $In_{0.53}Ga_{0.47}As$ and ~ 0.1 μm of superposed $Aℓ_2O_3$. The thickness of the $In_{0.53}Ga_{0.47}As$ layer was chosen so that, for $V_g = 0$, it is totally depleted and a positive $V_g$ is required for channel conduction. For $l_g = 10$ μm and a surface state density ~$8 \times 10^{11}/cm^2$-eV near mid-gap they obtained a $g_m = 17$ mS/mm. They also found that, in analogy to InSb MISFET, these have a DC drain current drift which has a logarithmic time dependence attributed to tunneling of charge carriers into traps in the gate insulator. Kaumans et al[29] investigated the nature of the semiconductor-dielectric interface of $In_{0.53}Ga_{0.47}As$ and $SiO_2$ using various post-deposition thermal annealing cycles and made non-optimized n-channel inversion-mode MISFET with a $\mu_{fe} < 100$ $cm^2/V$-s. Considerably better results were obtained by Gardner et al[30,31] using low temperature chemical vapor phase-deposited $SiO_2$ annealed for 16 hours at 300°C prior to gate metallization and annealed again for 15 min. following deposition of the gate metal. Their n-channel inversion-mode MISFET had $l_g$ of 1.5 to 3 μm and width of 150 μm. They obtained $\mu_{fe} = 2.5 \times 10^3$ to $4.6 \times 10^3$ $cm^2/V$-s. Such devices also exhibited a significant DC drain current drift with $I_{DS}$ decaying to ~ 50% of their initial value in ~ 100 s. However, similar self-aligned structures with ion-implanted source and drain electrodes, $l_g$ ~ 1 μm, width of 140 μm, gate insulator thickness of 0.9 μm, interface state density < $10^{11}/cm^2$-eV and a fixed oxide charge density ~ $4 \times 10^{10}/cm^2$ had a $g_m = 43$ to 64 mS/mm and the drain current instability was reduced to ~ 2% of its original value. More recently[32], by minimizing the gate overlap parasitic capacitance and by using a gate length of 1 μm they obtained a $g_m = 107$ mS/mm corresponding to an electron velocity of $2.5 \times 10^7$ cm/s. Depletion-mode MISFET of similar configuration made of n-type $In_{0.53}Ga_{0.47}As$ were found to have[33] an estimated $f_T = 20$ GHz, surface state densities in the range between $2.5 \times 10^{12}$ in accumulation to $2 \times 10^{10}/cm^2$-eV in inversion and a minimum noise figure of 3.4 dB with 9.4 dB associated gain at 4 GHz.

Gate-controlled galvanomagnetic measurements have been made on transistor-like five-terminal MISFET structures by Mullin and Wieder[34] using 0.25 μm thick $In_{0.53}Ga_{0.47}As$ epilayers grown on semi-insulating InP substrates with 0.1 μm thick $Al_2O_3$ gate insulators. Hall effect and resistivity measurements were made on such structures as a function of the applied gate voltage and the data were used to derive the density of surface states and their position within the bandgap. They found that while the surface Fermi level of virgin structures is pinned ~ 2 eV below the conduction band even mild thermal annealing at 120°C for 16 h reduces the density of interface states from ~ $5 \times 10^{12}/cm^2$-eV to $2 \times 10^{11}/cm^2$-eV at midgap and allows the surface potential to be displaced over most of the fundamental bandgap. Wieder et al[35] also demonstrated an enhancement-type MISFET based on surface accumulation of nearly semi-insulating $In_{0.53}Ga_{0.47}As$ which has its residual donors compensated by deep level $Fe^-$ acceptors. A non-optimized structure with an 8 μm long gate 240 μm in width with either $SiO_2$ or $Al_2O_3$, 0.12 μm thick gate insulators and $Al$ gate electrodes had a field-effect mobility of 793 $cm^2$/V-s. O'Connor et al[36] used a 10 to 25 nm thick silicon nitride gate insulator, made by reacting ammonia with silane at 300°C, for an $In_{0.53}Ga_{0.47}As$ MISFET, defined as an insulator-assisted Schottky gate FET. Such as device with a gate $l_g$ = 1.2 μm, width of 250 μm, gate capacitance, ~ 4.2 pF had a $g_m$ = 130 mS/mm and a current of ~ $2 \times 10^{-3}$ $A/cm^2$ for 10 V reverse bias. Subsequently Cheng et al[37] described a similar device with a 5 to 12 nm thick electron beam-evaporated $SiO_2$ gate insulator and a self-aligned recessed gate structure; they obtained a $g_m$ = 150 mS/mm, a $v_s$ = $2.4 \times 10^7$ cm/s and an estimated $f_T$ = 15 GHz.

## $In_{0.53}Ga_{0.47}As$ Homojunction Field-Effect Transistors

The feasibility of p-n homojunction gate field-effect transistor (JFET) was demonstrated by Leheny et al[38] using ~ 1 μm thick liquid phase epitaxially-grown, n = $2 \times 10^{16}/cm^3$, $In_{0.53}Ga_{0.47}As$. A 20 μm long 1 mm wide gate was made by Zn diffusion through a silicon nitride mask at 750°C for 70 s in a $ZnAs_2$ atmosphere. The resultant metallurgical junction at a depth of 0.5 μm required $V_g$ = - 4 V for channel pinchoff and had a $g_m$ = 1 mS/mm. A homojunction quasi-Schottky barrier gate diode employing a depleted, thin $p^+$-layer to reduce the gate leakage current while eliminating the problems associated with a hetero-inter-face gate was described by Chen et al[39]. Using MBE an $n^+$ $In_{0.53}Ga_{0.47}As$ layer, 0.5 μm thickness, was first grown on an(100)-oriented $n^+$ InP substrate followed by a 3 μm thick, Sn-doped, n = $10^{17}/cm^3$ layer and thereafter by an 8 nm Be-doped, p = $8 \times 10^{18}/cm^3$ layer. The effective barrier height rose from ~ 0.27 eV to 0.47 eV. The properties of JFET made by MBE were also investigated by Chang et al[40]. They deposited first a buffer layer of high resistivity $In_{0.52}Ga_{0.48}As$ on (100)-oriented semi-insulating InP followed by a relatively thin transition layer of $In_{0.53}Ga_{0.25}Al_{0.22}As$; thereafter, the undoped n-type channel, 0.7 μm in thickness, n = $2 \times 10^{16}/cm^3$ was deposited followed by an 0.8 μm thick Mn-doped, p = $10^{18}/cm^3$ gate layer. For $l_g$ = 2 μm they obtained a $g_m$ = 50 mS/mm and found that in the narrow gate devices there is an inflection point in the $I_{DS}$ vs $V_{DS}$ curves accompanied by a sharp rise in $I_{DS}$ attributed to an unfavorable channel length to depth ratio. High frequency JFETs were made by Chai et al[41] using MBE-grown n-type $In_{0.53}Ga_{0.47}As$ layers deposited directly, without any intermediate buffer layer, on SI InP substrates. The junction gate was made by Be ion implantation with a fluence of $10^{14}/cm^2$ at 30 keV followed by annealing at 675°C for 20 min. in flowing hydrogen with an activation ~ 20%. The $g_m$ = 86 mS/mm was considerably smaller than that expected of the measured $\mu_0$ = 5.5 to $6.5 \times 10^3$ $cm^2/V$-s and the $l_g$ < 1 µm. Further investigations revealed that $\mu_0$ decreases at the channel-substrate interface to ~ $10^3$ $cm^2/V$-s; this as well as a corresponding reduction in $v_s$, might be responsible for the smaller than expected $g_m$. Nevertheless, these JFETs with a 250 µm wide gate were found to have, at channel pinchoff, a source-drain breakdown strength ~ 20 V; they also had a 5.2 dB gain at 11 GHz with a power added efficiency of 14 %. Schmitt and Heime[42] have suggested that some of the problems encountered with ion-implanted ternary alloy JFETs might be due to the long range diffusion tails which follow high temperature annealing of the implanted acceptors compensating the donors in the channel and reducing the electron mobility. However, Selders et al[43] demonstrated that $SiO_2$-capped $In_{0.53}Ga_{0.47}As$ subjected to rapid thermal annealing of ion-implanted Be (800°C for 0.5 s) can produce JFETs with $g_m$ = 130 mS/mm and an $f_T$ = 15 GHz. Schmitt and Heime[42] have made ternary alloy JFETs by diffusing Zn at 600°C for 10 min. which had been earlier deposited from a "spin-on" solution. With the metal gate as a mask, photo-lithographic and etching techniques were used to fashion junction gates 1.2 µm in length and 300 µm in width to produce devices with $g_m$ = 100 mS/mm, $v_s$ = $2.3 \times 10^7$ cm/s and $f_{max}$ = 30 GHz.

A self-aligned JFET grown by MBE deposited on SI or $n^+$ InP was described by Cheng et al[44]; it was intended to make such transistors compatible with optoelectronic devices integrated on the same substrate. The vertical junction structure consists of an 0.05 µm thick $p^+$ = $2 \times 10^{19}/cm^3$ cap layer, a 0.4 to 0.5 µm thick, n = $10^{17}/cm^3$ channel and a Be-doped $p^+$ = $5 \times 10^{18}/cm^3$, 0.4 to 0.8 µm thick confinement layer all of $In_{0.53}Ga_{0.47}As$. Cr-Au gate metal patterns were photolithographically defined and self-aligned techniques were used to make the Au-Ge source and drain electrodes. A preferential, crystallographically selective etching solution was used to produce an undercut gate ~ 1 µm in length, 100 to 320 µm in width. Such devices have been made to operate in both depletion and enhancement. In depletion typical $g_m$ = 90 mS/mm while in enhancement $g_m$ = 60 to 70 mS/mm. An alternative means of channel isolation was also employed[45] by replacing the SI InP layer with a 1 µm thick undoped $In_{0.52}Al_{0.48}As$ layer grown by metal organic vapor-phase epitaxy (MOVPE) with no significant improvement in performance.

MOVPE was also used by Wake et al[46] for making JFET. They found, however, that considerable outdiffusion of the p dopant can occur during growth and this decreases the mobility and saturated velocity of the electrons in the channel as well as reducing its free electron density due to compensation of its donors. Cadmium is an acceptor and has a lower diffusion coefficient than Zn in InP; by using Cd in the buffer layer in concentration equal to or smaller than that of the channel they obtained high quality junctions. These were used with a preferential etching solution and self-aligned techniques to make JFETs with $l_g$ = 1.5 µm, width of 270 µm and $g_m$ = 210 mS/mm; the gate to source capacitance $C_{gs}$ = 0.5 pF for $V_g$ = 0, the calculated $v_s$ = $2.7 \times 10^7$ cm/s, close to the estimated theoretical maximum and the calculated $f_T$ = 18.5 GHz.

## Heterojunction Field-Effect Transistors

Among the reasons for the current research emphasis on ternary alloy hetero-junction gate field-effect transistors (HJFET), in particular, two-dimensional gas (2DEG) modulation doped transistors, is their potentially superior high speed and low noise characteristics in comparison with other FET. Initially, $In_{0.52}Al_{0.48}As$ with a fundamental bandgap, $E_g$ = 1.46 eV, whose lattice constant matches that of InP, was intended, primarily, as a heterojunction gate for raising the effective barrier height of $In_{0.53}Ga_{0.47}As$ while avoiding problems associated with interfacial misfit dislocations. It was also considered to be an adequate buffer layer between the active channel and its SI InP substrate. Ohno et al[47] have made such a depletion-mode HJFET using MBE to grow, sequentially, on a (100)-oriented InP substrate a 0.1 µm $In_{0.52}Al_{0.48}As$ buffer layer, a 0.125 µm $In_{0.53}Ga_{0.47}As$ layer with a superposed barrier-enhancing 0.06 µm thick $In_{0.52}Al_{0.48}As$ layer followed by an Al gate. Photolithographic techniques and a phosphoric acid/hydrogen peroxide etch were used to make devices with AuGe source and drain electrodes, 2.75 µm gate length, 3.5 µm source-gate separation, and 10 µm source drain separation. They obtained a saturated $g_m$ = 57 mS/mm, an apparent barrier height of 0.80 eV and a gate leakage current, for a reverse bias of 8 V, of 416nA/µm$^2$. A similar double heterostructure depletion mode HJFET was made, using MBE, by Barnard et al[48]. With a gate 0.6 µm in length, 0.65 µm wide and source-drain spacing of 2.9 µm they obtained, in the depletion-mode, a $g_m$ = 135 mS/mm. The gate leakage current was less than 62 nA/µm$^2$ for $V_g$ < 3 V rising to 310 nA/µm$^2$ for $V_g$ = 4 V. In the enhancement-mode these devices had a small $g_m$ (attributed to negatively charged interface states at the channel-buffer interface) which decreased with increasing positive bias.

In modulation-doped field-effect transistors (MODFET) very high charge carrier mobilities are obtained by separating, spatially, the conduction electrons in the channel layer from their ionized donor impurities which are present in an adjacent, larger bandgap layer. The carriers are confined in a 2DEG quantum well by the inter-facial band edge discontinuities between these layers.

The structure and configuration of a MODFET are subject to the following criteria which determine the thickness and doping of each layer:

(a)  The $In_{0.52}Al_{0.48}As$ cap layer need be n-doped to ~ $10^{17}/cm^3$; it should be thin enough so that it is nearly depleted by band bending from both its metal Schottky barrier and its heterojunction interface with the $In_{0.53}Ga_{0.47}As$ channel.

(b)  The $In_{0.53}Ga_{0.47}As$ layer containing the 2DEG usually has a background n-type electron density of $5 \times 10^{15}$ to $2 \times 10^{16}/cm^3$ when grown by MBE. The 2DEG is confined in an accumulation region at the heterojunction interface. It is desirable for the layer thickness to be small enough, typically, 0.1 to 0.2 µm, so that this background charge concentration represents less than 10% of the ~ $10^{12}/cm^2$ surface charge density in the 2DEG well.

(c) An undoped, essentially, SI $In_{0.52}Al_{0.48}As$ buffer layer should, preferably, isolate metallurgically the $In_{0.53}Ga_{0.47}As$ layer from possible outdiffusion of impurities or propagation of defects from the InP substrate.

Electrical and galvanomagnetic measurements made on MBE-grown modulation-doped $In_{0.53}Ga_{0.47}As$-$In_{0.52}Al_{0.48}As$ structures by Cheng et al[49] exhibited electron mobility enhancement by a factor of 2 at 300°K and a factor of 6 at 77°K in comparison with n-type $In_{0.53}Ga_{0.47}As$ with the same electron concentration. Chen et al[50] have described the construction and performance of a MODFET employing such a hetero-structure; $In_{0.53}Ga_{0.47}As$ 1.5 μm thick with $n = 2 \times 10^{15}/cm^3$ was first grown by MBE on SI (100)-oriented InP followed by an 9 nm undoped spacer layer of $In_{0.52}Al_{0.48}As$ and then by an 0.15 μm Si-doped $n = 10^{17}/cm^3$ layer of the same alloy. The MODFET was made with a gate $l_g = 5.2$ μm, 340 μm in width and a source-drain separation of 10.4 μm. Source and drain contacts were made by sequential deposition of 20 nm Ge, 0.1 μm Au-Ge, 20 nm Pt and 20 nm Au which were alloyed at 450°C for 1 min. The enhanced $\mu_0$ was credited for the measured $g_m = 31$ mS/mm at 300°K and 69 mS/mm at 77°K with the $g_m$ fairly constant for $V_g < 2$ V; an abrupt decrease of $g_m$ for $V_g > 2$V was attributed to the initiation of parasitic charge transport in the $In_{0.52}Al_{0.48}As$ layer. Chen et al[51] proposed a theoretical model for the dependence of $g_m$ on $V_g$ whose principal features are shown, schematically, in Fig. 9. If $0 > V_g > V_{FB}$ then $g_m$ is independent of $V_g$ and the band diagram of Fig. 9a is applicable; the flatband voltage, $V_{FB}$, is defined as the gate voltage required to quench the 2DEG, as shown in Fig. 9b; in this case the MODFET conducting channel is not completely pinched off due to the presence of background carriers and $g_m$ is to some extent a function of $V_g$. Figure 9c shows that as $V_g$ exceeds $V_{FB}$ the depletion edge extends into the undoped $In_{0.53}Ga_{0.47}As$ layer and $g_m$ becomes a strong function of $V_g$.

Pearsall et al[52] have made recessed gate depletion-mode MODFET with $l_g = 1.2$ μm, 125 to 250 μm in width and source-drain spacing of 8 μm. At 300°K they obtained a $g_m = 90$ mS/mm and at 77°K a $g_m = 200$ mS/mm. Their measurements suggested that a substantial fraction of the channel current is carried not in the 2DEG but in the rest of the $In_{0.53}Ga_{0.47}As$ layer and that this occurs as a consequence of real space transfer of moderately "hot" electrons out of the 2DEG potential well. It has been suggested by Chan et al[53] that an $In_{0.53}Ga_{0.47}As$/InP-based inverted modulation-doped structure might have a higher $v_s$ than one with a normal configuration, one in which the InP layer is on top of the $In_{0.53}Ga_{0.47}As$ channel while in an inverted structure it is below the channel. They found that in the inverted structure $I_{DSS}$ is consistent with the low field electron concentration and the bulk $v_s$, while in a normal structure $I_{DSS}$, is significantly smaller than expected and might be attributed to real space transfer of electrons into the InP layer. Seo et al[54] have used MBE to make an inverted $In_{0.53}Ga_{0.47}As$/$In_{0.52}Al_{0.48}As$ HJFET in which the active channel is a single quantum well of $In_{0.53}Ga_{0.47}As$ 10 to 40 nm in thickness. Using a recessed gate configuration they obtained with a 1.8 μm long and 60 μm wide gate a $g_m = 130$ mS/mm and a gate leakage current of 3 μA for $V_g$

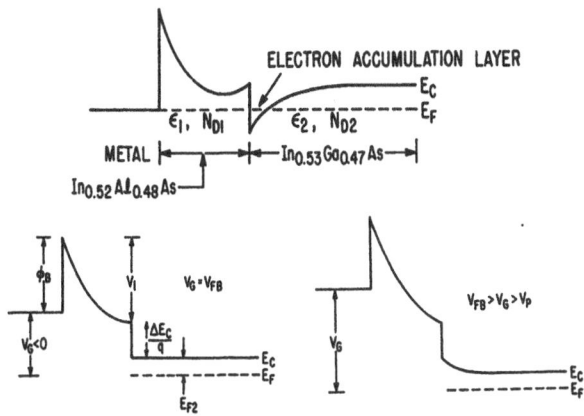

Figure 9. Band diagram of $In_{0.53}Ga_{0.47}As$ MODFET as a function of applied gate voltage: $E_1$, $N_{D1}$ and $E_2$ $N_{D2}$ are the respective fundamental bandgaps and donor doping densities of $In_{0.52}Al_{0.48}As$ and $In_{0.53}Ga_{0.47}As$ with $N_{D1} \gg N_{D2}$; a) with two-dimensional quantum well; b) at flatband; c) beyond flatband, depletion layer extends into $In_{0.53}Ga_{0.47}As$ (after Chen et al, ref. 51).

$= -3$ V. However, just as for the other MODFET, they found a strong dependence of $g_m$ on $V_g$ affected by real space charge transfer.

OVERVIEW

The low barrier height of narrow bandgap semiconductors which prevents their use for Schottky barrier gate FET has led to the search for alternatives. If compatible gate insulators with interfacial properties similar to those of the $Si-SiO_2$ system were available then insulated gate depletion and inversion-mode transistors analogous to MOSFETS might be feasible. Such devices might use the high electron mobility and saturated velocity of the narrow gap semiconductors while retaining the configurational simplicity of MOSFET. However, insulators presently available are far from ideal; only fragmentary information is available on insulator-semiconductor interfaces. MISFET as well as two-terminal MIS structures employing narrow gap semiconductors exhibit hysteresis and a logarithmic time dependent drift of their DC character-istics. This drift, attributed to tunneling of electrons from a semiconductor surface into traps located in the insulator within ~ 4 nm of its interface, is also present in $Si-SiO_2$ structures (it is smaller by two orders of magnitude in comparison with that of III-V semicon-ductors). Furthermore, electron scattering from charged interfacial centers as well as roughness scattering limit the surface channel mobility of MISFETs to less than 20% of their Hall mobilities. Fast and slow surface states also interpreted as spatial fluctuations in surface potential affect adversely the $g_m$ and $f_T$ of MISFETs. They depend on empirically-based semiconductor surface preparation, prior to deposition of the insulator, on the parameters of the deposition process itself, on the fundamental properties of the native oxides as well as on those of the synthetic dielectric layers.

Homojunction FET circumvent some of the problems associated with MISFETs. However, the gate voltage swing in forward bias must not exceed the built in contact potential difference between the ion-implanted or diffused p$^+$ gate and the n channel. Leakage currents limit the maximum applicable reverse gate bias and the relatively large junction capacitance of such structures reduce their $f_T$. Nevertheless, excellent results have been obtained with $In_{0.53}Ga_{0.47}As$ enhancement and depletion-mode JFET and a depleted p-doped layer under the metal gate of an MESFET has been used to raise its effective surface barrier height.

A MODFET is, in some respects, the analog of a MISFET with the insulator replaced by a doped but depleted large bandgap semiconducting layer adjacent to an undoped narrow bandgap 2DEG channel. Modulation doping implies that in such a heterostructure charge carriers are transferred from the large gap into the narrow bandgap channel where they are confined by the band offsets between them and are not subject to impurity scattering. Preliminary results obtained with $In_{0.53}Ga_{0.47}As/ In_{0.52}Al_{0.48}As$ MODFETs appear promising although at his stage of their development their performance is limited in part by the residual impurity concentration of $In_{0.53}Ga_{0.47}As$ and, if the gate voltage is large enough, by real space transfer out of the 2DEG well.

REFERENCES

1.  V. L. Frantz, Proc. IEEE, 53, 760(1965).
2.  F. C. Luo and M. Epstein, Proc. IEEE, 60, 997(1972).
3.  D. L. Lile and J. C. Anderson, Solid-State Electron. 12, 735(1969).
4.  A. VanCalster, Solid-State Electron. 22, 77(1979).
5.  T. P. Brody and H. E. Kunig, Appl. Phys. Lett. 9, 259(1966).
6.  W. B. Pennebaker, Solid-State Electron. 8, 509(1965).
7.  Y. Katayama, N. Kotera and K. F. Komatsubara, Proc. 10th Internat. Conf. Phys. Semicond. Cambridge, Mass.(1970), U.S. Atomic Energy Comm. Div. Tech. Info. pp. 464-468.
8.  Y. Katayama, N. Kotera and K. F. Komatsubara, Proc. 2nd Conf. Sol. State Dev. Tokyo, Suppl. Journal Japan. Soc. Appl. Phys. 40, 214(1971).
9.  J. Shappir, S. Margalit and I. Kidron, IEEE Trans. Electron. Dev. ED-22, 960(1975).
10. H. Fujisada, Japan. J. Appl. Phys. 24, L835(1985).
11. C-Y. Wei, K. L. Wang, E. A. Taft, J. M. Swab, M. D. Gibbons, W. E. Davern, D. M. Brown, IEEE Trans. Electron. Dev. ED-27, 170(1980).
12. T. Ohashi, D. P. Bour, T. Itoh, J. D. Berry, S. R. Jost, G. W. Wicks and L. F. Eastman, J. Vac. Sci. Tech. B4(2), 622(1986).
13. S. Margalit, J. Shappir and I. Kidron, J. Appl. Phys. 46, 3999(1975).
14. H. Fujisada and T. Sasase, Japan. J. Appl. Phys. 23, L162(1984).
15. H. Fujisada and M. Kawada, Japan, J. Appl. Phys. 24, L76(1985).
16. D. A. Baglee, D. K. Ferry, C. W. Wilmsen and H. H. Wieder, J. Vac. Sci. Tech. 17, 1032(1980).
17. B. T. Moore and D. K. Ferry, J. Vac. Sci. Technol. 17, 1037(1980).
18. S. Kawaji and Y. Kawaguchi, J. Phys. Soc. Japan V, 21, 336(1966).
19. R. K. Reich and D. K. Ferry, IEEE Trans. Electron. Dev. ED-27, 1062(1980).
20. S. R. Borrello, G. R. Pruett and J. D. Sawyer, Proc. 3rd Internat. Conf. Photocond. Pergamon Press (1971) Oxford, pp. 385-394.
21. C. A. Mead and W. G. Spitzer, Phys. Rev. 134, A713(1964).

22. A. Kolodny and I. Kidron, IEEE Trans. Electron. Dev. ED-27, 37(1980).

23. E. Bertagnolli, Thin Solid Films, 135, 267(1986).

24. A. V. Rzhanov and L. G. Nezvestny, Thin Solid Films, 58, 37(1979).

25. J. J. Rosenberg, "Germanium MISFET Utilizing a Germanium Nitride Gate Insulator", Ph.D. Dissertation, Columbia University, New York (1983).

26. H. H. Wieder, A. R. Clawson, D. I. Elder and D. A. Collins, IEEE Electron Dev. Lett. EDL-2, 73(1981).

27. A. S. H. Liao, R. F. Leheny, R. E. Nahory and J. C. DeWinter, IEEE Electron Dev. Lett. EDL-2, 288(1981).

28. K. Ishii, T. Sawada, H. Ohno and H. Hasegawa, Electron. Lett. 18, 1034(1982).

29. R. Kaumanns, J. Selders and H. Beneking, Inst. Phys. Conf. Ser. No. 63, Ch. 7, 329(1981).

30. P. D. Gardner, S. Y. Narayan, S. Colvin and Y-H. Yun, RCA Rev. 42, 542(1981).

31. P. D. Gardner, S. Y. Narayan and Y-H. Yun, Thin Solid Films, 117, 173(1984).

32. P. D. Gardner, S. G. Liu, S. Y. Narayan, S. D. Colvin, J. P. Paczkowski and D. R. Capewell, IEEE Electron. Dev. Lett. EDL-7, 363(1986).

33. P. D. Gardner, S. Y. Narayan, Y-H. Yun, J. Paczkowski, B. Dornan and R. E. Askew, Inst. Phys. Conf. Ser. No. 65, 399(1982).

34. D. P. Mullin and H. H. Wieder, J. Vac. Sci. Technol. b1, 782(1983).

35. H. H. Wieder, J. L. Veteran, A. R. Clawson and D. P. Mullin, Appl. Phys. Lett. 43, 287(1983).

36. P. O'Connor, T. P. Pearsall, K. Y. Cheng, A. Y. Cho, J. C. M. Hwang and K. Alavi, IEEE Electron. Dev. Lett. EDL-3, 64(1982).

37. C. L. Cheng, A. S. H. Liao, T. Y. Chang, E. A. Caridi, L. A. Coldren and B. Lalevic, IEEE Electron. Dev. Lett. EDL-5, 511(1984).

38. R. F. Leheny, R. E. Nahory, M. A. Pollack, A. A. Ballman, E. D. Beeb, J. C. DeWinter and R. J. Martubm UEEE Electron Dev. Lett. EDL-1, 110(1980).

39. C. Y. Chen, A. Y. Cho, K. Y. Cheng and P. A. Garbinski, Appl. Phys. Lett. 40, 401(1982).

40. T. Y. Chang, R. F. Leheny, R. E. Nahory, E. Silberg, A. A. Ballman, E. A. Caridi and C. J. Harrold, IEEE Electron. Dev. Lett. EDL-3, 56(1982).

41. Y. G. Chai, C. Yuen and G. A. Zdasiuk, IEEE Trans. Electron. Dev. ED-32, 972(1985).

42. R. Schmitt and K. Heime, Inst. Phys. Conf. Ser. No. 79, Ch. 11, 619(1985).

43. J. Selders, H. J. Wachs and H. Juergensen, Electron. Lett. 22, 313(1986).

44. J. Cheng, S. R. Forrest, R. Stall, G. Guth and R. Wunder, Appl. Phys. Lett. 46, 885(1985).

45. J. Cheng, R. Stall, S. R. Forrest, J. Long, C. L. Cheng, G. Ruth, R. Wunder and V. G. Riggs, IEEE Electron. Dev. Lett. EDL-6, 384(1985).

46. D. Wake, A. W. Nelson, S. Cole, S. Wong, I. D. Henning and E. G. Scott, IEEE Electron. Dev. Lett. EDL-6, 626(1985).

47. H. Ohno, J. Barnard, C. E. C. Wood and L. F. Eastman, IEEE Electron. Dev. Lett. EDL-1, 154(1980).

48. J. Barnard, H. Ohno, C. E. C. Wood and L. F. Eastman, IEEE Electron. Dev. Lett. EDL-1, 174(1980).

49. K. Y. Cheng, A. Y. Cho, T. J. Drummond and M. Morkoc, Appl. Phys. Lett. 40, 147(1982).

50. C. T. Chen, A. Y. Cho, K. Y. Cheng, T. P. Pearsall, P. O'Connor and P. A. Garbinski, IEEE Electron. Dev. Lett. EDL-3, 152(1982).

51. C. Y. Chen, A. Y. Cho, K. Alavi and P. A. Garbinski, IEEE Electron. Dev. Lett. EDL-3, 205(1982).
52. T. P. Pearsall, R. Hendel, P. O'Connor, K. Alavi and A. Y. Cho, IEEE Electron Dev. Lett. EDL-4, 5(1983).
53. W. K. Chan, H. M. Cox, S. G. Hummel, P. S. Davisson and R. F. Leheny, IEEE Electron. Dev. Lett. EDL-6, 247(1985).
54. K. S. Seo, P. K. Bhattacharya and Y. Nashimoto, IEEE Electron. Dev. Lett. EDL-6, 642(1985).

# THE PHYSICS OF THE QUANTUM WELL LASER

J. Nagle and C. Weisbuch

Laboratoire Central de Recherches, Thomson CSF
B.P 10
Orsay, 91401 France

## INTRODUCTION

The Quantum Well Laser (QWL) has now a few well-established advantages as compared to the usual double-heterostructure (DH) laser.[1-7] In spite of an abundant litterature, very few cases have been documented to the level of the profound knowledge that we now have of the DH laser.[8,9] In this contribution, we will single out the main features of the QWL operation, in order to provide some guidance for new materials choices, in particular for the infrared.

## PRELIMINARY : THE COMPARISON BETWEEN QW AND DH LASERS

The operation of QWL's originates in two large effects at variance from the DH Laser :

( i) the number of available quantum states for carriers in the active region is reduced, leading to significantly smaller quasi-2D Density Of States (DOS) to be inverted to reach threshold.

(ii) the overlap of the stimulated optical wave and the active material layer diminishes with active layer thickness. The wave-guiding properties of double heterostructure materials, quite limited due to the small difference of index of refraction between the well and barrier material, extinguish at vanishingly small thickness of active layer material. The confining factor $\Gamma$ defined by :

$$\Gamma = \int_{-d/2}^{d/2} |E\ (z)|^2\ dz\ /\int_{-\infty}^{-\infty} |E\ (z)|^{\ 2}\ dz \qquad (1)$$

where d is the active layer thickness centered at z = 0 and E the electric field of the optical wave, diminishes as $d^2$ for single quantum wells.[8,9] In order to retain optical wave confinement while using quantum wells for carrier confinement, Tsang[10,11] used the Separate Confinement Heterostructures (SCH) depicted in Figure 1c and 1d, where layers with intermediate composition are used in order to confine the optical wave. In such structures, $\Gamma$ varies merely as d.

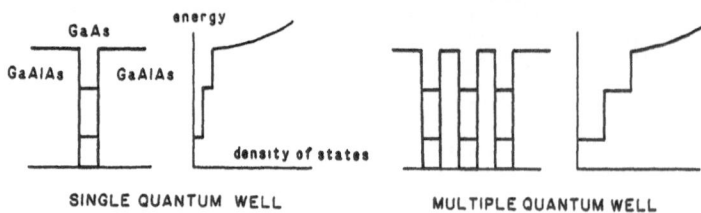

SINGLE QUANTUM WELL          MULTIPLE QUANTUM WELL

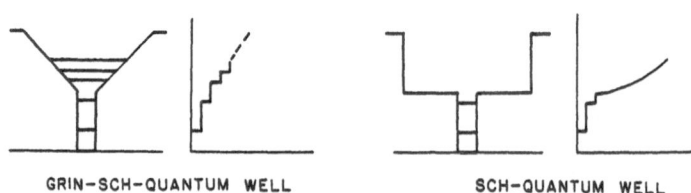

GRIN-SCH-QUANTUM WELL          SCH-QUANTUM WELL

FIGURE 1 : Various quantum well laser structures discussed in the
text schematically depicted by their conduction band
edge space variation and quantized energy levels (left-
side of each figure) and their 2D Density-Of-States(DOS)
(right side).
Top left : Single Quantum Well. Each quantized well state
introduces a 2D DOS equal to $m*/\pi\hbar^2$, while the onset of
3D states at the top of the well introduces a much larger
DOS.
Top right : Multiple Quantum Well (MQW) : Each quantized
state introduces a 2D DOS equal to $N\ m*/\pi\hbar^2$, N being the
number of wells.
Bottom left : Gradex-Index Separate Confinement Heterostructure (GRIN-SCH) : the Al graded- composition layers
adjacent to the quantum well provide good optical confinement. Note the ladder of quantum states in the graded
region, which correspondingly yields a ladder of $m*/\pi\hbar^2$
steps in the DOS.
Bottom right : Separate-Confinement Heterostructure
(SCH) : The intermediate-composition layers adjacent to
the quantum well provide good optical confinement, while
introducing a large DOS deleterious to laser operation.

These two effects of small number of states and $\Gamma$ tend to
cancel each other at first order. One has therefore to rely on
"second order" effects in order to discriminate the operating
features of the QWL when compared to the DH laser. These are :

( i) the quasi-2D density of states (DOS) that is associated to the
few $k_z$-quantum states confined in the well. This square DOS
helps to create finite gain even at low kinetic energy of
carriers. Therefore, when operating lasers at low temperatures
where band filling is small, one expects finite gain at the
bottom of the band for QWL's whereas the gain for DH lasers
tends to 0. Another advantage of the square DOS is that it
reduces the number of higher-lying states to populate to reach
a given population inversion at the bottom of the band. On the
other hand, the "price to pay" for the square DOS is a saturation of the gain value when bandfilling reaches unity,whereas
the gain in a DH laser increases with bandfilling due to the $\sqrt{E}$
increasing density of states.

(ii) the variation of $\Gamma$ with d in optimized SCH structures is slightly surlinear, due to the waveguiding effect of the quantum
well itself.

(iii) As will be discussed below, some subtle effects on matrix
elements and density of states can come into play : selection
rules at threshold of subbands increase the matrix elements
when compared to bulk crystals ; the splitting between heavy
and light holes diminishes the average density of states at the
top of the valence band.

In this paper, we will present the basic calculations in QWL's, then
the application to the GaAlAs and GaInAsP cases, finally we give
some hints of where the future might lie for narrow-gap
semiconductors.

BASIC CALCULATIONS IN QWL'S

Like in DH lasers[8,9], the basic quantities entering the operation of QWL's are the absorption coefficient $\alpha(E)$ (equivalent to a
gain $g(E)$ when the active medium is inverted) and the spontaneous
recombination rate $r_{sp}(E)$ :

$$\alpha(E) = \frac{\pi q^2 \hbar}{\varepsilon_0 m_0^2 c} \frac{1}{nE} \int_{-\infty}^{\infty} \rho_c(E')\rho_v(E'') |M(E',E'')|^2 [f_v(E'') - f_c(E')] \, dE'$$

(2)

$$r_{sp}(E) = \frac{4 \pi q^2}{\varepsilon_0 m_0^2 c^3 \hbar^2} nE \int_{-\infty}^{\infty} \rho_c(E')\rho_v(E'') |M(E',E'')|^2 f_c(E')[1-f_v(E'')] \, dE'$$

(3)

$$E'' = E' - E$$

where q is the electron charge, $m_0$ the free electron mass, $\rho_c(E)$ and
$\rho_v(E)$ are the density of states for electrons and holes, $f_c(E)$ and
$f_v(E)$ the Fermi-Dirac distribution functions for electrons and holes
with quasi-Fermi levels $F_e$ and $F_h$. $M(E',E'')$ is the interband dipole
matrix element. In equations (2) and (3) a summation over the various
allowed transitions between different bands must be made. We can
point out now some of the major factors entering the calculated
$g(E)$ and $r_{sp}(E)$.

Energy quantization

For quantum well lasers, the finite-height quantum well
calculations using the envelope wavefunction approximation yield
excellent agreement with experiment, given the uncertainties in
energy level determinations in QWL's. The QW wavefunctions are
therefore given by

$$\Psi_{e,h}(\vec{r}) = e^{i\vec{k}\perp \vec{r}\perp} \phi_{env}(z)\, u_{c,v,\vec{k}}(\vec{r}) \tag{4}$$

where $\vec{k}_\perp$ represents the transverse carrier momentum, $\vec{r}_\perp$ the transverse position, $u_{c,v,\vec{k}}$ is the usual rapidly varying Bloch wavefunction, $\phi_{env}(z)$ the envelope wavefunction (i.e. slowly varying on the unit cell scale) determined by the Schrödinger-like equation

$$\frac{-\hbar^2}{2m^*} \Delta \phi_{env}(z) + V(z)\phi_{env}(z) = E_{conf}\,\phi_{env}(z) \tag{5}$$

where $V(z)$ represents the potential energy of the band extremum under consideration. It is well established now that the boundary conditions are in the present case the continuity of $\phi_{env}(z)$ and $(m^*)^{-1}.d\,\phi_{env}/dz$ at the interfaces. In that approximation, the energy levels $E_{conf}$ are given by the implicit equation

$$tg\left(\frac{d\sqrt{2m_1 E_{conf}}}{2\hbar}\right) = \sqrt{\frac{m_1}{m_2}\left(\frac{\Delta V - E_{conf}}{E_{conf}}\right)} \tag{6}$$

where $m_1$ and $m_2$ are the effective masses on either side of the heterojunction and $V$ is the band extremum discontinuity.

Density of States and Occupation Factor

The usual 2D density of states $m^*/\pi\hbar^2$ leads to a 2D number of confined electrons given by

$$n_{2D}(T_e, F_e) = \int \rho_c(E) f_c(E) dE = \sum_i \frac{m^*(E_i) k\, T_e}{\pi\hbar^2} \ln\left[1 + \exp\frac{F_e - E_i}{k\, T_e}\right] \tag{7}$$

where $m^*(E_i)$ is the energy-corrected electron effective mass, $i$ represents the $i^{th}$ confined electron band, $F_e$ is the electron quasi Fermi level, $T_e$ is the electron effective temperature, $E_i$ the bottom of the $i^{th}$ band. One usually uses an energy dependent effective mass deduced from effective mass theory.

Similar equations hold for the two sets of hole bands, heavy and light. The situation is however strongly complicated due to the non-trivial band-mixing between hole states at non-zero kinetic energy. Such effects are usually neglected, although they can play a significant role in density of states and matrix elements. As density-of-states hole masses for transverse kinetic energy, we use the simplest Luttinger[13] approximation with no band-mixing, i.e.

$$m_{hh,t} = (\gamma_1 + \gamma_2)^{-1}\ ;\ m_{lh,t} = (\gamma_1 - \gamma_2)^{-1} \tag{8}$$

where $\gamma_1$ and $\gamma_2$ are the usual Luttinger coefficients. One usually assumes equal temperatures for electrons and holes.

## Interband matrix elements

The optical matrix elements M (E', E") are calculated in the usual manner where the slowly-varying envelope wavefunction can be factorized in the dipole matrix element and, being normalized to unity, does not contribute to the matrix element. It is then equal to the matrix element of the periodic part of the wavefunction.

Several Japanese[14-17] teams have evaluated the interband matrix element M (E', E"). They have shown that in the simple Kane approximation with no hole band mixing, the dipole matrix elements can be calculated against the usual direction-averaged bulk matrix element $< |M_o| >^2 = M^2$. For laser emission along the quantum well plane, they showed that the interband matrix elements are :

$$\text{Heavy-hole band} \qquad \text{TE} \quad M^2 = \frac{3}{4} M^2 (1 + \frac{E_{conf}}{\varepsilon}) \tag{9}$$

$$\text{TM} \quad M^2 = \frac{3}{2} M^2 (1 - \frac{E_{conf}}{\varepsilon})$$

$$\text{Light-hole band} \qquad \text{TE} \quad M^2 = \frac{1}{4} M^2 (5-3 \frac{E_{conf}}{\varepsilon}) \tag{10}$$

$$\text{TM} \quad M^2 = \frac{1}{2} M^2 (1+3 \frac{E_{conf}}{\varepsilon})$$

where $E_{conf}$ is given by equation (6) and $\varepsilon$ represents the total energy of a confined particle from the bottom of the bulk band, i.e.

$$\varepsilon = E_{conf} + \hbar^2 k_\perp^2 / 2 \, m^*$$

The values reported in equations (9) and (10) are due to the averaging of the relative orientations of the electron and hole wavefunction relative to the direction of light propagation, and should be strongly affected by the hole band mixing discussed above. Near confined heavy-hole band-edge, one deduces from (9) that the transition only occurs in the TE mode and has 1.5 times the oscillator strength of the bulk.

## Calculation of the current-density and gain-density relations

We have recently shown that the optical cavity of separate confinement structures may be significantly populated. Therefore, we inject in equation (3) (and its mirror for holes) an additional 3D band with the usual 3D density of states

$$\rho_{3D} (E) \approx 2^{\frac{1}{2}} m^{*3/2} \pi^{-2} \hbar^{-3} E^{\frac{1}{2}} \tag{11}$$

We note $n_{2Dwell}$, $n_{2Dcav}$ (resp. $n_{3Dwell}$, $n_{3Dcav}$) the 2D densities (resp. equivalent 3D densities) of electrons in the well and in the cavity and use similar notations for hole populations.

We use throughout the calculation $n_{2Dwell}$ as an input parameter. From it we deduce $F_e$ and $n_{total} = n_{2Dwell} + n_{2Dcav}$. Assuming electrical neutrality we can calculate $F_h$. Equations (2) et (3) yield $g(E)$ and $r_{sp}(E)$. We can then obtain the total radiative recombination rate R and the radiative current $J_{rad}$ required to maintain the electron density $n_{2Dwell}$ in the active material :

$$J_{rad} = qR = q \int r_{sp}(E)dE \qquad (12)$$

If necessary, other non-radiative recombination mechanisms depending on the electron density can be added to evaluate the injection current such as the Auger effect or the recombination current in the cavity. If $C_A$ and B are the Auger and cavity recombination coefficients respectively, these current are then given by :

$$J_A = qd\, C_A n_{3Dwell}\, p^2_{3Dwell} \quad \text{and} \quad J_{cav} = qd\, Bn_{3Dcav}\, p_{3Dcav}$$

## Laser threshold calculation

The laser threshold condition is defined as[8,9]

$$\Gamma g_{th} = \alpha + \frac{1}{L} \log \frac{1}{R} \qquad (13)$$

where $g_{th}$ is the material threshold gain (defined in $cm^{-1}$) for optical waves propagating along the waveguide, L is the cavity length and R the reflectivity of the laser facets (supposed equal here). The quantity $\alpha$ represents the loss due to all mechanisms in the structure which tend to attenuate the optical wave. It can be written as

$$\alpha = \Gamma\alpha_a + (1 - \Gamma)\, \alpha_c + \alpha_s + \alpha_e \qquad (14)$$

where $\alpha_a$ is the absorption loss in the active layer due for instance to free carrier absorption, $\alpha_c$ is the absorption loss in the confining and cladding layers, $\alpha_s$ is the optical scattering loss due to the imperfections of the waveguide and $\alpha_e$ represents the loss due to external layers when the optical wave extends beyond the cladding layer, which is normally not the case. A typical value for $\alpha$ is 10 $cm^{-1}$ in GaAs-GaAlAs lasers. The second term in equation (13) is typically 37 $cm^{-1}$ for a cavity length 300 µm and facet reflectivity R = 0.3.

In standard DH lasers[8,9], the detailed gain calculation from the quantum-mechanical evaluation of the stimulated emission rate leads to a rather linear relation between the maximum value of the gain curve $g_{max}$ and the injection current

$$g_{max} = \beta\, (J_{nom} - J_o) \qquad (15)$$

where $\beta$ is the so-called gain factor defining from material parameters the relation between carrier injection and gain assuming unit internal quantum efficiency, $J_{nom}$ is the nominal current density for

1 μm thick active layer and $J_o$ the value of $J_{nom}$ for which $g_{max}$ is extrapolated to zero. The threshold condition can therefore be written as

$$\Gamma g_{max,th} = \beta\Gamma \ (J_{th} - J_o) = \alpha + \frac{1}{L} \log \frac{1}{R} \qquad (16)$$

and, for an active layer thickness d expressed in microns and an internal quantum efficiency η representing the ratio of radiative to non-radiative recombination at threshold

$$J_{th} = \frac{J_o \ d}{\eta} + \frac{\alpha d}{\eta\beta\Gamma} + \frac{d}{\eta\beta\Gamma} \frac{1}{L} \log \frac{1}{R} \qquad (17)$$

This formula is well documented by the abundant experimental results in DH lasers. The common factor d leads to a slow decrease of the threshold with active layer thickness for large thicknesses while the fast decrease of Γ below ≈ 1000 Å leads to an increase of $J_{th}$ at lower values of d. It can be shown that, at small thicknesses, Γ takes the approximate value :

$$\Gamma = (\frac{d}{\lambda_o})^2 \ (2_\pi)^2 \ n \ \Delta n \qquad (18)$$

where $\lambda_o$ is the vacuum wavelength, n and Δn are the average and difference in the index of refraction of materials. In the case of GaAs imbedded in $Ga_{0.7}Al_{0.3}As$ confining layers, n = 3.6 and Δn = 0.2, which yields for d = 100 Å, Γ ≈ 0.004. When compared to the value of Γ = 0.28 for a 1000 Å thick DH laser, this means that the medium gain g in equation (2) must be 100 times larger in a single QWL than in a similar DH laser. If a separate confinement heterostructure is used, it can be shown that the confinement factor varies linearly as

$$\Gamma_{SCH} \approx 3.10^{-4} \ d(Å) \qquad (19)$$

for a $GaAs/GaAl_{0.18}As/GaAl_{0.40}As$ SCH structure, chosen as to optimize the value of the optical wave at the center of the waveguide where the quantum well is placed. In the case of a 100 Å QW, $\Gamma_{SCH}$ ≈ 0.03, significantly larger than in a SQWL. Analysis of the Graded-Index SCH laser gives similar improvement of Γ over SQWL's.

The simple relation (15) between gain and injection current exists in QWL's only in a limited range of current injection. Numerical calculations evidence strong non-linearities due to the saturating behaviour of the QW-DOS. This points out that great care must be exercised in the evaluation of various experimental data in quantum wells : under modes of operation which appear as rather "weak excitation" in the usual 3D case, one actually deals with very large filling effects in the 2D-QW case. The gain curve is shown in figure 2 for SQW lasers and multiple QW lasers. The main features of the 2D behaviour of QWL's appear clearly :

- the modal gain $\Gamma g$ is significant even at extremely low injection.
- as $\Gamma \approx 0.03$, one can deduce an approximate value of $\beta \approx 0.11$ cm$^{-1}$/A.cm$^{-2}$.$\mu$m assuming a linear (!) variation of g with the current injection . This is to be compared with $\beta \approx 0.045$ cm$^{-1}$/A.cm$^{-2}$.$\mu$m in standard DH lasers. This is due to the fact that gain increases quickly with injected carriers for smaller active regions.
- One clearly observes the saturation of the gain for the single quantum well. The simplest way to increase the saturated gain value is to use multiple quantum well structures as depicted in Fig. 1b, either in the straight MQW form represented or in a SCH configuration with MQW's located inside a confining cavity for optical waveguiding, so-called Modified MQW (MMQW). In first order the N-Well MQW structure has a DOS which is N times that of the SQW, i.e. $N(m^*/\pi h^2)$. The saturated gain value is increased by that amount, but it is also the case for the confining factor and for the number of quantum states to be inverted at threshold. This last effect increases the $J_0$ factor in equation (15), whereas the two compensating effects of $\Gamma$ and DOS would lead to equivalent $\beta$'s (eq.(15)) for SQW's and MQW's. Band filling effects (i.e. saturation) actually yield higher $\beta$'s for MQW's, as can be seen in fig.2.
- depending on the required modal gain at threshold, one therefore needs a SQW (at low loss) or MQW to optimize the threshold current as shown in figure 2.
- the steepness of the square DOS leads to larger differential gain (dg/dn) and smaller linewidth enhancement factors. Those two factors should potentially lead to QWL's with higher-modulation bandwidth and narrower spectrum than DH lasers.[22]

A main operating parameter of semiconductor lasers is the temperature dependance of their threshold current, usually approximated by the formula $J_{th}(T) = J_{th}(T_0) \exp (T/T_0)$. The $T_0$ in the standard theory developed in equations (2) - (3) originates in the change of the quasi-Fermi levels with temperature : to reach the threshold gain at a higher temperature, one requires a higher carrier density, as the Fermi-Dirac distribution is smoothened out. The calculation yields $T_0$ = 220 K in usual DH heterostructures, which is actually what is measured. The square DOS in 2D QW lasers tends to improve the $T_0$ factor, as the high energy tail of the distribution function creates less population (unused in the gain increase) because of the constant DOS instead of the $\sqrt{E}$ increasing DOS in the 3D case. This however is reversed in the SCH laser with a thin well where the quasi-Fermi level for electrons can approach the confining layer level : then a small thermally-induced smoothing of the distribution function leads to a large increase in the population of the confining layer. Such lasers have then a poor $T_0$. The GRIN-SCH laser is of course much better in that respect due to the gradual increase of the DOS with energy.

Several authors [16-24] have defined more or less simplified models to carry out the optimization of threshold current and/or $T_0$ as a function of number of wells, well thickness, cavity length ... As will become clear below, however, we still lack enough detailed knowledge of a variety of phenomena (non-radiative recombination processes, material variations, band tailing, etc.) in order to produce useable QWL design rules.

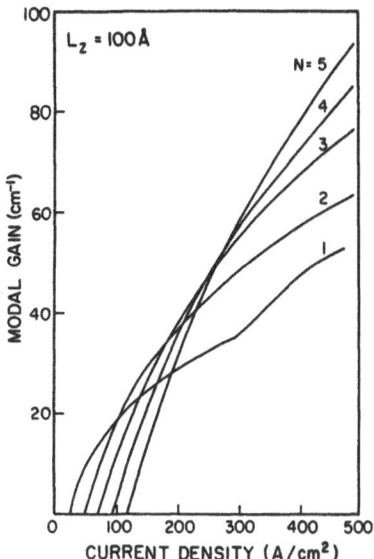

FIGURE 2 : The modal gain  (= $\Gamma g$) variation as a function of the injected  current  density  J with the various number of GaAs  quantum wells N. In  this  case, the quantum well thickness d is assumed to be 100 $\overset{\circ}{A}$ (From Arakawa and Yariv[22] )

THE GaAs QUANTUM WELL LASER

The GaAs QWL's are now the most widely documented[1-7, 25,26] as such lasers  are superior to DH lasers in  threshold  current,  $T_o$, power handling, operating lifetime etc.

A variety of multiple quantum wells have been studied,  with separate cavities or not.  It appears that,  although MQW's are not the optimal choice for threshold considerations,  .they have a very high power handling capability and the highest operating frequency, as in that latter case one uses short cavity lasers with short photon lifetimes and as high differential gain dg/dI as possible. Figure 3 represents measured differential gains and evidences the advantage of MQW's over SQW's and of the latter over DH lasers.

The anisotropy of the gain has been discussed by several authors [14-17] along formulas ( 9 ) and (10),  see figure 4.  The spectral analysis of gain curves below threshold has been performed along the method of  Hakki and Paoli[18] : one  measures  the electro-luminescence of the laser diode at high spectral resolution.  The modulation of the spontaneous emitted light along the cavity direction is due to the cavity modes,  the amplitude being dependent on the attenuation of the spontaneous light as it travels back and forth in the cavity.  Hakki and Paoli showed that :

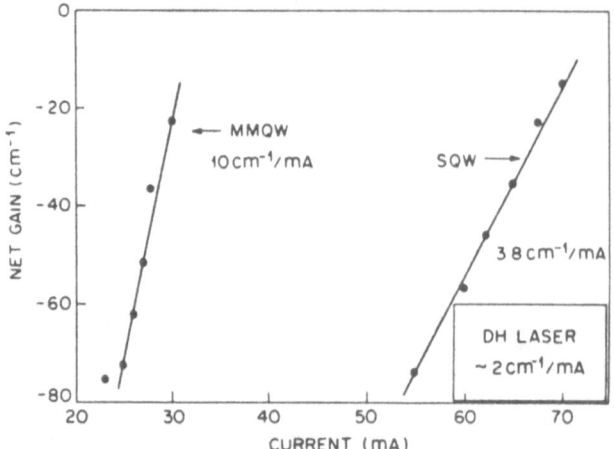

FIGURE 3 : The net optical gain ($\Gamma g - \alpha - (1/R)(\ln 1/R)$) as a function of injection current for a single quantum well and
a modified multiquantum well heterostructure laser. The
insert recalls the value for usual DH lasers, i.e.
2cm-1/mA (From Tsang[1] ).

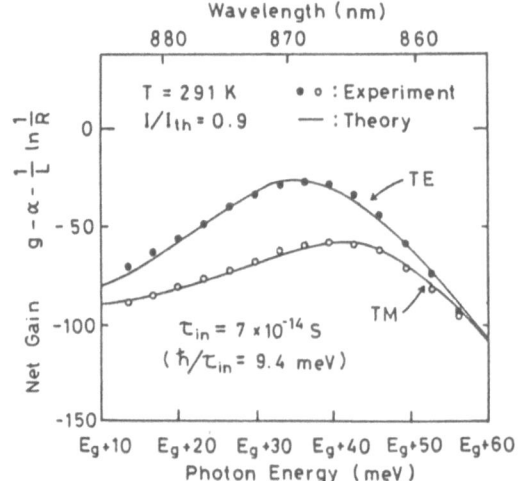

FIGURE 4 : Comparison between the calculated and the measured net
gain profiles. The calculated profiles coincide well with
the experimental data when broadening of the gain curve
due to a relaxation time in the range $(0.7 \approx 1.0) \times 10^{-13}$ s
is introduced (From Yamada et al.[17] ).

$$\Gamma g(E) - \alpha - \frac{1}{L} \ln \frac{1}{R} = \frac{1}{L} \ln \frac{\sqrt{I_{max}(E)} - \sqrt{I_{min}(E)}}{\sqrt{I_{max}(E)} + \sqrt{I_{min}(E)}} \qquad (20)$$

where $I_{max}(E)$ and $I_{min}(E)$ are the maximum and minimum intensities in the Fabry-Perot modulated spectrum.

- Such measurements are shown in figure (5) for a series of SQW GRINSCH lasers, along with the calculated gain curves. The theory reproduces well experiment, i.e. wider gain curves and larger high-energy slopes with diminishing well thickness. These effects, in contradiction with the early descriptions of quantum well laser operation, are due to the increasing population of excited quantum well states.

Several authors have reported that minimum thresholds are obtained with GRIN-SCH and SCH structures. This has been recently explained by the very low number of carriers required to fulfill equation (16), while the superiority of the GRIN-SCH QWL was shown to be due to the smaller DOS of its light-confining cavity than in the SCH.[27]

The QWL structure with its ultra-thin active layers allows complete thermally-induced atomic species interdiffusion.[28] It has been thoroughly shown that a number of impurity species activate this interdiffusion. Through localized impurity implantation and rather low temperature annealing, it has been possible to completely interdiffuse the implanted regions while retaining the high-quality of the unmodified material. Such a treatment appears highly promising for low-cost processing of gain-guided buried heterostructure lasers thanks to the lower refractive index of the interdiffused region (alloy) as compared to the high GaAs index of the active layer. The interdiffusion scheme has already been used in window lasers to protect the mirror facets from recombining electron-hole pairs and thus prevent catastrophic damage at high powers.

QWL's allow to operate at short wavelengths by reducing the QW thickness. However, at the shortest possible wavelengths allowed in the GaAs-GaAlAs material system, the quantum well width becomes very small, of the order of 20 Å. Then, the interface fluctuations (usually at least one monolayer i.e. $\approx$ 3 Å) induce an important inhomogeneous broadening mechanism which reduces QW laser gain. Furthermore, the first confined level approaches the cavity level implying large cavity recombination currents. Saku et al.[29] have shown that a good compromise between 2D effects and quantum well disorder can be obtained by using GaAlAs quantum wells imbedded in higher-Al content GaAlAs barriers. The result of their optimization is shown in figure 6.

The GaInAsP/InP QUANTUM WELL LASER

Whereas the superiority of SQW's and MQW's is now well established in the GaAs/GaAlAs system, the opposite is true in the GaInAsP/InP system. A major issue in QWL's is therefore to understand this outstanding difference : up to now, there are only two reported instances of laser action in a GaInAs SQWL,[30,57] and all the reported MQWL's have features which are not better than those of DH lasers.[31-35]

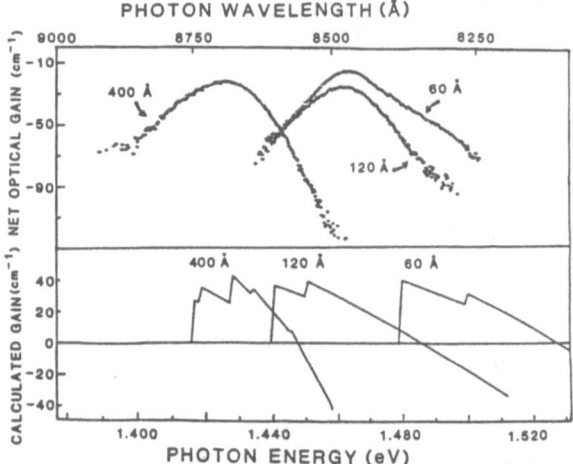

FIGURE 5 : Measured net optical gain curves (top) and corresponding calculated modal gain ($\Gamma g$) curves (bottom) for
three GRIN-SCH GaAlAs-GaAs quantum well lasers with
active layer thickness 400,120 and 60 Å. No broadening
parameter has been introduced, which explains the absence
of low-energy tails in calculated curves. The relevant
features are the high-energy slopes.(From Nagle et al.[27])

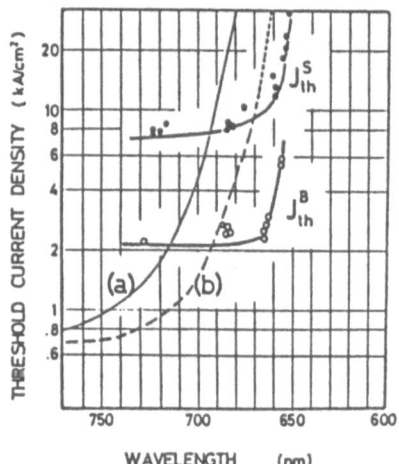

FIGURE 6 : Threshold current density at 300 K as a function of
lasing wavelength for AlGaAs laser diode. Curve (a)
is the lowest $J_{th}$ realized so far by conventional DH
laser diodes with $Al_x Ga_{1-x}As$ active layer. Curve (b)
is the result of MOCVD grown SQW laser with 40-60 nm
QW layer. GaAlAs SCH-MQW results with well Al content from 0.15 to 0.35 are indicated by open and
closed circles. Open circles broad area diodes.Closed
circles : stripe geometry diodes (From Saku et al.[29]).

The case of the GaInAsP/InP standard DH laser has been the source of numerous studies[36-43] : although state of the art GaInAsP DH lasers have threshold current densities quite similar to those of GaAs DH lasers, their $T_O$ is much worse. The $T_O$ question in GaInAsP/InP lasers has long been a matter of controversy but there seems now to be a consensus about the following points[39-41]:

- for optimum lasers ($d \approx 0.15$ µm), the threshold current can be as low as 670 A.cm$^{-2}$. The typical values of $T_O$ are 60-80 K. The density of carriers at threshold is $\simeq 2.10^{18}$cm$^{-3}$
- the threshold current has a radiative component and several non-radiative components. The radiative component is $\approx$ 60 % at 300 K.[39] The main non radiative component is due to Auger recombination, with smaller contributions due to inter-valence band light absorption and thermal carrier leakage from the active region. The Auger recombination process is of the CHSH type, i.e. an electron and a heavy-hole recombine by exciting a heavy-hole to the spin-split-off valence band
- the Auger coefficient value is $C_A$= 4.10$^{-29}$cm$^6$s$^{-1}$ and 2.10$^{-29}$cm$^6$s$^{-1}$ for GaInAs and 1.3 µm GaInAsP material respectively.

We have performed gain measurements in SCH-SQW lasers[44]. The astonishing result is the diminishing maximum gain with increasing injection current which prevents QWL operation. At the same time the gain curve broadens. Observing the smoothened-out luminescence (widened spectrometer slits), we observe a heating of the carriers (as seen from the high energy slope of the 1.5 µm emission line) and a significant population of the optical confinement cavity at 1.3 µm (figure 7). Eventually, lasing is obtained at 1.3 µm for a current of 450 mA. Such experimental data can only be explained if we assume carrier heating with increasing injection, which has been observed in some DH lasers[38], but was recently disputed[39]. We however cannot explain otherwise the decrease in gain with increased injection.

A central issue is the value of the Auger coefficient in quantum wells. There is an abundant literature on the theoretical evaluation of the various possible processes in 3D[45], but there seems to be a consensus now about the dominance of the CHSH process. At some point, the Auger effect in QW's has been predicted to be much weaker than in 3D[47] but consensus is now that it should be comparable[48,49], as is also the experimental result by Sermage et al.[50]

Sugimura[49] and Asada et al.[16] have produced detailed calculations where they can optimize separately $I_{th}$ and $T_O$ for the various types of QW lasers in the GaInAsP/InP system. However, the values found for the threshold current are significantly lower than those obtained up to now in MQW InGaAsP/InP lasers.

EXTENSION TOWARD NEAR-INFRARED-LASERS (2 - 10 µm)

The spectral region 2 - 10 µm has recently become much more widely studied than before. Early studies in the IV-VI compounds have shown that the lead chalcogenides could give efficent low-temperature lasers in the 4 - 12 µm range. Such lasers have been commercially available for a while and have been mainly used in

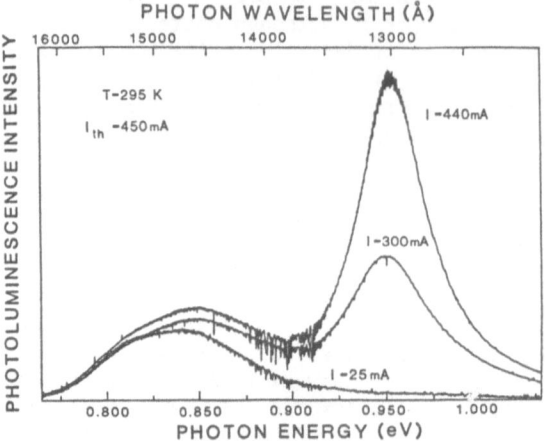

PHOTON WAVELENGTH (Å)

T-295 K

$I_{th}$ -450mA

I -440mA

I ~300mA

I ~25mA

PHOTON ENERGY (eV)

FIGURE 7 : Spontaneous emission of GaInAs SCH-QWL (both
polarisations). Relative intensities are significant. Accidents of the curves around 0.9
eV due to atmospheric water absorption. Quantum
well width d = 50 Å. Optical cavity width 3000
Å with lattice-matched GaInAsP at 1.3 μm (From
Nagle et al.[44]).

high-resolution gas spectroscopy and pollution monitoring. Recently,
the development of low-loss infrared fiber optics beyond 2 μm opens
the market of very-high distance repeaterless data communication.

The various materials which should allow laser emission in
the 2 - 10 μm range are III-V's (In and Sb-related materials), IV-
VI's (Pb-related materials) and II-VI's (Hg-related materials). The
field of homostructure and DH lasers in that range has been recently
excellently reviewed by Horikoshi.[51] The main result of his thorough
study is shown in figure 8 , where one sees that the threshold
currents increase so dramatically with temperature that no laser can
now operate C.W. at room temperature beyond 2 μm. As the detailed
investigations reported by Horikoshi tend to point to the intrinsic
character of this limit, it is of utmost interest to evaluate whether
quantum well heterostructures would improve the situation. We have
not yet performed a case by case quantitative analysis of the
possibilities, but we give some hints of what parameters are of
importance in order to select a materials system.

Let us first point out the basic effects in the GaAlAs and
GaInAsP systems. Due to the asymmetry between electron and hole
bands the electron quasi-Fermi level is high in the band. The
electron inversion is therefore always high and the inversion
required for net gain originates in the precise position of the hole
quasi-Fermi level around the hole band extremum. The high quasi-
Fermi level for electrons induces significant populations in the 3D

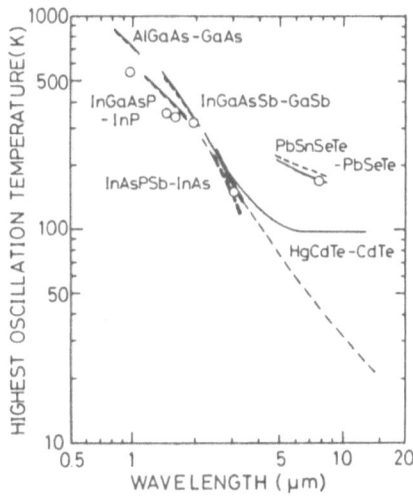

FIGURE 8 : High-temperature limits for laser oscillation
in DH lasers with various materials under
pulsed operation ; curves represent the
temperatures at which the calculated internal
quantum efficiency falls to 2.5 % for each DH
material due to non-radiative processes :
results (---) for nondoped active-region lasers
and (—) for lasers with p-type active regions
($p = 2 \times 10^{17} cm^{-3}$). Experimentally observed
highest lasing temperatures with 100-nsec-long
current pulses are given for DH lasers made
from various materials (From Horikoshi[51]).

DOS's of confining layers. One therefore has to take into account
with precision the relative carrier masses and band discontinuities
in order to carry out quantitative performance evaluations. The good
performance of 2D GaAlAs lasers is therefore due to :

- the 2D DOS
- the good confining factor of GRIN-SCH's
- the large quantum yield of QW structures,(although this univer-
  sally observed factor is hard to evaluate quantitatively)
- the larger matrix elements in QW's

and some other effects more difficult to quantify such as the
reduced free carrier absorption due to small confinement factor,
reduced scattering loss by interface disorder etc.

All these effects lead to 2D electron population at threshold
smaller by $\approx$ 50 % in the optimal 60 Å GRIN-SCH-QW laser than in the
best GaAs DH laser.

In GaInAsP QWL's, the confinement factors are smaller as are
the electron effective masses. Band filling effects are therefore
more important and the equivalent 3D population is high. Carrier

heating, presumably by the Auger effect, then leads to a thermal runaway effect whereby additional injected carriers heat up the carrier distribution so efficiently that the gain diminishes and the confining layer gets heavily populated. The saturated value of the 2D gain is not the limiting factor as it can be large enough to overcome losses as evidenced by the excellent lasing characteristics of GaInAsP SQW lasers at low temperatures.[44]

One must be aware that the positive effects of the bidimensionality may be lost as soon as new components of the current arise. For example radiative or non radiative recombination of carriers in the cavity of separate confinement structures may become important due to high band filling (this effect will be most noticeable for small wells and/or for systems with low effective mass or small band discontinuities). Another recombination path is non-radiative Auger process especially in small-gap materials. Auger recombination current may become dominant when the equivalent 3D carrier populations at threshold are large i.e. for smaller wells again or for lasers with high losses. As already mentioned, Auger effect can have some further detrimental effect by heating the carrier distribution, therefore also contributing to band filling and raising of the threshold population.

The strong departures from ideal 2D temperature dependence are mostly due to those strongly temperature-dependent extra-components of the injection current. One way to reduce them is to use MQW instead of SQW, somewhat lowering band filling and threshold population.

This analysis of the GaAlAs and GaInAsP cases shows the delicate interplay of various factors, which require for quantitative evaluation the detailed knowledge of many semiconductor parameters. Those are often not now available in the IR domain, as is the case for bandgap discontinuities for instance. We can tentatively list the various factors entering the operation of IR QW lasers with indicated level of plus or minus character (indicated on a self explanatory scale from +++ to ---) :


+++ In equations (2) and (3) the nE factors play in opposite direc- tions. Therefore one expects an improvement proportional to $(nE)^2$, which can be $\approx 40$.

--- The Auger effect in small gap materials is shown to be much more important than in wider-gap materials[52-54]

-(-) The electron masses being small, band-filling effects can be more important ; the saturated-gain value is smaller, especial- ly when the hole mass is also small (chalcogenides).

 + This effect of gain saturation is partly compensated for chal- cogenides by the existence of many valleys that can help to obtain a significant DOS

+(-) The matrix element can be as large as for large gap materials (III-V's), but can also be smaller (chalcogenides)

 - As the confinement factor depends on the wavelength (eq. 18) all dimensions scale like it in the calculation of the confine- ment factor. Therefore, if differences of indices of refraction are similar, one will need thicker active layers where quantum effects might diminish, in spite of the lighter masses.

 ++ Bandgap discontinuities can be large, which helps to approach real 2D density of states

As can be seen from the preceeding list, only a careful selection of materials and design parameters could lead to optimal QW laser design in the IR range. In view of the extreme importance of the Auger effects, efforts must certainly be focused on the minimization of the equivalent 3D density in order to diminish Auger losses.The excellent results of Partin[55] tend to prove that in the PbTe case 2D operation is beneficial. An additional possibility of QW structures is to induce size-quantized (or strain induced in the case of Strained Layer Multi-Quantum Wells)modifications of the band structure which can lead to extremely significant diminutions of Auger processes as proposed by Adams.[56]

## ACKNOWLEDGEMENTS

The authors wish to thank M. RAZEGHI, S. HERSEE, T. WEIL, B. VINTER, B. de CREMOUX for useful discussions.

## REFERENCES

1 - W.T. TSANG, IEEE J. Quantum Electronics QE-20, 1119 (1984)

2 - L.J. VAN RUYVEN, J. Luminescence 29, 123 (1984)

3 - R.D. BURNHAM, W. STREIFER and T.L. PAOLI, J. Crystal Growth 68, 370 (1984)

4 - N. HOLONYAK Jr., R.M. KOLBAS, R.D. DUPUIS and P.D. DAPKUS, IEEE J. Quantum Electron. QE-16, 170 (1980)

5 - B. de CREMOUX, Proc. of ESSDERC 85, Studies in Electrical and Electronic Engineering 23, Solid State Devices 85, Ed. by P. BALK and O.G. FOLBERTH, Elvesier (1986), p. 83

6 - W.T. TSANG, in Semiconductors and Semimetals, Vol. 22A, eds R.K WILLARDSON and A.C. BEER, Academic, New York (1985), p. 95

7 - N. HOLONYAK Jr. and K. HESS in Synthetic Modulated Structures, L.L. CHANG and B.C. GIESSEN eds., Academic, New York (1985)

8 - H.C. CASEY Jr. and M.B. PANISH, Heterostructure lasers, Academic, New York (1978)

9 - G.H.B. THOMPSON, Physics of Semiconductor Laser Devices, Wiley, New York (1980)

10 - W.T. TSANG, Electron. Letters 16, 939 (1980)

11 - W.T. TSANG, Appl. Phys. Lett. 39, 134 (1981)

12 - D.A. BROIDO and L.J. SHAM, Phys. Rev. B 31, 888 (1985)

13 - D.S. CHEMLA, Helvetica Physica Acta 56, 607 (1983)

14 - H. KOBAYASHI, H. IWAMURA, T. SAKU and K. OTSUKA, Electron. Lett. 19, 166 (1983)

15 - H. IWAMURA, T. SAKU, H. KOBAYASHI and Y. HORIKOSHI, J. Appl. Phys. 54, 2692 (1983)

16 - M. ASADA, A. KAMEYAMA and Y. SUEMATSU, IEEE J. Quantum Electron. QE-20, 745 (1984)

17 - M. YAMADA, S. OGITA, M. YAMAGISHI and K. TABAKA, IEEE J. Quantum Electron. QE-21, 640 (1985)

18 - N.K. DUTTA, J. Appl. Phys. 53, 7211 (1982)

19 - M.G. BURT, Electron. Letters 19, 210 (1983)

20 - D. KASEMSET, CHI-SHAIN HONG, N.B. PATEL and P.D. DAPKUS, IEEE J. Quantum Electronics QE-19, 1025 (1983)

21 - A. SUGIMURA, IEEE J. Quantum Electronics QE-20, 336 (1984)

22 - Y. ARAKAWA and A. YARIV, IEEE J. Quant. Electron. QE-21, 1966 (1985)

23 - P.W.A. McILROY, A. KUROBE and Y. UEMATSU, IEEE J. Quantum Electronics QE-21, 1958 (1985)

24 - B. SAINT-CRICQ, F. LOZES-DUPUY and G. VASSILIEFF, IEEE J. Quantum Electronics QE-22, 625 (1986)

25 - S.D. HERSEE, M. BALDY, P. ASSENAT and B. de CREMOUX, Electron. Letter 18, 870 (1982)

26 - S.D. HERSEE, M. RAZEGHI, R. BLONDEAU, M. KRAKOWSKI, B. de CREMOUX and J.P. DUCHEMIN, IEDM 83 Technical Digest, p. 288 ; see also M. BALDY, S. HERSEE and P. ASSENAT, Rev.Techn.THOMSON CSF 15,5 (1983)

27 - J. NAGLE, S. HERSEE, M. KRAKOWSKI and C. WEISBUCH, (to be published)
28 - See e.g. ref. 3 and further references therein.
29 - T. SAKU, H. IWAMURA, Y. HIRAYAMA, Y. SUZUKI and H. OKAMOTO,
     Jap. J. Appl. Phys. $\underline{24}$, L 73 (1985)
30 - J. NAGLE, M. RAZEGHI and C. WEISBUCH, unpublished
31 - H. TEMKIN, K. ALAI, W.R. WAGNER, T.P. PEARSALL and A.Y. CHO,
     Appl. Phys. Letters $\underline{42}$, 845 (1983)
32 - T. YANASE, Y. KATO, I. MITO, M. YAMAGUCHI, K. NISHI,
     K. KOBAYASHI and R. LANG, Electronics Letters $\underline{19}$,701(1983)
33 - N.K. DUTTA, S.G. NAPHOLTZ, R.T. YEN, R.L. BROWN, T.M. SHEN
     and N.A. OLSSON, Electronics Letters $\underline{20}$, 727 (1984)
34 - Y. SASAI, N. HASE and T. KAJIWARA, Jap. J. Appl. Phys. $\underline{24}$,
     L 137 (1985)
35 - N.K. DUTTA, S.G. NAPHOLTZ, R. YEN, T. WESSEL, T.M. SHEN
     and N.A. OLSSON, Appl. Phys. Lett. $\underline{46}$, 1036 (1985)
36 - See e.g. the reviews by R.J. NELSON and N.K. DUTTA, in
     Semiconductors and Semimetals, eds. R.K. WILLARDSON and
     A.C. BEER, vol. 22C, volume editor W.T. TSANG,
     Academic, Orlando (1985), p. 1 ; Y. HORIKOSHI, in GaInAsP
     Alloy Semiconductors, ed. T. P. PEARSALL, Wiley, New York
     (1982) p. 379.
37 - N.K. DUTTA and R.J. NELSON, J. Appl. Phys. $\underline{53}$, 74 (1982)
38 - M. ASADA and Y. SUEMATSU, IEEE J. Quantum Electronics
     $\underline{QE-19}$, 941 (1983)
39 - C.H. HENRY. R.A. LOGAN, H. TEMKIN and F.R. MERITT, IEEE
     J. Quantum Electronics $\underline{QE-19}$, 941 (1983)
40 - C.H. HENRY, B.F.LEVINE, R.A. LOGAN and C.G. BETHEA, IEEE
     J. Quantum Electronics $\underline{QE-19}$, 905 (1983)
41 - C.H. HENRY, R.A.LOGAN, F.R. MERITT and J.P. LUONGO, IEEE
     J. Quantum Electronics $\underline{QE-19}$, 947 (1983)
42 - R. OLSHANSKY, C.B. SU, J. MANNING and W. POWAZINIK, IEEE
     J. Quantum Electronics $\underline{QE-20}$, 838 (1984)
43 - A.P. MOZER, S. HAUSSER and M.H. PILKUHN, IEEE J. quantum
     Electronics $\underline{QE-21}$, 719 (1985)
44 - J. NAGLE, S. HERSEE, M. RAZEGHI, M. KRAKOWSKI, B. de CREMOUX
     and C. WEISBUCH, Surface Science, to be published
45 - B. ETIENNE, J. SHAH, R.F. LEHENY and R.E. NAHORY, Appl. Phys.
     Letters $\underline{41}$, 1018 (1982)
46 - See e.g. ref. 36 and further references therein.
47 - L.H. CHIN and A. YARIV, IEEE J. Quantum Electronics
     $\underline{QE-18}$, 1406 (1982)
48 - R.I. TAYLOR, R.A. ABRAM, M.G. BURT and C. SMITH, IEE
     Proc. J., Optoelectronics, $\underline{132}$, 364 (1985)
49 - A. SUGIMURA, IEEE J. Quantum Electronics $\underline{QE-19}$, 932 (1983)
50 - B. SERMACE, D.S. CHEMLA, D. SIVCO and A.Y. CHO, IEEE
     J. Quantum Electronics $\underline{QE-22}$, 774 (1986)
51 - Y. HORIKOSHI, in Semiconductors and Semimetals, eds. R.K.
     WILLARDSON and A.C. BEER, vol. 22C, volume editor by
     W.T. TSANG, Academic, Orlando (1985), p. 93
52 - A. SUGIMURA, IEEE J. Quantum Electronics $\underline{QE-1}$, 352 (1982)
53 - B.L. GELMONT, Z.N. SOKOLOVA and I.N. YASSIEVITCH, Fiz. Tekh.
     Poluprovodn. 16, 592 (1982)  Sov. Phys. - Semicond. $\underline{16}$, 382
     (1982)
54 - B.L. GELMONT and Z.N. SOKOLOVA, Fiz. Tekh. Poluprovodn. $\underline{16}$,
     1670 (1982)  Sov. Phys. - Semicond. $\underline{16}$, 1670 (1982)
55 - D.L. PARTIN, Appl. Phys. Letters $\underline{45}$, 487 (1984)
56 - A.R. ADAMS, Electronics Lett. $\underline{22}$, 250 (1986)
57 - M.B. Panish, H. TEMKIN and S. SUMSKI, J. Vac. Sci. Technol.
     $\underline{B3}$, 657 (1985)

RAMAN SCATTERING AT INTERFACES

Gerhard Abstreiter

Physik-Department, Technische Universität München
D-8046 Garching
Federal Republik of Germany

## INTRODUCTION

Inelastic light scattering is a powerful probe of various properties
of semiconductors, heterostructures, and superlattices. Three types of
excitations have been applied. Allowed phonon scattering is used to obtain
information on composition, structure, orientation, periodicity and built-
in strain in semiconductor thin layer structures. "Forbidden" LO- phonon
scattering in polar semiconductors is sensitive to internal electric fields
and barrier heights. The study of the formation of semiconductor heterostructures from clean surfaces to overlayers with a thickness of several
hundred Angstroms is possible. The most widely studied properties of low

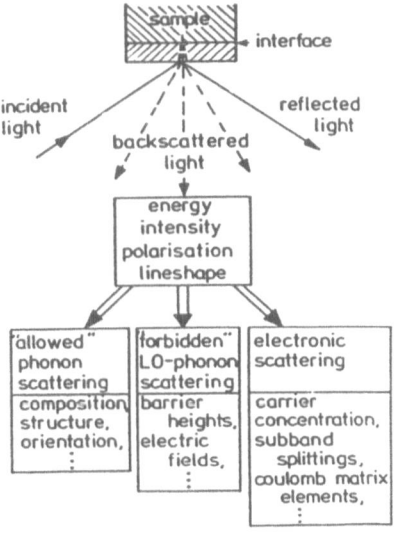

Fig. 1.    Schematics of inelastic light
scattering

dimensional systems are the elementary excitations of two-dimensional electron and hole gases at semiconductor interfaces and in superlattices[1]. Light scattering became a multi-purpose experimental technique which goes far beyond its power as a spectroscopic tool that yields just the energies of electronic or vibronic excitations. The essential information which can be extracted from the analysis of back scattering light is shown schematically in the block diagram of Fig. 1. In the present paper we present a comprehensive overview of the various possible excitations in low-dimensional systems. Emphasis is put on electronic excitations of two-dimensional carrier systems. The possibilities of applying this technique to narrow-gap semiconductors is briefly discussed. The paper ends with a concise description of specific phonon aspects in low dimensional systems. For more details the reader is referred to the many review articles which appeared recently in the literature for each subject.

ELECTRONIC LIGHT SCATTERING

It has been shown that electronic light scattering in semiconductors under resonance conditions is sensitive enough to observe the elementary excitations of two-dimensional carrier systems which are confined at semiconductor interfaces or surfaces [2,3]. For resonances the photon energies are close to certain band gap energies, where optical transitions are associated by states occupied with free carriers. The first observations of resonant light scattering in high mobility two-dimensional electron gases were published for GaAs-Al$_x$Ga$_{1-x}$As heterostructures [4,5]. This pioneering work was followed by a large variety of experiments which are reviewed in [1,6,7,8]. The light scattering mechanisms found for two-dimensional systems are similar to those of three dimensional systems [9]. Large resonance enhancements are observed in GaAs around the E$_o$ and the E$_o$+$\Delta_o$ energy gap. Their energies are about 1,5 eV and 1,9 eV. In quantum well structures these energy gaps are modified slightly due to shifted subband energies. Carriers are quantized in one direction and are free, within the effective mass approximation, in parallel direction. The dispersion of two subbands in k$_{\shortparallel}$ direction and possible excitations are shown schematically in Fig. 2. The scattering wave vector is determined by the scattering geometry and the photon energy. In back-scattering geometry, which is usually applied in opaque semiconductors, $q \approx 4\pi n/\lambda$ , where $n$ is the refractive index. The high refractive index of the studied semiconductors usually leads to $q_\perp \gg q_{\shortparallel}$, where $q_\perp$ is the component of the scattering wave vector normal to the layers and $q_{\shortparallel}$ is the in-plane component. The ratio is varied by the angle of incident and scattered photon wave vector with respect to the surface. In ideal back-scattering geometry only intersubband excitations are observable. A coupling to intrasubband (in-plane) excitations is only possible with finite $q_{\shortparallel}$. This is shown schematically in Fig. 3, where the energy is plotted versus wave vector $k_{\shortparallel}$. Inter- and intrasubband excitations split into two types, which are either of single-particle or collective nature. We first consider only the inter-subband excitations.

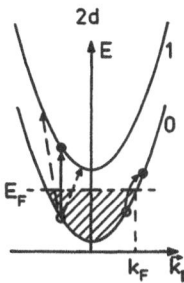

Fig. 2.    Electronic inter- and intrasubband excitations in a two-dimensional system (schematically)

Excitations from a lower occupied subband to a higher unoccupied subband reveal approximately the subband splitting (see Fig. 2). With increasing carrier density such excitations are screened by the collective behaviour of the total carrier system. The measured excitations are shifted to higher energies which is often referred to as depolarization shift. It reflects the three dimensional nature due to the finite extension of the carrier system in the direction of quantization. Band structure effects make it possible to observe single-particle excitations also in the high density case. Optical transitions, which involve for example the spin-orbit split-off valence band can lead to scattering via spin density fluctuations. They are observed as depolarized spectra. The polarizations of incident and scattered light are perpendicular to each other. The spectra are proportional to the imaginary part of the dielectric function of the electron gas. Neglecting small corrections due to excitonic effects, peaks appear at energies which correspond to the bare subband splittings of the two-dimensional system ($\omega_{01}$ in the example shown in Fig. 3).

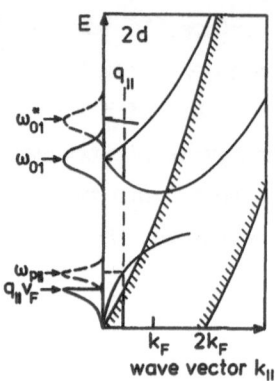

Fig. 3.  Energy versus wavevector of single
         particle and collective excitations
         in two-dimensional carrier systems

The collective, depolarization shifted spectra are observed in parallel
polarisations. The upward shift is given by an effective plasma frequency
$\omega_P^*$ such that $\omega_{01}^{*2} = \omega_{01}^2 + \omega_P^{*2}$. $\omega_P^*$ depends on the difference in carrier occupation of the two subbands $n_o \cdot n_1$, the bare energy separation $\omega_{01}$ and the
Coulomb integral of the wavefunctions of the two subbands involved. The
investigation of single-particle and collective inter-subband excitations
consequently leads to direct information on Coulomb matrix elements in
two-dimensional carrier systems. In polar semiconductors like GaAs, the
longitudinal plasma oscillations are coupled to the LO-phonons leading
to coupled phonon-plasmon modes which are determined by the zeros of the
total dielectric function. For a two subband system this is shown schematically in Fig. 4. The behaviour is more complicated when more levels are involved. This has been studied by several authors based on the work of [10].

Electronic intersubband excitations have been studied in various semiconductor heterostructures and multilayer systems. Examples of the most
widely studied electron systems in GaAs structures are shown in Fig. 5.
Collective excitations are observed for parallel polarisations (z(yy)z),
Also shown are typical conduction band diagrams and possible intersubband
transitions. The potential wells are achieved by certain doping layer
sequences. The used laser lines are close to the $E_o + \Delta_o$ energy gap of
GaAs. Similar spectra have been obtained by electrons in InP [11] and Ge [12].
Two-dimensional hole systems have been studied in accumulation layers of

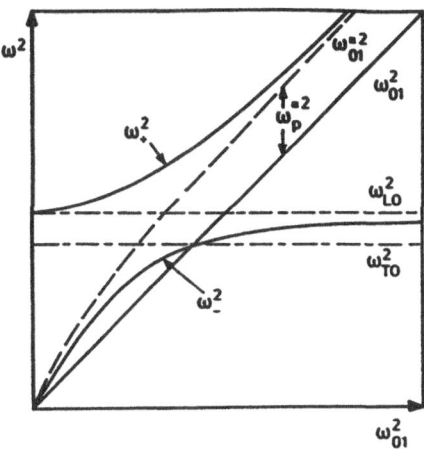

Fig. 4.    Dependence of the single particle
and coupled LO-phonon intersubband
frequencies on subband splitting
(schematically)

Si [13] and in p-type modulation doped GaAs-Al$_x$Ga$_{1-x}$As quantum wells [14]. The
spectral lineshapes for hole excitations are vastly different from those of
electron gases. This is due to the complex valence band structure which
leads to subbands with entirely different dispersion in $k_{\parallel}$. Resonance
enhancement is only achieved when the top of the valence band is involved
in the optical transitions. Therefore photon energies close to the E$_O'$ gap
in Si have been chosen in these experiments.

Resonant photon  energies involving carrier occupied states are not easily
useable for narrow gap semiconductors. The small energy gaps involved are
not accessible with conventional Raman spectrometers. Consequently no results of single particle excitations are reported so far in narrow gap
semiconductors. However, there exist some reports on the observation of
collective LO-phonon-intersubband excitations in InAs metal -insulator-
semiconductor structures /15, 1/. Those experiments were performed with
laser energies close to the E$_1$ gap of InAs. At this resonance direct optical interband transitions involve no carrier occupied states in the conduction band. Consequently carrier-density mechanisms are negligible
and usually not observable. The coupling to the LO-phonons, however, leads
to resonant enhancement of the collective excitations due to other scattering
mechanisms as for example deformation potential or Fröhlich intraband
electric field induced mechanisms. Resonance enhancement by the LO-phonon
scattering mechanisms occurs at all optical gaps. Fig. 6 shows a differential
Raman spectrum of InAs which exhibits the carrier induced changes of the
spectrum. The positive signals at $\omega_-$ and $\omega_+$ are the collective intersubband
modes, the negative signal at $\omega_{LO}$ shows the reduced intensity of the LO-
phonon due to the two-dimensional electron gas at the interface. The

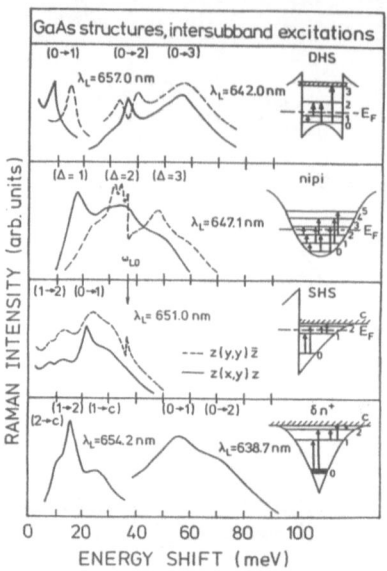

Fig. 5.   Raman spectra of single-
particle and collective
intersubband excitation
in various potential wells
achieved in different GaAs
quantum well and superlattice
structures (from Ref. 20)

experiments with InAs are the only ones performed so far where collective
excitations of the surface electron system are observed, even though the
optical transitions do not involve carrier occupied states.

As mentioned already, a finite $q_{\shortparallel}$ can be realized by certain scattering
configurations which deviate from ideal back-scattering. Under such conditions also in-plane excitations are observed which are either two-dimensional plasmons or single particle excitations. The value of $q_{\shortparallel}$ can be
changed by turning the sample with respect to incident and scattering light
directions. Thus the dispersion of the intrasubband excitations can be
studied experimentally. Such experiments have been performed recently [16,17].
The results for plasmons and single particle excitations are in good agreement with theoretical predictions.

In summary, electronic excitations exhibit a large variety of information
on various aspects of two-dimensional carrier systems. Under resonance

conditions light scattering is sensitive to carrier concentrations as low as $10^{11}$ cm$^{-2}$. However, the application to narrow gap semiconductors is limited due to experimental problems.


## PHONON ASPECTS


A discussion of Raman scattering at semiconductor interfaces is not complete without mentioning at least shortly the usefulness of vibronic excitations with respect to the analysis of interfaces and superlattices. In polar semiconductors like GaAs and back-scattering from (110) surfaces only the TO-phonon is symmetry allowed. Under resonance conditions, however, forbidden, electric field induced light scattering by LO-phonons can be as strong or even stronger than the allowed scattering. The intensity of the forbidden LO-phonon is directly proportional to the square of the electric field and consequently to the barrier height at the surface or interface. This optical method for the determination of barrier heights can be used from clean, ultra-high -vacuum cleaved surfaces up to coverages of several hundred Angstroms. It has been used to study the formation of Ge-GaAs and $Si_xGe_{1-x}$-GaAs interfaces /19, 20/. Examples are shown in Fig. 7. Forbidden LO-phonon scattering can be sensitive to coverages of less than a thousandth of a monolayer. It is an important tool to study the formation of heterostructures and Schottky barriers.

As mentioned already in the introduction, allowed phonon scattering is often used to get a variety of information on thin films and superlattices. An essential quantity in strained layer superlattices is the built in strain in such systems. It is directly reflected by the shift of the phonon lines to smaller or larger energies depending on the sign of the strain. In Fig. 8 a series of phonon Raman spectra is shown which were obtained from a

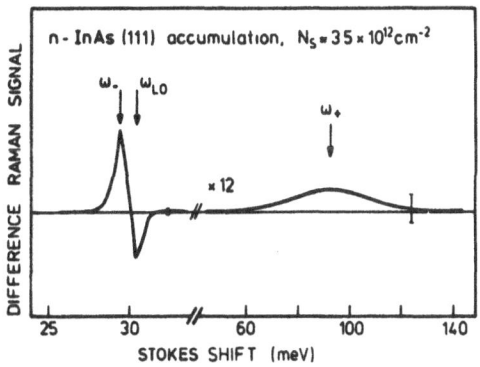

Fig. 6:

Differential Raman spectrum obtained with a chopped gate voltage (between falt band and positive voltage which corresponds to a carrier concentration of $3,5 \times 10^{12}$ cm$^{-2}$) at a n-InAs surface (from Ref. 1).

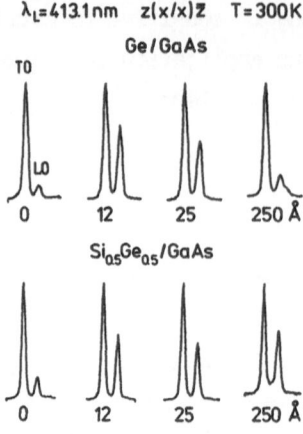

λ<sub>L</sub>=413.1nm   z(x/x)Z   T=300K

$\lambda_L=413.1\,nm$   $z(x/x)\bar{z}$   $T=300K$

Ge/GaAs

TO

LO

0        12        25        250 Å

Si$_{0.5}$Ge$_{0.5}$/GaAs

0        12        25        250 Å

Fig. 7.    Raman spectra of TO and forbidden
LO phonon scattering in GaAs for
Different epitaxial overlayers of
Ge and Si$_{0.5}$Ge$_{0.5}$. (from Ref. 20)

GaAs cleavage surface with different Si$_{0.5}$Ge$_{0.5}$ overlayers. The freshly,
ultrahigh vacuum cleaved sample exhibits only the TO-phonon line of GaAs.
The characteristic three Si$_{0.5}$Ge$_{0.5}$ modes appear already at layer thicknesses of the order of 1 nm. They are shifted downwards which is caused
by the built-in strain due to the 2% lattice mismatch. For larger thickness the three modes shift to the positions characteristic for the unstrained situation. Formation of dislocation lines after a critical thickness is responsible for the strain relaxation. These selected examples
demonstrate the multi-purpose of Raman scattering for the analysis of
semiconductor heterostructures and superlattices.

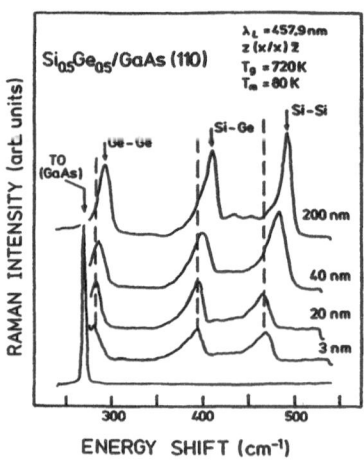

Fig. 8.    Raman spectra of strained and
unstrained Si$_{0.5}$Ge$_{0.5}$ overlayers
on GaAs. (from Ref. 20)

# References

1. G. Abstreiter, M. Cardona, and A. Pinczuk, "Light Scattering by Free Carrier Excitations in Semiconductors", in Light Scattering in Solids IV, Topics in Applied Physics 54, eds. M. Cardona and G. Güntherodt, Springer-Verlag Berlin, Heidelberg (1984) p. 5

2. E. Burstein, A. Pinczuk and S. Buchner, "Resonance Inelastic Light by Charge Carriers at Semiconductor Surfaces", in Physics of Semi-conductors 1978, ed. B.L.H. Wilson, London, The Institute of Physics, (1979), p. 1231

3. A. Pinczuk, G. Abstreiter, R. Trommer and M. Cardona, "Resonance Enhancement of Raman Scattering by Electron-Gas Excitations of n-GaAs, Solid State Commun, $\underline{30}$, 429 (1979)

4. G. Abstreiter and K. Ploog, "Inelastic Light Scattering From a Quasi Two-dimensional Electron System in GaAs-Al$_x$Ga$_{1-x}$As Hetero-junction ", Phys. Rev. Letters, $\underline{42}$, 1308 (1979)

5. A. Pinczuk, H.L. Störmer, R. Dingle, J.M. Worlock, W. Wiegmann and A.C. Gossard, "Observation of Intersubband Excitations in a Multi-layer Two-Dimensional Electron Gas", Solid State Commun. $\underline{32}$, 1001 (1979)

6. G. Abstreiter, "Light Scattering in Semiconductor Heterostructures" in Molecular Beam Epitacy and Heterostructures, eds. L.L. Chang, and K. Ploog, Dordrecht, Martinus Nijhoff Publishers (1985), p. 425

7. A. Pinczuk and J.M. Worlock, "Light scattering by Two-Dimensional Electron Systems in Semiconductors", Surface Science $\underline{113}$, 69, (1982)

8. G. Abstreiter, R. Merlin and A. Pinczuk, "Inelastic Light Scattering by Electronic Excitations in Semiconductor Heterostructures", in IEEE, special issue on: Quantum Well Structures: Physics and Applications, in press

9. E. Burstein, A. Pinczuk, and D.L. Mills, "Inelastic Light Scattering By Charge Carrier Excitations in Two-Dimensional Plasmas: Theoretical Considerations", Surface Science $\underline{98}$, 451, (1980)

10. D.A. Dahl and L.J. Sham, "Electrodynamics of Quasi-Two-Dimensional Electrons", Phys. Rev. B $\underline{16}$, 651 (1977)

11. G. Abstreiter, R. Huber, G. Tränkle, and B. Vinter, "Subband Energies in Accumulation Layers on InP", Solid State Commun. $\underline{47}$ 651 (1983)

12. R. Merlin, A Pinczuk, W.T. Beard and C.E.E. Wood, J. Vac. Sci. Technol. $\underline{21}$, 516 (1982)

13. M. Baumgartner, G. Abstreiter, and E. Bangert, "Hole Subbands in Silicon Surfaces", J. Phys, C $\underline{17}$, 1617 (1984)

14. A. Pinczuk, H.L. Störmer, A.C. Gossard and W. Wiegmann, "Energy Levels of Two-Dimensional Holes in GaAs-(AlGa)As Quantum Well Structures" in Proc. of the 17$^{th}$ Int. Conf. on the Physics of Semiconductors, eds. J.D. Chadi and W.A. Harrison, New York, Springer Verlag (1985), p. 329

15. L.Y. Ching, E. Burstein, S. Buchner and H.H. Wieder, "Resonant Raman Scattering at InAs Surfaces in MOS Junctions" Proc. of the 15$^{th}$ Int. Conf. on the Physics of Semiconductors, eds. S. Tanaka and Y. Toyozawa, Kyoto, (1980), J. Phys. Soc. Japan 49 951 (1980)

16. D. Olego, A. Pinczuk, A.C. Gossard and W. Wiegmann, "Plasma Dispression in a Layered Electron Gas: A Determination in GaAs-(AlGa)As Heterostructures", Phys. Rev. B 25, 7867, (1982)

17. G. Fasol, N. Mestres, H.P. Hughes, A. Fischer, and K. Ploog, "Raman Scattering by Coupled-Layer Plasmons and In-Plane Two-Dimensional Single-Particle Excitations in Multi-Quantum-Well Structures", Phys. Rev. Lett. 56, 2517 (1986)

18. G. Abstreiter, "Inelastic Light Scattering in Semiconductor Heterostructures" in Festkörperprobleme XXIV, ed. P. Grosse, Vieweg Braunschweig (1984), p. 291

20. G. Abstreiter, "Light Scattering in Novel Layered Semiconductor Structures", in Festkörperprobleme XXVI, ed. P. Grosse, Vieweg Braunschweig (1986), in press

# SOURCES AND DETECTORS FOR PICOSECOND/FEMTOSECOND SPECTROSCOPY

W. Sibbett

Department of Physics
University of St. Andrews
St. Andrews, KY16 9SS
Scotland

Time-domain spectroscopy in the picosecond and femtosecond regimes has been demonstrated as a useful diagnostic tool in the study of the kinetics of nonequilibrium charge carriers and phonons in semiconductors. The applicability of the technique is essentially determined by the excitation source and by the detection system involved. For this reason, a review will be presented where the emphasis will be directed towards (i) the generation of frequency-tunable ultrashort coherent laser pulses and (ii) the types of linear or nonlinear optical processes that provide quantitative measurements with adequate time resolution, detection sensitivity and dynamic range.

# HIGH PRESSURE TECHNIQUES FOR RESEARCH IN SEMICONDUCTORS: A REVIEW

Ian L. Spain

Department of Physics
Colorado State University
Fort Collins CO, 80523

## I. INTRODUCTION

High pressure studies of semiconductors have given valuable insights to their room-pressure properties and have also been used to generate new phases. It is important to note that hydrostatic pressure acts as a perturbation on the electronic properties without a change of symmetry within a single, homogeneous phase. Accordingly, the changes in optical or electronic properties can be interepreted in a straightforward manner, at least in principle. The effects of uniaxial or shear stresses will not be dealt with in this review. However, shear stresses are generated in epi-layers or thin films of different compressibility than the substrate, even if compressed by a hydrostatic fluid.

The changes in physical properties resulting from the application of hydrostatic pressure can be roughly subdivided as follows:- Firstly, at relatively accessible pressures, say up to 1 GPa, properties such as the conductivity change smoothly with pressure, mainly due to modification of the electron energy band gap, $\varepsilon_g$, which typically changes at the rate $|d\varepsilon_g/dP| \lesssim 150$ meV/GPa. Sizeable changes can occur in intrinsic carrier densities,

$$\frac{1}{n_i} \frac{dn_i}{dp} \sim \frac{1}{kT} \frac{d\varepsilon_g}{dP}$$

For instance, the intrinsic carrier, density $n_i$, decreases by 40% at room temperature with an increase of only 0.1 GPa in pressure if $d\varepsilon_g/dP = 100$meV/GPa. Extrinsic carrier densities on a molar, or per atom basis, normally remain constant, provided the temperature is in the exhaustion region. The carrier effective masses and mobilities usually change at relatively slow rates (eg $\lesssim 1\%$ for $\Delta P \sim 0.1$ GPa). Experimental information about the variation of mobility with pressure can give valuable information about carrier scattering mechanisms, sometimes with surprising results (see Lancefield, Adams and Gunney, 1984, for example).

Secondly, relatively higher pressures (eg 1-10 GPa) can be used to change the electronic energy levels sufficiently that electronic transitions can occur. A good example of this is the cross-over of the conduction band minimum at the $\Gamma$ point with minima near the X-point in GaAs at $\sim$4GPa. This occurs because the $\Gamma$ point energy increases rapidly ($dE_\Gamma/dP \sim 110$ meV/GPa) while the X-minima slowly decrease ($dE_X/dP \sim -10$ meV/GPa). Note that a pressure of 10 GPa, which is easily attained in a diamond anvil cell, can cause the

Γ level to shift by more than 1eV. The properties of the X-electrons can be studied directly above the transition. Although a similar transition can be induced by alloying with Al, the disorder introduces an additional complexity in the interpretation of results.

Structural transitions can also be induced by application of hydrostatic pressure. Transition pressures generally increase with band gap (eg 2.3 GPa for InSb, 12GPa for Si, 17GPa for GaAs, 27GPa for GaP etc.) and metallic phases result. There is also the possibility of producing metastable phases when the pressure is released from the metallic phase. There has been increased theoretical activity aimed at predicting high pressure and metastable phases, with encouraging successes,which will be discussed later in this paper.

The present paper will briefly review the types of apparatus which are useful for experimentation with semiconductors, with particular emphasis on the diamond anvil design. Examples of research will be given to illustrate the techniques. It is impossible for the review to be exhaustive, and only representative areas of interest can be mentioned. Review literature is cited whenever possible, enabling the reader to gain access to a far more extensive list of references than would be feasible here.

## II  TYPES OF HIGH PRESSURE APPARATUS USED FOR SEMICONDUCTOR RESEARCH

The apparatus employed for semiconductor research can be divided conveniently into several basic types:

### 1)  Cylindrical vessel fed from compressor

A typical apparatus is illustrated in fig.1. The pressure range is controlled by the maximum safe operating pressure of the tubing, which is typically 0.5 mm inner diameter, half-hard 316 stainless steel, which is weakly paramagnetic. Commercial tubing with outer diameter of 4.7 mm or 3.2 mm is available with safe operating pressure of 1.5 GPa (Paul and Warschauer, 1956). Fittings suitable for high magnetic fields and/or low temperature need to be machined of suitable alloys (eg beryllium copper) but their design can be the same as commercial fittings, which are constructed of tool steel.

The apparatus illustrated in figure 1 is for electrical measurements with leads passing from the sample along the capillary tubing to an external seal. This seal can be effected with epoxy resin or frozen oil. A Bridgman unsupported area seal is used for the end plug (see Bridgman, 1952). A good example of this type of apparatus is described by Schirber (1970). Optical windows can be incorporated, and a modern design for optical measurements on semi-conductors in the infrared region is described by Wasilewski, Stradling and Porowski (1985).

Cryogenic pressure vessels are typically constructed of beryllium-copper     (Paul, Benedek and Warschauer, 1959), which can be heat-treated (precipitatation hardening mechanism) to give a 0.2% yield strength of ∿1GPa. Non-magnetic grades can be obtained,  (Telcon 250, for example) in which the concentration of cobalt alloying ingredient is minimized. Mechanical properties of the alloy are slightly enhanced at low temperature. Single-walled vessels designed with care can typically be used to ∿ 1.0GPa, and double-walled, or auto-frettaged single-walled up to 1.5 GPa( for a review of stresses in pressure vessels, see Crossland and Spain, 1983. Paureau (1977) has reviewed low temperature apparatus).

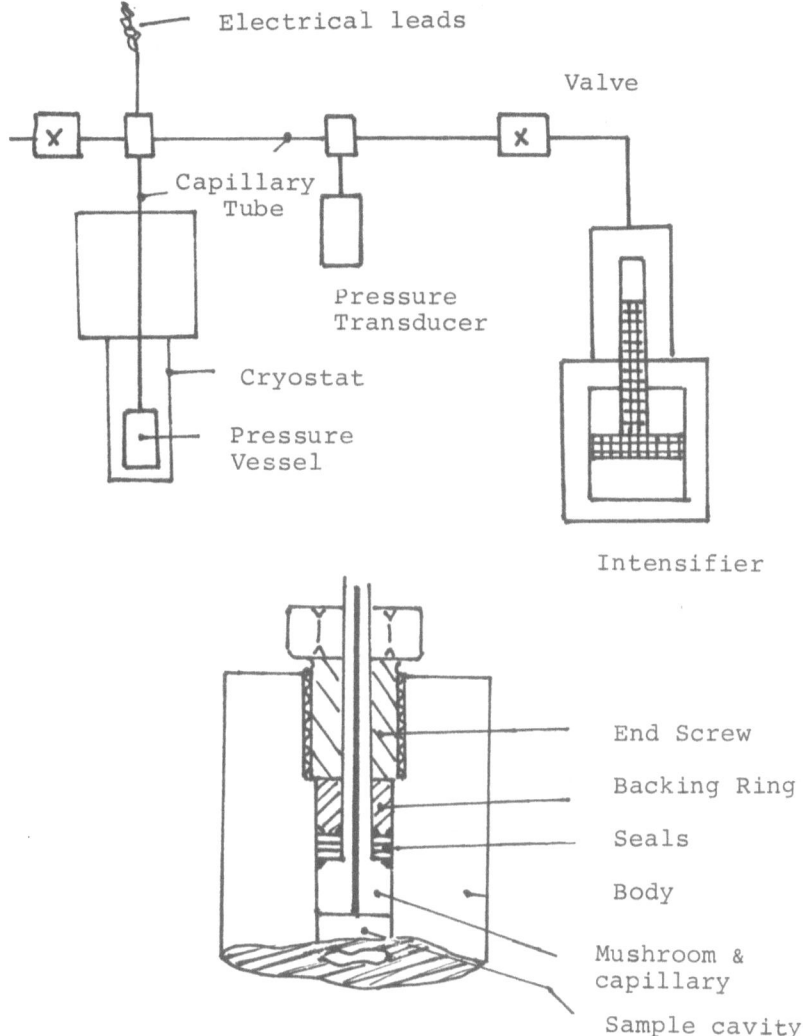

Figure 1     Schematic of apparatus employing tubing to conduct pressurizing
fluid to the sample cavity. Inset shows the Bridgman unsupported area seal.

## 2) Piston - Cylinder Apparatus

The requirement of higher pressures (>1.5 GPa) precludes the use of
tubing, so that pressures need to be generated within the vessel itself.
Many different designs of piston - cylinder apparatus have been reported,
some reaching pressures as high as 8GPa with multiwalled vessels, and
active mechanical support of critical components. (see for example,
Bradley, 1969 and Crossland and Spain, 1983 for reviews). Multi-walled
vessels with tungsten carbide pistons can be simple in design, and useful
for 5GPa. However, they are bulky and difficult to use for studies at
low temperature, or in magnetic fields, although small coils can be
incorporated inside the sample volume (Liftschitz and Maines 1979).

A very useful design incorporates a piston - cylinder configuration with
a clamping device (Fig.2) (see Fujiwara et al, 1980 and Wasilewski et al,

1986). Pressure is generated by advancing the piston in a hydraulic press. Then a clamping nut is tightened to hold the piston in position. The vessel can then be removed from the press and placed in a cryostat or magnetic field. Typical designs use similar materials and seal designs to the vessels described in Section II-1, but can be operated to 2-3 GPa. (For a review, see Paureau (1977).

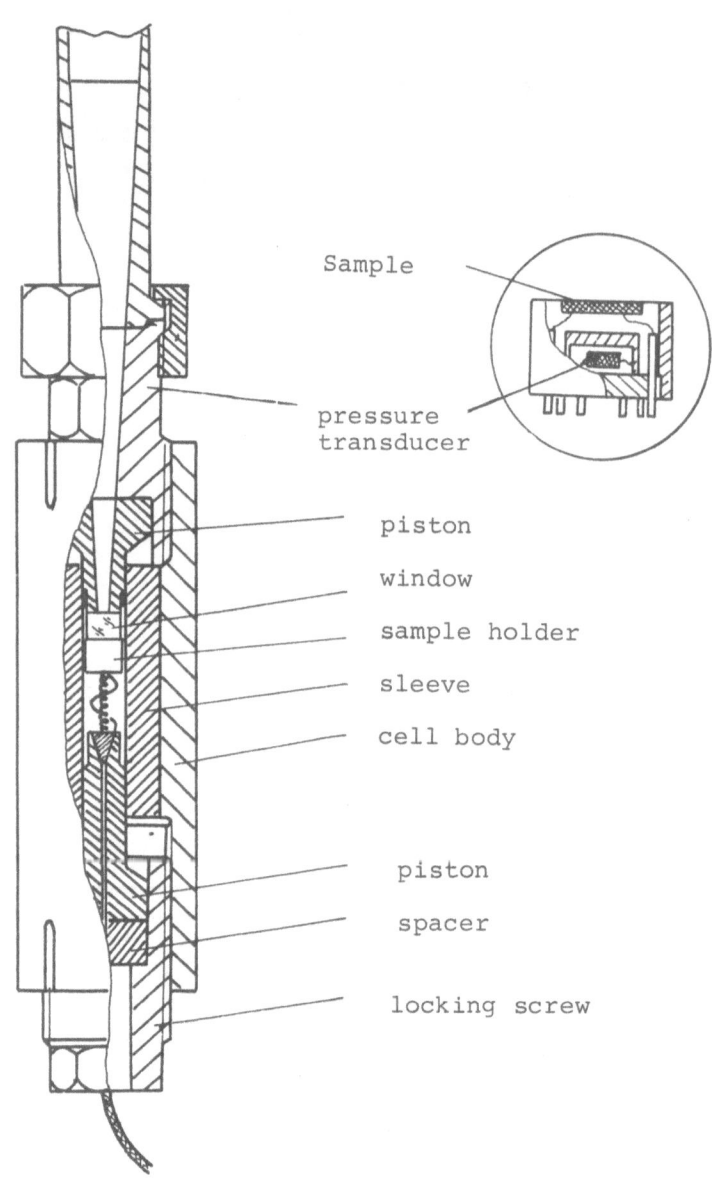

Sample

pressure transducer

piston

window

sample holder

sleeve

cell body

piston

spacer

locking screw

Figure 2    Schematic diagram of a cylindrical pressure vessel with a clamping device useful for electrical and optical studies at low temperature (Figure adapted from Wasilewski, Porowski & Stradling 1986).

## 3) Anvil Devices

Bridgman found that higher stresses could be produced in thin samples than the compressive strength of structural materials, if tapered anvils were used (see Bridgman, 1952). Since that time, many anvil configurations have been devised for synthesis of semiconducting materials, and measurements of these physical properties (see for example Bradley 1969 and Crossland and Spain 1983 for a discussion).

An opposed anvil device of hardened tool steel, suitable for Hall measurements on semiconductors has been described by Pitt (1968) in which the sample was embedded in MgO/epoxy mixture. Magnetic fields up to 1.5T were generated using the anvils and backing blocks as the poles of an electromagnet. Vyas et al (1973) used this apparatus cooled to 120K to obtain data on GaAs. (see also Bandyopadhyay et al 1980). A similar sample and anvil geometry was used to generate pressures up to ∼1.5 GPa using titanium anvils at room temperatures and high magnetic fields (9T) (Pitt et al, 1973). This allowed magnetophonon oscillations to be studied in GaAs and InP. Optical fibres have been fed through the anvils for photoconductivity measurements (Gunney et al 1982).

The most exciting results on semiconductors using Bridgman anvils have been obtained with the diamond anvil cell, which will be discussed at greater length in Section IV.

## III PRESSURE MEASUREMENT

Bourdon, or other gauges based on mechanical strain, can be used with apparatus in which the fluid is fed to the sample cavity (see Scaife and Peggs 1983 for a review). Care must be taken to ensure that there is no pressure differential between the sample and the manometer. Measurement sensitivity is typically ∼0.1% and absolute accuracy ∼0.25 - 0.5% with these gauges, when used to ∼1GPa.

The variation of the electrical resistance of a metal coil is typically used in piston-cylinder devices (see Peggs and Wisniewski 1983 for a review). Manganin is often used because its temperature coefficient of resistance is close to zero at room pressure. However, the resistance-temperature curve passes through a maximum there, and this maximum shifts with pressure (Beavitt, 1969). The pressure coefficient is small $(\frac{1}{R}\left(\frac{\partial R}{\partial P}\right)_T \sim 2.3 \times 10^{-2}/\text{GPa})$, so that precise temperature control is necessary to ensure that the resistance change from thermal effects is negligible compared to that from pressure effects. Temperature control $\pm 0.01°C$ is typically needed to ensure an accuracy of $\pm 0.25\%$ up to 1GPa. The resistance change is nearly linear with pressure, but corrections become more important above ∼0.5 GPa. The gauge is sensitive to non-hydrostatic stress and must be in an electrically insulating fluid medium.

Other types of alloy have been considered and are reviewed by Peggs and Wisniewski (1983). A particularly useful gauge for measurements to low temperature and high pressure is based on heavily-doped n-InSb. (Konczykowski et al 1977), used by Wasilewski et al(1986) in the apparatus depicted in fig.2.

Anvil devices have most frequently been calibrated by plotting ram hydraulic pressure against "known" fixed point transitions (eg in bismuth) (see Bean, 1983). Unfortunately, it now appears that this technique led to results which have not been substantiated by more reliable measurements. (see for example Piermarini and Block 1975). Pressure measurement in the diamond anvil cell has usually been carried out using the ruby fluorescence technique (see IV).

The importance of carrying out measurements under hydrostatic stress conditions must be emphasized. Application of non-hydrostatic stresses results in a change of symmetry of the semiconductor, which in turn lifts degeneracies of electronic states. The attainment of hydrostatic stress conditions is particularly troublesome at high pressure, and/or low temperatures, where all materials solidify. In such cases the use of rare gases, such as $He^4$, Ne, Ar or $N_2$ is recommended (see Section IV).

IV  THE DIAMOND ANVIL HIGH PRESSURE CELL

A side view of two diamond anvils, the gasket and sample is shown in figure 3.  The diamonds are often brilliant cut gem-stones with points removed to form a culet, or anvil, but those shown in the figure are of a more recent design which withstands higher stresses (Seal 1985). The outer diameter of the stone is typically ∿4mm and the anvil diameter 0.5 - 1mm. The metallic gasket is usually of a tough metal such as Inconel 718, beryllium copper, or 316X stainless steel. A sheet of initial thickness ∿0.5 mm is indented by the diamonds, after which the sample cavity is drilled. The sample is incorporated inside it, together with a ruby chip (typically 30μm diameter) and the pressure transmitting fluid. Pressure is then built up by applying force to the back surfaces of the diamonds, which are separated from the hard backing plate by a thin (eg 12μm) sheet of softer metal (eg Zr or Al).

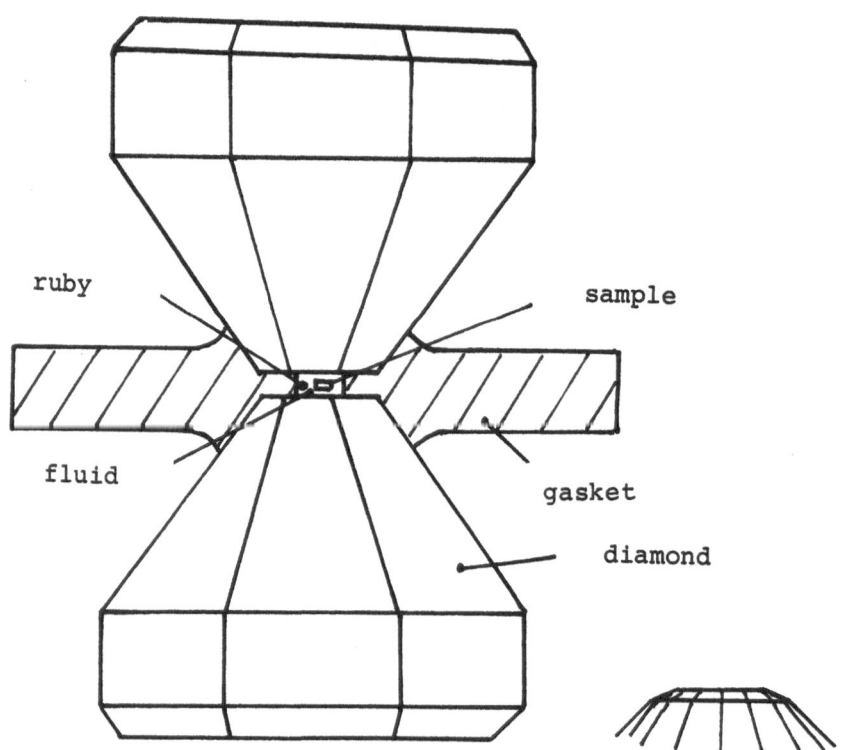

Figure 3    Side section of two diamond anvils, gasket and sample with typical dimensions. Inset shows shaped anvil tips for attainment of pressures in excess of 200 GPa.

A review of diamond anvil cell measurement techniques has been given by Jayaraman (1983), who illustrates several techniques for applying force to the diamonds. The cell illustrated in fig 4 is small and simple, used by us for low temperature studies. Force is applied by tightening the end screw transmitted via a set of spring washers to the piston (fig 4). All parts are machined of Be/Cu, and the cell has been used to 30 GPa with 0.5 mm culet diameter, and 10 GPa with 0.8 mm diameter. Cells often use a hardened steel, or tungsten carbide backing plate behind the diamonds, since the stresses in this area can be considerable. Sapphire can also be used, and is useful where large optical aperture is needed (Hirsch and Holzapfel 1981).

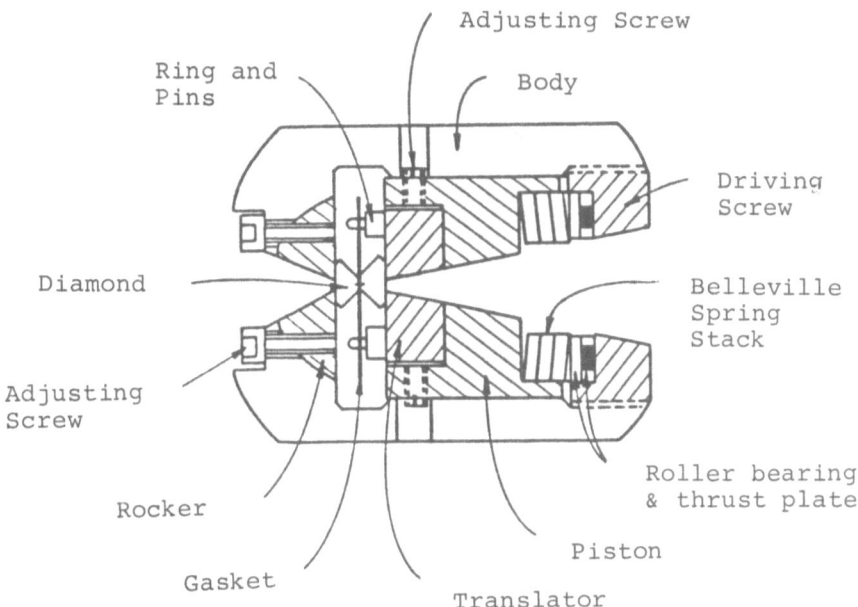

Figure 4    Cross section of a typical diamond anvil cell used in the author's laboratory.

Provision must be made for the diamonds to be aligned, so that anvil faces share a common axis, and are parallel.  The method employed with this cell is to align axially by translating the backing plate in the piston via four screws, then for parallelism by the "hemispherical" rocker via four rear screws. The parallel alignment is normally made to better than one optical fringe.

The dimensions of the anvils and sample cavity depend on a number of factors. Very small, thin, samples (eg 30μm diameter x 10μm thick) are used to reach pressures in excess of 400 GPa with specially shaped anvil tips (Goettel et al 1985 and Moss et al 1986). However, these samples are clearly under non-hydrostatic pressure. Pressures of 50 GPa can be reached with close to hydrostatic conditions using anvil diameters of 500 μm, sample cavity diameters of 150μm and initial thickness of 80μm, reducing to ∿30μm at high pressure. More recent measurements have been made by us on semiconductors up to ∿ 10 GPa using 800 μm diameter anvils, 350 μm diameter sample cavity and initial thickness 200 μm which can decrease to 100 μm at high pressure. Close to hydrostatic pressures can be achieved with $He^4$, Ne, or Ar (Bell and Mao 1980) or $N_2$ (Le Sar et al, 1979) as the pressure-trasmitting media. These substances can be loaded in several ways. One basic method is to force fluid into the cavity at high pressure, the other is to load liquid at atmospheric pressure but low temperature. The former method is more reliable but requires more expensive equipment (Besson and Pinceaux 1979, Mills et al 1980).

The ruby fluorescence technique is widely used for pressure measurement (Barnett et al 1973, Piermarini et al 1975). Below about 30 GPa a linear wavelength-pressure relationship is suitable, with $d\lambda/dP = 0.365$nm /GPa. At higher pressures a non-linear expression is applicable (Mao et al, 1978).

$$P = 380.8 \left[ \left( \frac{\Delta\lambda}{694.2} + 1 \right)^5 - 1 \right]$$

The fluorescence peaks broaden with non-hydrostatic stress, but severe effects of non hydrostatic stress can be observed in semiconductors well before the broadening in the fluorscence lines becomes apparent (Adams, Appleby and Sharma, 1976). At low temperature the wavelength of the R1 fluorescence peak shifts slightly from its zero pressure value ($\sim$694nm, depending on sample), but the pressure coefficient remains constant within experimental error (Noach and Holzapfel, 1977).

The transparency of the diamond anvils over a wide range of photon energies makes them particularly useful for physical property measurements. IR and optical photons up to $\sim$5eV ($\sim$250 nm) and X-ray photons above 10 keV($\sim$1.25Å) are transmitted, although IR absorption bands can be troublesome (Adams and Sharma 1977, Seal 1985). Diamonds are selected to be defect-free up to typically 15x magnification and are normally slightly yellow (Type I) due to nitrogen impurities. Lower IR absorption is found in white diamonds (IIa) and these can be selected to have relatively low fluorescence background, of particular importance in Raman and Brillouin spectroscopy (Adams and Sharma 1977).

## V  SOME REPRESENTATIVE MEASUREMENTS

### V.I  Lattice Parameters and Phase Transitions

Most physical phenomena are best interpreted as a function of interatomic separation rather than pressure. Two basic techniques can be used to measure the compression of samples, firstly single crystal or powder x-ray diffraction (see Jayaraman 1983 for a review), or, secondly, measurement of sample dilation using optical observations (Brasch, 1980, Tanaka and Maeda, 1986). X-ray diffraction can be carried out quickly using synchrotron sources (Baublitz et al, 1981, Skelton et al, 1983) but a typical exposure may take several days using a conventional laboratory source.

An example of the work on semiconductors can be gained from silicon, where x-ray diffraction experiments have obtained the volume up to 12 GPa for the cubic (diamond) phase with an absolute accuracy of $\sim$0.5% (Hu et al 1986). The compression agrees with the prediction of the Murnaghan equation

$$P = \frac{B_o}{B_o'} \left( \left( \frac{a_o}{a} \right)^{3B_o'} -1 \right)$$

Using values of the bulk modulus $B_o$ and its pressure derivative, $B_o'$ obtained at 1 atmosphere pressure, from precision elastic constant data. This equation can be used to predict the variation of the lattice parameter, a, for all cubic semiconductors up to $P \sim 0.1 - 0.2\ B_o$. (For a review see Bolsaitis and Spain 1984).

At 12 GPa a structural transition occurs to the $\beta$-Sn phase (Jamieson, 1963) then at 13.5 GPa to simple hexagonal (Olijnyk, Sikka and Holzapfel, 1984, Hu and Spain 1984) and at $\sim$ 43 GPa to a hexagonal close-packed phase (Olijnyk et al 1984, Hu et al 1986). The simple hexagonal phase has not been observed in any other element. Theoretical calculations of the high pressure structural behaviour are in excellent agreement with experiment (McMahan and Moriarty 1983, Needs and Martin, 1984, Chang and Cohen 1985).

Metastable phases are found on release of pressure. A body-centred structure with 8 atoms per unit cell is found on slow release (Wentorf and Kasper 1963, Kasper and Richards 1964), and tetragonal structures on fast release ($\lesssim$0.1s) (Zhao et al, 1986). We have found recently that GaAs also forms a metastable phase on release of pressure, using x-ray diffraction techniques, while Weinstein (1986) has observed that metastable phases can result when pressure is released from superlattice structures.

## V.2 Vibrational Properties

It should be possible to obtain the elastic constants of materials in the diamond cell, using GHz ultrasonic techniques, but to date the only measurements of these parameters have used Brillouin spectroscopy (Whitfield et al 1976). The technique is difficult because of the weak signals from samples, and are restricted to optically transparent materials at the laser frequency. A good example of the use of the technique is represented by measurements of Shimizu et al (1981) who obtained data on the longitudinal and transverse velocities of $H_2$ up to 20GPa and used them to estimate the equation of state.

A number of diamond anvil studies of the Raman spectra of group IV (Ge, Si), III-V(GaAs, InP, GaP) and II-VI (ZnS, Zn Te, ZnSe) semiconductors have been published (see Jayaraman 1983 and Aoki et al 1985, for reviews) following pioneering work of Brasch et al (1968) who developed the technique. Raman spectra in these materials are relatively simple to understand because of the high symmetry of the structure. The first order spectrum of Ge and Si consists of one line, because the LO + TO phonons are degenerate at the zone centre. Data of $\omega(P)$ have been used to obtain the Grüneisen parameter $\gamma = -\frac{d\ln\omega}{d\ln v}$. The LO & TO phonons energies for the III-V and II-VI compounds are non-degenerate, and the splitting can be interpreted to give the variations of the effective change transfer between the ions (Carlone et al 1981).

Second-order Raman spectra exhibit critical points which can be interpreted to give the shift of phonon energies in other regions of the Brillouin Zone than the centre. Of particular interest is the softening of the TA mode at the zone boundaries, related to the onset of a phase transition (Weinstein and Piermarini 1975).

## V.3 Band Gap Variation with Pressure and Defect Levels

The variation of band gaps with pressure can be measured in several ways. Firstly, Welber (1976, 1977) developed micro-techniques for optical absorption and he and others have obtained data for several semiconductors, such as Ge (Welber et al 1977) GaAs (Welber et al 1975) InP (Muller et al 1980, Kobayashi et al 1981, Menoni et al 1986). These date typically probe the direct band gaps (zone centre), which increase strongly with pressure. It had already been established from lower pressure data (eg $\lesssim$ 1GPa) on optical and electrical transport measurements that the principal conduction band minima behave similarly in different group IV and III-V semiconductors (Camphausen et al 1971) (see fig.5). The $\Gamma$-level typically moves upwards in energy relative to the conduction band at $\sim$100 mev/GPa, the L-level upwards at $\sim$50 meV/GPa and the X-level downwards ($\sim$-10meV/GPa). The diamond anvil cell measurements allowed the non-linearity of the energy - pressure curves to be investigated. It was found that the direct band gap varied nearly

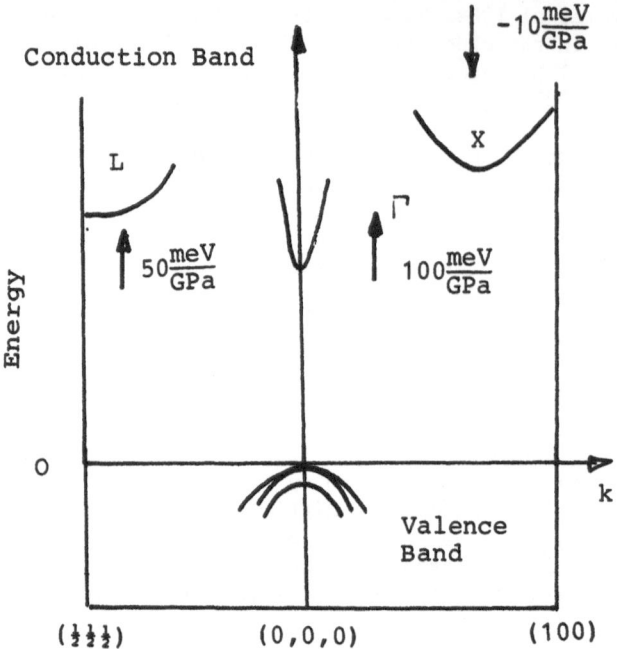

Figure 5   Schematic of the dispersion relationship for group IV and
III-V semiconductors and the variation of conduction band minima with
pressure. Band gaps depend strongly on material, but pressure
coefficients only weakly.

linearly with lattice parameter, a, and that deviations could be related
to band structure features.

More detailed information has been obtained from luminescence
measurements. Figure 6 summarizes data on GaAs obtained by Wolford and
Bradley, 1985 and LeRoux, Neu and Verie, 1986. Note that the $\Gamma$ and
X-bands cross at $\sim$ 4GPa. The position of the L band was not determined
from their data, but is included for completeness. Note also that a deep
level associated with nitrogen impurity and the X-band ($N_x$) passes into
the band-gap just above 2GPa. These measurements were carried out on
GaAs at 20K using argon as the compressing medium. The fact that the
$N_x$ doublet could be observed with a width of 3meV without splitting was
adduced as evidence that the pressure was close to hydrostatic.

Band gaps can also be measured from photoconductivity experiments
(see for example Gonzalez, Besson and Weill, 1986) but experiments in
which leads are attached to samples within the diamond anvil cell will
be discussed in the following section.

V. 4.   Electrical Measurements in The Diamond Anvil Cell

Electrical leads can be attached to samples in the diamond anvil cell.
Two-point methods (Block et al 1977, Sakai et al 1982) have been super-
ceded by 4-point techniques, (Walling and Ferrarro 1978, Mao and Bell,
1981, Reichlin 1983, Tozer and King 1985) enabling electrical resistivity,
photoconductivity (Gonzalez et al 1986) magneto-resistance and Hall
effect (Patel et al 1986) measurements to be carried out. An illustration

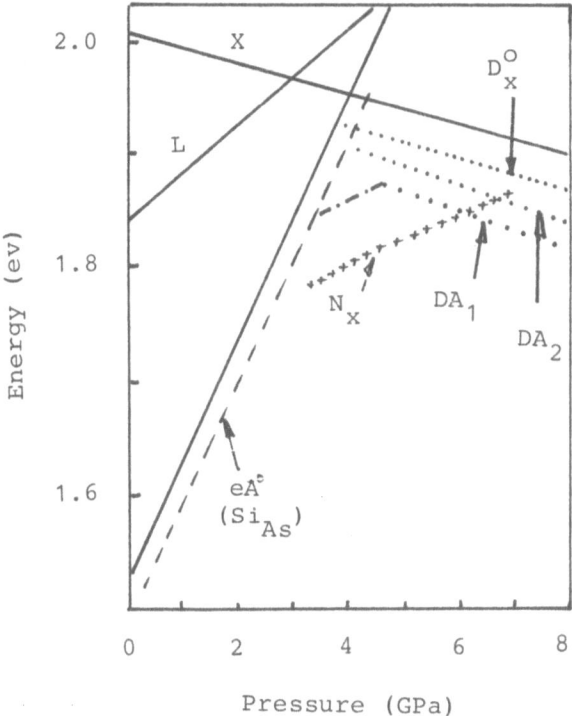

Figure 6  Variation of conduction band energy minima with pressure for GaAs together with some impurity levels (from Wolford & Bradley, 1985 and Leroux et al 1986).

of the technique used in the present laboratory is illustrated in figure 7 . Fine (12.5 um diameter) leads are insulated from the metallic gasket by a thin insulating layer. The guide pins through the gasket enable it to be positioned precisely, so that lead breakage is minimized. Also, the diamonds are rounded at the edges, or bevelled. Other methods of introducing leads use methods differing in detail from the above, but the principles are similar.

Figure 8 illustrates the results of Hall and resistivity measurements on n–GaAs up to ∿6 GPa (Patel et al 1986). The peak in the Hall coefficient occurs at ∿3.7 GPa when the contributions to the conductivity of Γ and X-electrons are equal. The band crossover deduced from a two-band model occurs at 4.1 GPa in good agreement with photoluminescence measurements (Wolford and Bradley, 1985). The transition from Γ- to X-electron conduction can be seen clearly in the Hall mobility.

Gonzalez Besson and Weill (1986) have observed photoconductivity in n-type GaAs up to 7.0 GPa, obtaining pressure coefficients of the direct band transition ($\Gamma_v$-$\Gamma_c$) of 120 meV/GPa. The value from luminescence data is 107 meV/GPa (Wolford and Bradley 1985) and from photo-absorption is 125 meV/GPa (Welber et al 1975). Differences between these values can be partially attributed to the fact that the techniques do not probe the direct band-edge transition, but involve impurity or exciton levels.

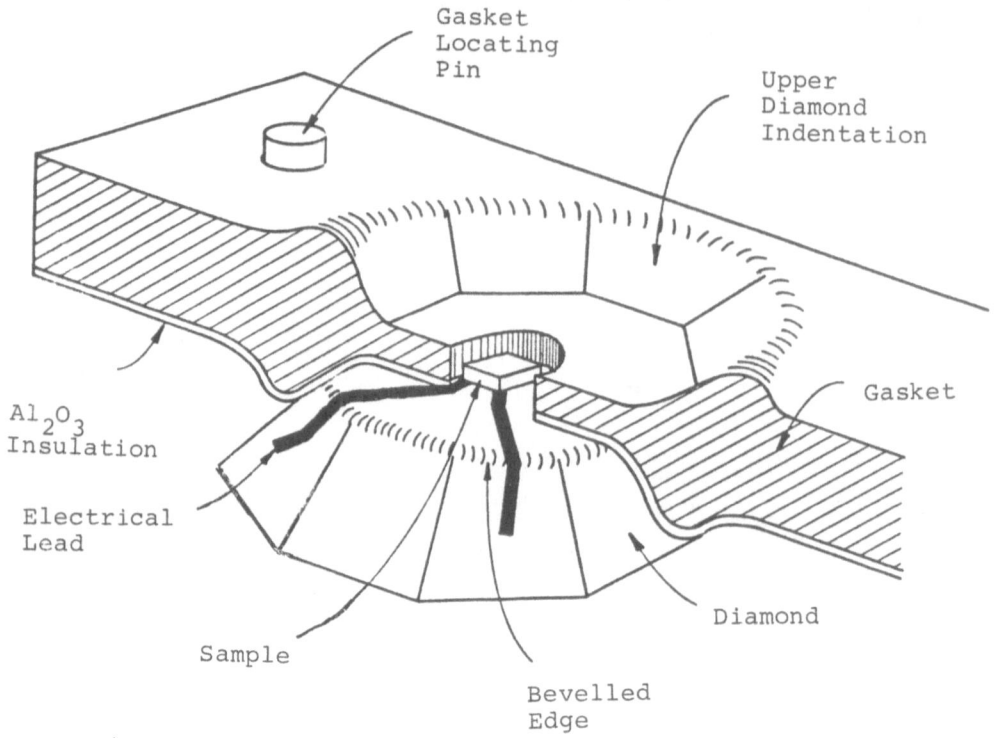

Figure 7 Sketch of the diamond anvil, gasket and sample used in Hall and restivity measurements (figure from Patel et al 1986)

These measurements can be carried out in principle to low temperatures and high magnetic fields, so that a variety of interesting experiments should be possible, such as Shubnikov-de Haas, magneto phonon and cyclotron resonance.

## V.5  Photo luminescence from Quantum-Well Bound States

As the final example of diamond anvil studies, the recent work of Wolford et al (1986) on luminescence from GaAs/Al$_x$Ga$_{1-x}$As quantum-well bound states will be presented. These measurements were carried out at 8K using a pressurizing medium in which the maximum deviation from hydrostaticity had been determined to be less than 1 part in 600 (Wolford and Bradley 1985). Pressure was used to lift the $\Gamma$ states above the X-states, so that transitions could be observed as illustrated in figure9. Accordingly, the valence band offset, $\Delta E_v$, could be determined directly for the first time with a resolution of meV ($\Delta E_v$ = 0.032 $\pm$ 0.02 eV at $\sim$ 3GPa.).

This study gives an interesting example of the way in which the pressure variable can be used to give unique information on semiconductor interfaces.

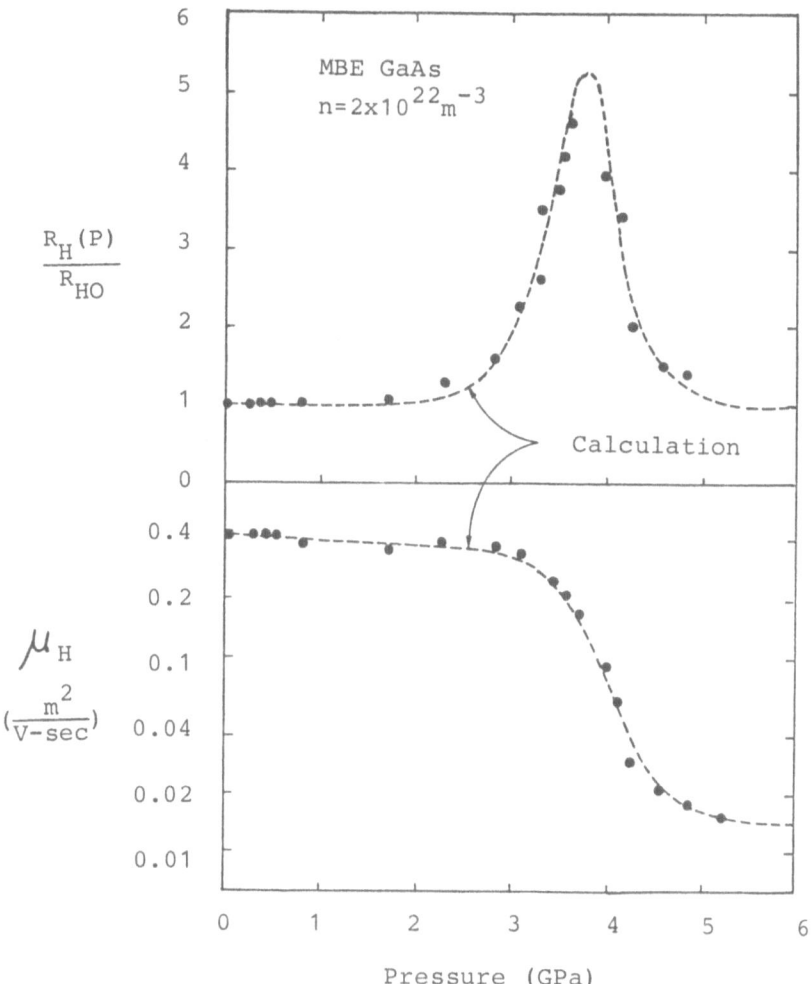

Figure 8 Variation of the Hall coefficient and electron mobility for an n-type sample of GaAs (from Patel et al 1986)

## VI  SUMMARY AND CONCLUSIONS

A brief review of high pressure techniques applicable to semiconductor research illustrates the rich variety of problems that have been addressed. Although stress has been laid on diamond anvil techniques, the complete laboratory includes a battery of different high pressure apparatus. One of the major drawbacks for instance, of the diamond anvil cell is the small size of the sample cavity, so that electrical leads cannot be applied to standard samples, or even geometries. The use of tungsten carbide anvils of larger diameter is worthwhile in this case. However, experience with diamond anvils has shown how these experiments can be carried out with hydrostatic stress conditions. Also, the incoroporation of optical fibres through the tungsten carbide anvils (Gunney et al 1982)

allows ruby fluorescence techniques to be used for pressure measurement .
Thus, technology transfer between these different apparatus is helping
the state-of-the-art to advance as a whole.

The next few years will see many exciting developments in both
techniques and applications. Dum vivimus vivamus!

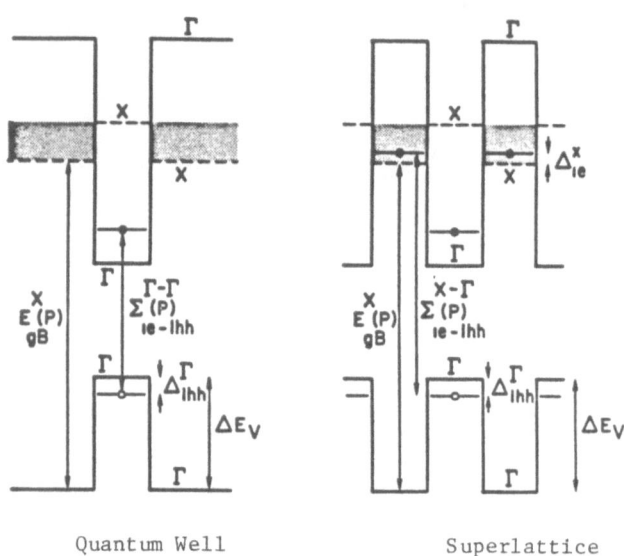

Quantum Well                    Superlattice

Figure 9    Schematic band diagram for GaAs/AlGaAs heterostructures.
Energies necessary for determining  offsets are indicated. Minima found
in X-states by valence offset are shaded.(Figure from Wolford et al (1986).

Acknowledgements

Many people have helped me at various times to gain knowledge and
experience of this field, and all are thanked. Profs R A Stradling and
A R Adams gave me kind hospitality during the writing of this paper, and
Dr David Dunstan offered many helpful criticisms. Financial support was
given by the Army Research Office, Triangle Park, N.C. (DAAG29-84-0049).

"Note added in proof: A very useful review of diamond anvil cell techniques
has appeared (A.Jayaraman (1986) (Rev.Sci.Inst. 57,1013))."

# REFERENCES

Aoki, J, Anastasskis, E and Cardona, M.(1985) p81 in Solid State Physics under Pressure (ed S.Minomura: Terra Scientific Publ.Co.(Dordrecht, Boston, Lancaster)).

Adams, D.M., Appleby, R., Sharma, S.K. (1976) J.Phys.E9 1140.

Adams, D.M., and Sharma, S.K. (1977) J.Phys. E10, 680.

Bandypadhyay, A.K., Nalini, A.V., Gopal, E.S.R., Subramanyam, S.V. (1980), Rev.Sci. Inst. 51 136.

Barnett, J., Block, S., Piermarini, G.J. (1973), Rev.Sci.Instr.44,1.

Bean, V.E. (1983) "Fixed Points for Pressure Metrology" Chapter 3, p93, in "High Pressure Measurement Techniques", (ed G.N. Peggs: Applied Sci.Publ., London and New York).

Beavitt, A.R. (1969) J.Phys.D2, 1675.

Bell, P.M. and Mao, H.K., (1980), Geophysical Laboratory Yearbook p404.

Besson, J.M. and Pinceaux, J.P. (1979) Science 206, 1073.

Block, S. and Forman, R.A., Piermarini, G.J. (1977) p503 in "High Pressure Research Application in Geophysics (ed.M.H.Manghnani and S.Akimoto: Academic Press, NY).

Bolsaitis, P. and Spain, I.K. (1977) p477 in "High Pressure Technology" (ed. I.L. Spain and J.Paauw: Marcel Dekker, NY).

Bradley, C.C. (1969), "High Pressure Methods in Solid State Research", Plenum Press, New York.

Brasch, J.W. (1980), Rev.Sci.Instr.51,1358.

Brasch, J.W., Melveger, A.J., Lippincott, E.R.(1968), Chem.Phys.Lett.2,99.

Bridgman, P.W. (1952), "The Physics of High Pressure" (Bell and Hyman, London).

Camphausen, D.L., Connell, G.A.A., Paul, W. (1971),Phys.Rev.Lett.26,184.

Carlone, C., Olego, D., Jayaraman, A. and Cardona, M.(1981),Phys.Rev. B22, 3877.

Chang, K.J. and Cohen, M.L. (1985), Phys.Rev.31,7819.

Crossland, B, and Spain, I.L. (1983), "High Pressure Generation and Containment" Chapter 8, p.307 in "High Pressure Measurement Techniques", (ed. G.N.Peggs: Applied Sci.Publ., London and New York).

Fujiwara H., Kadomatsu, H., Tohma, K. (1980), Rev.Sci.Instr.51,1345.

Gonzalez, J., Besson, J.M. and Weill, G.(1986), Rev.Sci.Instr.57,106.

Goettel K.A.., Mao H.K., Bell P.M. (1985), Rev.Sci.Instr.56,1420.

Gunney, B.J., Patel, D., Tatham, H.L., Hayes, J.R. and Adams, A.R.(.982), p.481 in "High Pressure in Research and Industry" (ed. C-M Backman, T Johanisson, and L Tegner: Arkitektkopia, Uppsala).

Hirsch, K.R. and Holzapfel, W.B. (1981), Rev.Sci.Instr.52,52.

Hu, J.Z., Menoni, C.S., Merkle, L.D. and Spain, I.L. (1986), to be published in Phys.Rev.B.

Hu, J.Z. and Spain, I.L. (1984), Sol.St.Comm.51, 263.

Jamieson, J.C. (1963), Science 139, 762.

Jayaraman, A. (1983), Rev.Mod.Phys.55, 65.

Konczykowski, M., Baj M, Szafrankiewicz, E., Konczewicz, L and Porowski,S. (1977), Proc.Int.Conf. on High Pressure and·Low Temperature Physics p.124, (edited C.W.Chu and J.A. Woollam).

Lancefield, D., Adams, A.R., Gunney, B.J. (1984). Appl.Phys.Lett.$\underline{45}$,1121.

Le Sar, R., Ekburg, S.A., Jones, L.N., Mills, R.L., Schalbe, L.R. and Shiferl, D. (1979), Sol.State.Comm.$\underline{32}$,131.

Leroux, M., Neu, G., Verie, C (1966), Sol.St.Comm.$\underline{58}$,289.

Lifshitz, N. and Maines, R.G.(1979), Rev.Sci.Instr.$\underline{50}$,608.

Mao, H K. and Bell, P.M.(1981), Rev.Sci.Instr.$\underline{52}$,615.

Mao, H.K., Bell, P M., Shaner, J.W., Steinberg, D.J.(1978),J.Appl.Phys. $\underline{49}$, 3276.

Menoni, C.S. and Spain I.L. (1983), "Ultra-high Pressure Measurement", Chapter 4 p.125 in "High Pressure Measurement Techniques", ed. G.N. Peggs, Appl.Sci. Publ. London and New York.

Menoni, C.S., Spain, I.L., Hochheimer, H.D., (1986), Phys.Rev.$\underline{B33}$,5896.

Mills, R.L., Liebenberg, D.N., Bronson, J.C., Schmidt, L.C.(1980), Rev.Sci.Instr.$\underline{51}$, 891.

Moss, W.C., Hallquist, J.O., Reichlin, R., Goettel, K.A. and Martin,S. (1986), Appl.Phys.Lett.$\underline{48}$,1258.

Muller, H., Trommer, R., Cardona, M., Vogl, P. (1980), Phys.Rev.$\underline{B21}$,2641.

Noach, R.A. and Holzapfel, W.B. (1977), p748 in Vol.I of "High Pressure Sci and Tech" (ed. K.D.Timmerhaus and M.S.Barker: Plenum Press, London and New York).

Needs, R.J., and Martin, R.M. (1984), P-ys.Rev.$\underline{B30}$,5390.

Olijnyk, J., Sikka, S.K., Holzapfel, W.B.(1984), Phys.Lett $\underline{103A}$,137.

Patel, D., Crumbaker, T., Sites, J.R. and Spain, I.L. (1986), (accepted by Rev.Sci.Instr.).

Paul, W., Benedek, G.B., Warschauer, D.M. (1959), Rev.Sci.Instr.$\underline{30}$,874.

Paul, W. and Warschauer, D.M. (.956), Rev.Sci.Instr.$\underline{27}$, 418.

Paureau, J., (1977) J.Phys.$\underline{E10}$,1093.

Wasilewski, Z., Stradling, R.A., Porowski, S., (1985), Sol.Stat.Comm. $\underline{57}$,123.

Peggs, G.N. and Wisniewski, R.(1983), "Electrical Resistance Gauges", Chapter 6, p215 in "High Pressure Measurement Techniques", (ed. G.N.Peggs: Appl.Sci.Publishers, London and New York).

Piermarini, G J. and Block, S.(1975), Rev.Sci.Instr.$\underline{46}$,973.

Piermarini, G.J., Block, S., Barnett, J.D., Forman, R.A.(1975), J.Appl.Phys.$\underline{46}$,2774.

Pitt, G.D.(1968), J.Phys.$\underline{E1}$,915.

Pitt, G.D., Lees, J., Hoult, R.A., Stradling, R.A. (1973), J.Phys.$\underline{C6}$,3282.

Reichlin, R.(1983), Rev.Sci.Instr.$\underline{54}$,1674.

Sakai, H., Kajiwana, T., Twuji, K., Minomura, S.(1982), Rev.Sci.Instr.$\underline{53}$, 499.

Scaife, W.G., and Peggs, G.N.(1983), "Pressure Transducers Based on Various Physical Effects", Chapter 5, p.179 in "High Pressure Measurement Techniques", ed. G.N.Peggs, Applied Sci.Publ.London and New York.

Schirber, J.E. (197o), Cryogenics 10,418.

Seal, J. (1984), High Temp.High Press 16,573.

Schimizu, H.E., Brody, E.M. Mao, H.K., Bell, P.M. (1981), Phys.Rev.
    Lett. 47, 128.

Skelton, E.F., Qadri, S.B., Webb, A.W., Lee, G.W.(1983), Rev.Sci,Instr.
    54,403.

Tanaka, J. and Maeda, J. (1986), Rev.Sci.Instr.57, 500.

Tozer, S.W., and King, H.E.. (1975), Rev.Sci.Instr.56,260.

Vyas, M.K.P., Pitt, G.E., Hoult, R.A.(1973), J.Phys.C6,285.

Walling, La P., Ferrarro, J.R. (1978), Rev.Sci.Instr.49,1557.

Wasilewski, Z., Porowski, S., Stradling, R.A.(1986), J.Phys.E19,480.

Weinstein, B.A., (1986) to be published.

Weinstein, B.A. and Piermarini, G.J. (1975), Phys.Rev.B12,1172.

Welber, B.(1976), Rev.Sci.Instr.47,183.

Welber, B.(1977), Rev.Sci.Instr.48,395.

Welber, B., Cardona, M., Kim, C.K., Rodriguez, S.(1975), Phys.Rev.B12,5729.

Welber, B., Cardona M., Tsay, Y.F., Bendow, B.(1977), Phys.Rev.B15,875.

Whitfield, C.H., Brody, E.M., Bassett, W.A. (1976), Rev.Sci.Instr.47,942.

Wolford, D.J. and Bradley, J.A. (1985), Sol.St.Comm.53,1069.

Wolford, D.J., Kuech, T.F., Bradley, J.A., Gell, M.A., Ninno, D.,
    Janos, M. (1986), Proc.of Int.Conf.on Physics and Chemistry of
    Semicondustor Interfaces, Pasadena, California, to be published
    in J.Vac.Sci.Tech.

Yin, M.T. and Cohen, M.L. (1982), Phys.Rev. B26, 5668.

Zhao, Y-Z., Buehler, F., Sites, J.R. and Spain, I.L. (1986), to be
    published in Sol.St.Comm.

# HIGH PRESSURE TRANSPORT  EXPERIMENTS
# IN 3 DIMENSIONAL SYSTEMS

Roger-Louis Aulombard, Abderrahmane Kadri, and
Karima Zitouni

Groupe d'Etudes des Semiconducteurs, U.S.T.L.
Place E. Bataillon, 34060 Montpellier Cedex, France

## INTRODUCTION

For more than twenty years hydrostatic pressure has been used as a
variable in experimental measurements in semiconductors. The investigation of its effect on the properties of 3D systems is clearly important
for a basic understanding of band structure, conduction processes, etc.
As early as 1968, W. PAUL (1) gave an extensive plenary paper on this
subject, and since then many authors have reviewed further developments
R.A. STRADLING (2) published one of the latest papers on the subject in
the Advances in Solid State Physics (1985).

In this paper we deal with the effects of hydrostatic pressure on
transport properties in 3D semiconductors.

It is well known that in semiconducting materials energy
band gaps and their associated electronic states (hydrogenic behavior) or
not purely electronic states (deep level behavior) may depend on external perturbations of the sample. Among these perturbations is pressure
which, if applied uniformly and before a phase transition sets in, pre-
serves crystal symmetries while the semiconducting properties change over
the interatomic distances.

Before discussing the influence of pressure on band structure, impurities, etc, we briefly comment on the experimental techniques used in
high-pressure studies.

EXPERIMENTAL TECHNIQUES

Very high pressures are easy to achieve in very small volumes. This is the case of diamond anvils, mainly used in optical (3) or X-ray (4) experiments. They can be used to obtain hydrostatic pressures of up to 80 kbar (Pressures are hydrostatic to better than 1 part in 500 according to WOLFORD and BRADLEY (5) ).

Experiments requiring more space can be performed in a high pressure Be-Cu clamp cell (see example in Fig 1-b) or in a Be-Cu cell connected to a gas compressor (Fig 1-a). Helium gas is used as a pressure transmitting medium and the experimental cell is connected to the compressor by a flexible high pressure capillary tube.

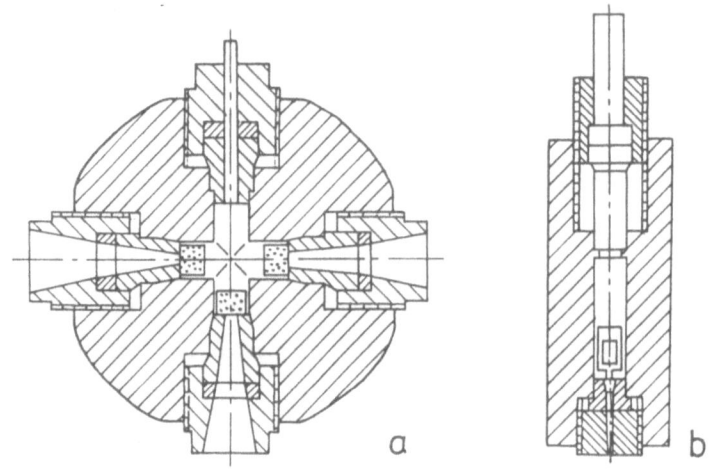

Fig 1 After S. POROWSKI et al (6) High Pressure "UNIPRESS" cells
        a) optical cell connected to a gas compressor by a capillary tube.
        b) clamp cell with liquid for pressures of up to 27 kbar.

Here the maximum pressure usually does not exceed 25 kbar but this technique is very flexible, and the temperature range available for investigations can be very broad (from T < 4.2K to T > 1000K). Moreover high magnetic fields can be easily applied.

THE EFFECT OF PRESSURE ON BAND STRUCTURE

When hydrostatic pressure is applied, the various electronic energy bands shift relative to one another. The main effects are to increase the value of the direct ($\Gamma$) energy gap (typically at $\simeq$ 10 ~ 15 meV/kbar) and

the indirect bandgap at the L point ($\simeq$ 5 meV/kbar), and to decrease the
indirect bandgap at the X point, ($\simeq$ -1~-5 meV/kbar). (Fig 2)

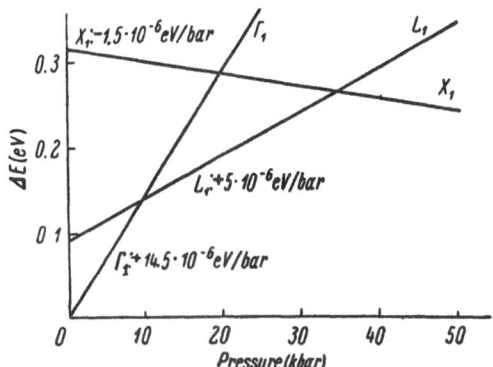

Fig 2 Shift with pressure of the three lowest conduction-band extrema of
GaSb, relative to the valence band maximum (after W. PAUL (1) ).

Due to the different variations with pressure of the conduction band
minima, the band structure can change from a direct to an indirect configuration and carrier tranfer may occur. One of the most significant results has been obtained by WOLFORD and BRADLEY in gallium arsenide (5)
from an analysis of photoluminescence and PL-excitation under pressure.
They obtained precise $\Gamma_{1c}-\Gamma_{15v}$ and $X_{1c}-\Gamma_{15v}$ gap dependences on pressure
(10.73 and -1.34 meV/kbar respectively) and deduced new indirect band-
gap energies at atmospheric pressure.

Transport experiments provide an opportunity to analyze the pressure-
induced electron redistribution between the different conduction band minima.   For example this technique was used (7-8) to specify the band structure   of the ternary system $Ga_{1-x}Al_xSb$. Despite encouraging reports the
band structure of the $Ga_{1-x}Al_xSb$ alloy was still insufficiently known. A
survey of the literature indicated scattered results. Consequently the
Shubnikov-de Haas oscillations(Fig 3) and the Hall effect were investigated as a function of pressure. Using the experimental results the distribution of the carriers between $\Gamma$ and L valleys were calculated and the

301

distance between these two minima was determined. Finally a value was proposed for x at which direct-indirect transition occurs.

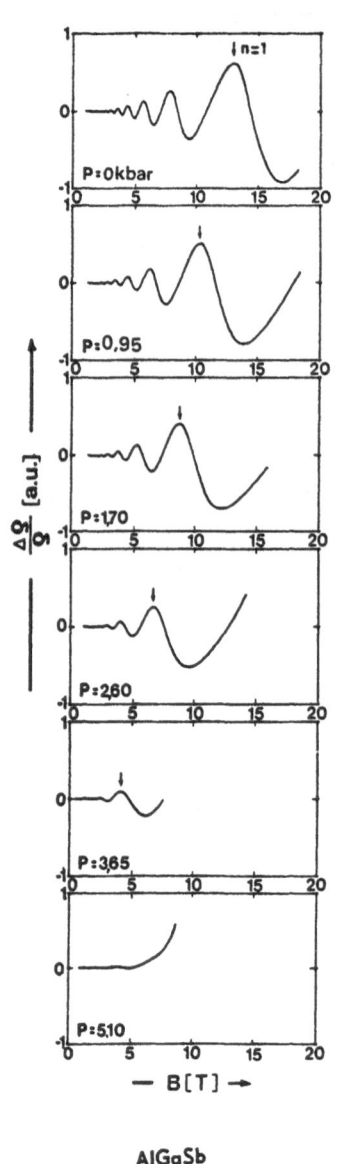

Fig 3 Oscillatory portion of the longitudinal magnetoresistivity at T = 4.2°K for $Ga_{0.93}Al_{0.07}Sb$.

$Ga_{1-x}Al_{x}Sb$ alloy is an attractive candidate for optoelectric devices at long wave lengths in the spectral regions of 1.3 - 1.55 $\mu$m where the fused silica fibers exhibit the lowest dispersion and loss. Consequently, precise knowledge of the band structure is a prerequisite for evaluating its expected domain of application.

As can be seen in the above example, transport experiments under hydrostatic pressure can give fundamental information about semiconductor band structure.

Since impurity states govern most of the properties of crystals, their investigation is of great significance for Solid-State Physics. Practically all experiments that provide information about impurity states can be performed under hydrostatic pressure. Shallow donors have been found to follow their "associated" conduction band minima as they shift pressure. In contrast, deep levels move toward or away from the nearest conduction band edge and are sometimes located too far below this band to be covered by the effective-mass approximation, even taking into account effective-mass and the dielectric constant versus pressure variations. Moreover, some of these deep states may exhibit metastability or persistent photoconductivity at low temperatures (DX center).

In this section, the above effects are explained using examples of transport experiments results in various materials. We start with the hydrogenic and non hydrogenic states in pure InSb and we discuss how pressure can be used to vary free electron density in the same sample through metastable state occupation. We then give data for pure n-InAs and some doped binary and ternary compounds.

## Undoped InSb

Great interest has been shown in the low free carrier density of n-type InSb. In the purest available samples, residual impurity still introduces three states (9 to 12) ; a shallow level and two deep levels.

The first (the hydrogenic level) has been extensively studied. It is located within or very near ( <0.6 meV) the conduction band. The problem of the dependence of the binding energy on the magnetic field has been discussed in detail and many authors have looked into the more specific problem of magnetic-field-induced metal-nonmetal transition in n-type InSb (13). In the purest sample ($n < 810^{13}$ $cm^{-3}$) there is evidence that the activation energy of the hydrogenic level corresponds to the effective Rydberg (0.7 meV below the conduction band).

It is known that hydrogenic states are dominated by a long-range impurity potential described by the effective mass theory. Therefore such impurity states shift with pressure along with their associated minima. If, as is often the case, the metal-nonmetal transition is studied as a function of magnetic field, high pressure shifts the field at which the transition occurs (Fig 4). S. POROWSKI (6) gives the following law to describe the pressure variation of the hydrogenic level relative to the conduction band minimum :

$$E_{\Gamma h} = 0.6 + 0.001 \, P \text{ (meV), with P in MPa.}$$

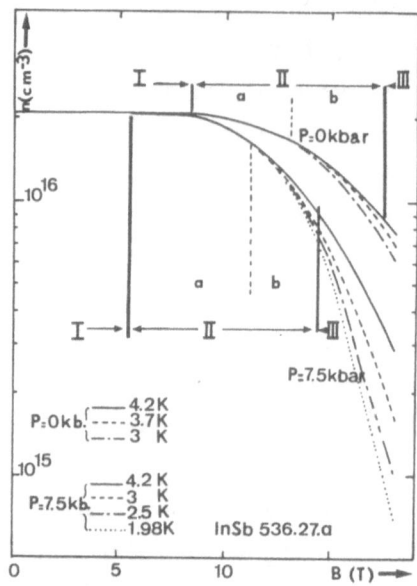

Fig 4 Magnetic field dependence of the carrier density at different tem-
    peratures of a InSb sample with and without pressure (after
    J.L. ROBERT et al (13) ).

In pure n - InSb with extrinsic concentrations below $10^{15}$ $cm^{-3}$ at
pressures above 7 kbar, a deionization of the other two donor levels was
observed. At ambiant pressure the levels lie above the bottom of the Γ band
(Fig 5) but their energies $E_{dL}$ and $E_{dX}$ relative to the Γ band decrease
linearly with pressure :

$$E_{dL} = -85 + 0.105 \ P \ (meV)$$
$$E_{dX} = -140 + 0.2 \ P \ (meV)$$

It can be seen that with increasing pressure, the impurity levels
are driven into the fundamental gap. A surprising result is that the
pressure coefficients are close to those expected for the X and L minima.
Nevertheless the separations are far to great for the levels to be described by simple effective mass theory. There is, however, a fundamental difference between the respective behaviors of these two levels.

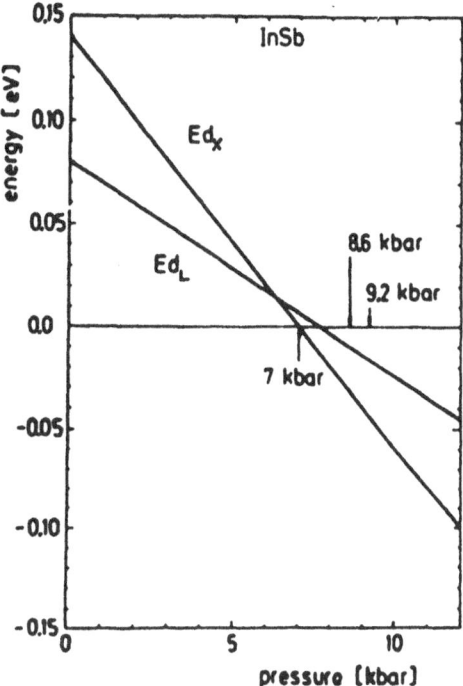

Fig 5 Energies of the $E_{dL}$ and $E_{dX}$ states relative to the $\Gamma$ conduction band edge as a function of pressure.

Whereas the $E_{dL}$ level has equal thermal (11) and optical (12) ionization energies ; this is not the case of the $E_{dX}$ level. An important point for more extensive $E_{dX}$ level studies is the evidence of metastability effects obtained by changing the pressure conditions for sample cooling. These effects are clearly correlated with the pressure induced changes in the thermal ionization energies, which determine the distribution of carriers between the conduction band and the impurity level. The existence of a potential barrier to the transitions between the impurity level and the conduction band, and the long persistence times of the off-thermodynamic equilibrium as the temperature is lowered toward 77K are

all consistent with the expected behavior of impurity centers with large
lattice relaxation (14). This means that the resonant impurity levels
responsible for the observed effects are not purely electronic, but are
affected by a strong electron-lattice coupling, i.e. any change in the
charge configuration of the impurity centers is accompanied by a drastic
rearrangement of the local atomic environment. This implies that the impurity states cannot be regarded as effective-mass-like but arise from a
strong and short-ranged potential. To describe such impurity centers, a
large lattice relaxation model is needed (Fig 6).

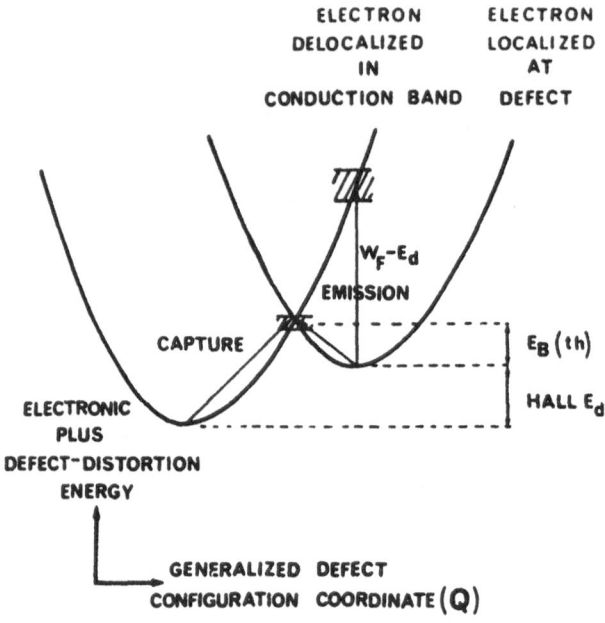

Fig 6 Configuration coordinate diagram for the $E_{dX}$ state in pure
   n-InSb.

Some workers (16) performing experiments on n-type InSb at 77K have
shown the possibility of obtaining low electron concentrations and high
mobilities by freezing electrons in this metastable state. By varying
free electron density in the same sample from $10^{11}$ $cm^{-3}$ to $10^{14}$ $cm^{-3}$,
using the lattice coupled defect states as a "tank" of electrons, the
metal-nonmetal transition was observed (17-18). Interplay between corre-
lation and localization effects at the metal-insulator transitions has
been shown (19). Moreover, using this high pressure freeze-out method the

static dielectric constant    could be measured from typical capacitance measurements on high resistant InSb (21).

As a concluding remark about energy levels in pure n-type InSb, it should be noted that the $E_{dL}$ states originate from the same impurity centers as some of the shallow impurity states (12,20). These experiments have also evidenced a level crossing interaction between $E_{dL}$ and $E_{Th}$ when the energies of these two levels approach one another. However, because of the small  value of the interaction energy, it is difficult to detect this crossing interaction in transport experiments (11).

### Undoped InAs

The investigation of impurity states in InAs is similar to that described above in n-type InSb.

Due to the small effective mass in the $\Gamma$ minimum, a strong effect of the magnetic field on the shallow donors has been observed in magnetic freeze out experiments (22). In the nominally undoped samples (n~ 1-2 x $10^{16}$ $cm^{-3}$) at zero magnetic field, the hydrogenic donor level has been found to be degenerate with the continuum of the conduction band. Room temperature electrical experiments under pressure (23-24) have shown that the electrons are trapped in two resonant-impurity levels (one located at 60 meV above the $\Gamma$ band edge and the other at 340 meV above it).

We have observed (25) a pressure dependence of the shallow impurity level (-0.077 meV/kbar) and a crossing between this level and the resonant level (located at 68 meV above the $\Gamma$ conduction band and shifting with pressure at a rate of -4 meV/kbar with respect to the minimum). Moreover, an extra-deepening of the shallow-donor level is observed when the pressure and the magnetic field are high enough to induce the occupation of the resonant states (Fig 7).

Although an $E_{dX}$ level, characterized by a metastable occupation at low temperatures,  may exist in InAs (level located at 340 meV) it cannot be observed, because it remains too high in energy to be analyzed at pressures lower than 20 kbar.

### Observation of donor states in some doped materials

GaAs material is not considered in this section. Information on this subject can be obtained in the special issue of semiconductors and semi-metals (26) which summarizes the most interesting GaAs results. We will here give a brief review of results concerned with resonance donor states introduced by impurities in some binary or ternary compounds.

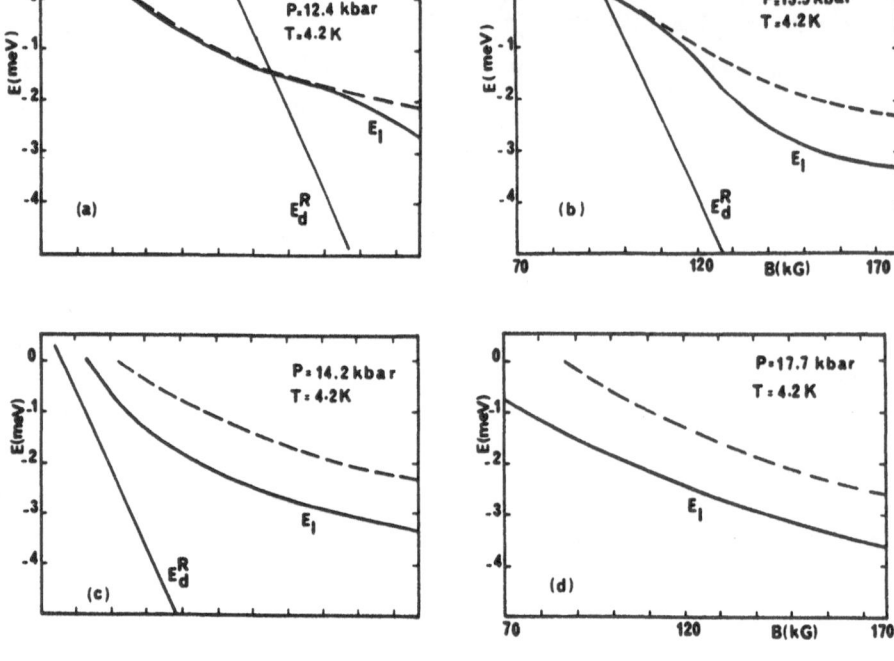

Fig 7 Variations of the resonant ($E_d^R$) and shallow ($E_I$) levels versus magnetic field at different pressures. The dashed line is the energy of the shallow donor level extrapolated from the low pressure results.

S and Se doped InSb

In doped InSb impurities such as sulfur and selenium introduce highly localized donor states. At normal pressures the resonance states have been found to lie above the bottom of the $\Gamma$ conduction band (27) and their energies with respect to this minimum are $E_S = 0.55$ eV in S-doped samples and $E_{Se} > 0.6$ eV in Se doped samples. Moreover, the energetic distance from the L and X minima is at least one order of magnitude greater than the ionization energy of the hydrogenic levels of L and X subbands. The one site-one band Koster-Slater model has been used (28) to describe the highly localized potential of these levels.

Ternary compounds

Examples of impurity states with large lattice relaxation have been given in $Ga_{1-x}Al_xSb$ : Te (29), $Ga_xIn_{1-x}Sb$ : S (30), $Ga_xIn_{1-x}Sb$ : Se (30) and $Ga_{1-x}Al_xAs$ : doped with Si (31), Sn, Te, S or Ge. The following table summarizes the positions of the levels introduced into the gap by donors in these materials, and their resulting potential barriers. The table suggests that Te is the shallowest of the non-$\Gamma$ donors followed by Si, Se, S and Ge in order of increasing depth.

Table 1  Summary of the levels introduced into the gap by donors in some
ternary compounds. The zero energy is taken at the conduction
band minimum.

For $Ga_{1-x}Al_xAs$ materials see STRADLING (2).

| Material | Donor | energetic position E (meV) | potential barrier (meV) |
|---|---|---|---|
| $Ga_{0.78}In_{0.22}Sb$ | S | 130 | 180 |
| $Ga_{0.78}In_{0.22}Sb$ | Se | 180 | 240 |
| $Ga_{0.7}Al_{0.3}Sb$ | Te | -46 | 260 |

COMPARISON BETWEEN THE EFFECTS OF ALLOYING AND HYDROSTATIC PRESSURE

Alloying a direct-gap semiconductor with another of wider gap (e.g.
InSb with GaSb to form $Ga_xIn_{1-x}Sb$ or GaSb with AlSb to form $Ga_{1-x}Al_xSb$)
has an effect similar to pressure in changing the relative positions of
the bands. Consequently, the investigation of the electrical properties
of ternary or quaternary systems with respect to the alloy composition is
of current interest in the field of superlattice growth.

The following table gives coefficients demonstrating the similarity
between pressure and alloying in opening up the direct band-gap of some
III-V compounds (from narrower to wider gaps).

Table 2 Pressure and "alloying" coefficients in some III-V ternary compounds.*Very nonlinear because of strong bowing.

For example in $Ga_xIn_{1-x}Sb$  The variation law for $E_\Gamma(x)$ is :
$$E_\Gamma(x) = 0.235 + 0.161\ x + 0.415\ x^2\ (32).$$

| Material | $\dfrac{dE_\Gamma}{dP}$ (meV/kbar) | $\dfrac{dE}{dx}$ (meV/atomic %) |
|---|---|---|
| $Ga_xIn_{1-x}Sb$ | 15 | 1.7 to 10 * (x = 0 to x = 1) |
| $Ga_{1-x}Al_xSb$ | 14,7 | 11 to 19 * (x = 0 to x = 1) |
| $Ga_{1-x}Al_xAs$ | 11.5 | ~ 15 |
| $In\ Sb_{1-x}As_x$ | 15 | ~ 2 * |

Moreover, alloying has an effect similar to pressure in introducing non levels into the bandgap. Consequently, the determination of the positions and pressure variations of the impurity states in particular alloy compositions will be important in the design of devices using such superlattices. In Figs 8 and 9 we show typical variations of energy bands and energy levels in $Ga_xIn_{1-x}Sb$ as a function of pressure and alloy composition.

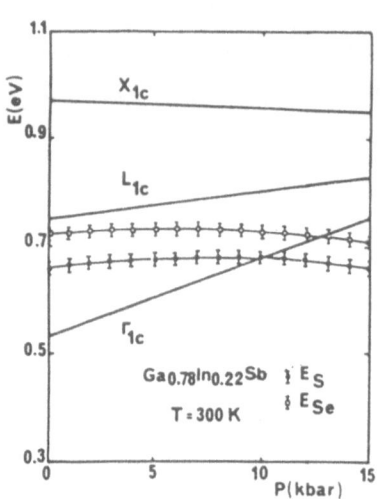

Fig 8 S- and Se-related energy levels and conduction band minima versus pressure in $Ga_{0.78}In_{0.22}Sb$ (with respect to the top of the valence band).

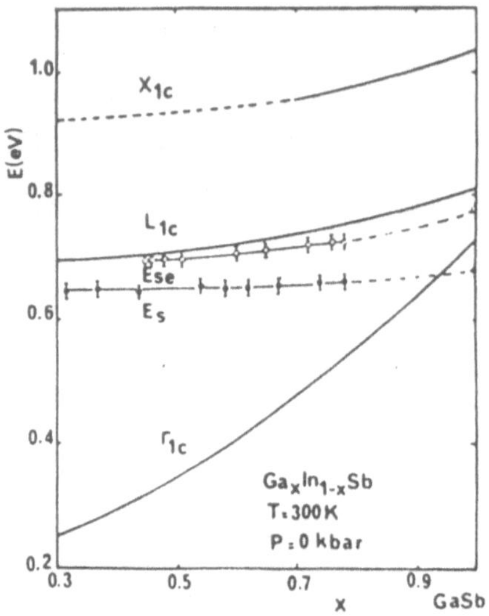

Fig 9 Conduction band minima and S- and Se-related energy levels as a function of mole percent of GaSb (with respect to the top of the valence band).

ACKNOWLEDGEMENTS

The authors would like to thank Professor J.L. ROBERT (G.E.S. - U.S.T.L. - Montpellier FRANCE) for many helpful discussions and we are very grateful to L. KONCZEWICZ and S. POROWSKI (UNIPRESS - Warsaw POLAND) for introducing us to high pressure technology and for helpful discussions.

# REFERENCES

1. W. PAUL, Proc. IX Int. Conf. Physics Semicond. Moscow, vol 1, p 16 (1968).

2. R.A. STRADLING, Advances in Solid State Physics, Edited by P. GROSSE (1985).

3. G.J. PIERMARINI and S. BLOCK, Rev. Sci. Instrum 46,33 (1975).

4. B. WELBER, Rev. Sci. Instrum. 47, 183 (1976).

5. D.J. WOLFORD and J.A. BRADLEY, Solid State Com. 53, 12 (1985).

6. S. POROWSKI and W. TRZECIAKOWSKI, Phys. Stat. Sol (b) 128, 11 (1985).

7. R.L. AULOMBARD, C. BOUSQUET, J.L. ROBERT, L. KONCZEWICZ, E. LITWIN-STASZEWSKA, S. POROWSKI, Proc. Int. Symp. on GaAs and related compounds, Vienne (1980).

8. R.L. AULOMBARD, L. KONCZEWICZ, A. KADRI, A. JOULLIE and J.C. PORTAL Proc. Int. Symp. on GaAs and related compounds, Biarritz (1984), (IOP, Bristol, 1985);

9. S. POROWSKI, M. KONCZYKOWSKI and J. CHROBOCZEK, Phys. Stat. Sol. (b) 63, 291 (1974).

10. L. DMOWSKI, M. KONCZYKOWSKI, R. PIOTRZKOWSKI and S. POROWSKI, Phys. Stat. Sol (b) 73, K131 (1976).

11. S. POROWSKI, L. KONCZEWICZ, A. RAYMOND, R.L. AULOMBARD, J.L. ROBERT and M. BAJ, Lect. Notes Phys. 177, 357 (1983).

12. Z. WASILEWSKI, A.M. DAVIDSON and R.A. STRADLING, Proc. XVI Int. Conf. Phys. Semicond. Montpellier (1982) p 89.

13. J.L. ROBERT, A. RAYMOND, R.L. AULOMBARD and C. BOUSQUET, Phil. Mag. B 42, 6 (1980).

14. D.V. LANG and R.A. LOGAN, Phys. Rev. Letters 39, 635 (1977).

15. D.V. LANG, R.A. LOGAN and M. JAROS, Phys. Rev. B 19, 1015 (1979).

16. E. LITWIN, W. SZYMANSKA and R. PIOTRZKOWSKI, Proc. 4th Int. Conf. Phys. narrowgap semiconductors, LINZ, AUSTRIA (1981).

17. A. KADRI, R.L. AULOMBARD, C. BOUSQUET, A. RAYMOND and J.L. ROBERT, Proc. XVI Int. Conf. Phys. Semicond. Montpellier (1982) p 235.

18. A. KADRI, M. BAJ, K. ZITOUNI, R.L. AULOMBARD, C. BOUSQUET, J.L. ROBERT and L. KONCZEWICZ, Rev. Phys. Appl. 19 (1984).

19. A. KADRI, K. ZITOUNI and R.L. AULOMBARD, X[th] Int. AIRAPT Conf. Amsterdam (1985).

20. Z. WASILEWSKI, A.M. DAVIDSON, P. KNOWLES, S. POROWSKI and R.A. STRADLING, Proc. 4th Int. Conf. Phys. narrow gap semiconductors, LINZ, AUSTRIA (1981).

21. J.C. THUILLIER, L. KONCZEWICZ, R.L. AULOMBARD, A. KADRI, X[th], Int. AIRAPT Conf. Amsterdam (1985).

22. L.A. KAUFMANN and L.J. NEURINGER, Phys. Rev. B 2, 1840 (1970).

23. D.G. PITT and M.K.R. VYAS, J. Phys. C 6, 214 (1973).

24. A.N. EL SABBAHY and A.R. ADAMS, Proc. of the XIV Int. Conf. on Phys. Semicond. Edinburgh (1978).

25. A. KADRI, R.L. AULOMBARD, K. ZITOUNI, M. BAJ and L. KONCZEWICZ, Phys. Rev. B 31, 12 (1985).

26. SEMICONDUCTORS AND SEMIMETALS Vol 20, Semi-insulating GaAs, Edited by R.K.A. WILLARDSON and A.C. BEER (1984).

27. S. POROWSKI, L. KONCZEWICZ, J. KOWALSKI, R.L. AULOMBARD and J.L. ROBERT, Phys. Stat. Sol. (b) 104, 657 (1981).

28. L. KONCZEWICZ and W. TRZECIAKOWSKI, Phys. Stat. Sol. (b) 115, 359 (1983).

29. L. KONCZEWICZ, E. LITWIN, S. POROWSKI, A. ILLER, R.L. AULOMBARD, J.L. ROBERT and A. JOULLIE, Proc. XVI Int. Conf. Phys. Semicond. Montpellier, FRANCE (1982).

30. K. ZITOUNI, A. KADRI and R.L. AULOMBARD, Phys. Rev. B 34, 4 (1986).

31. J.C.M. HENNING, J.P.M. ANSEMS and A.G.M. de NIJS, J. Phys. C 17, L915 (1984).

32. D. AUVERGNE, J. CAMASSEL, H. MATHIEU and A. JOULLIE, J. Phys. Chem. Solids 35, 133 (1974).

# HIGH PRESSURE TRANSPORT EXPERIMENTS IN 2D SYSTEMS

J.L. Robert, A. Raymond and C. Bousquet

Groupe d'Etude des Semiconducteurs, UA 357
Université des Sciences et Techniques du Languedoc
34060 - Montpellier-Cédex, France

Hydrostatic pressure can be used as a variable to introduce deep levels in the forbidden gap of III-V compounds. This technique is applied in the case of GaAs/Ga$_x$Al$_{1-x}$As heterojunctions to vary the density $n_s$ of the 2D electron gas. We show that one can reduce $n_s$ to values lower than $5.10^{10}$cm$^{-2}$ even for highly doped samples. As a result, several physical effects can be examined on the same sample such as:

- the $n_s$ dependence of the mobility
- the effect of screening on the width of the quantum Hall plateaux
- the magnetic field induced metal non metal transition.

All these effects are discussed in this paper.

## INTRODUCTION

The development of thin film growth techniques such as M.B.E. and MOCVD during the past decade has made it possible to produce abrupt semiconductor-semiconductor interfaces. A quasi two-dimensional electron gas is obtained, located at the interface in a quantum well, characterized by the existence of subbands due to electric quantization. In this potential well, the motion of electrons perpendicular to the interface is quantized in discrete eigen-states whereas the motion parallel to the interface is free. Unusual properties have been observed in selectively doped heterostructures. To reduce impurity scattering, Esaki [1] suggested spatially separating the free electrons and their parent impurities. This concept of modulation doping has been successfully put into concrete form in the GaAs-GaAlAs system and peak electron mobilities have recently exceeded $2.10^6$ cm$^2$ /Vs. The M.D. heterostructures usually consist of a nominally undoped GaAs layer followed by an undoped GaAlAs layer, a Si-doped GaAlAs layer and a thin cap layer to facilitiate ohmic contact formation. The basic structure is shown in Figure 1.

The smaller gap GaAs (material 1) has a higher electron affinity ($\chi_1$) than GaAlAs (material 2, $\chi_2$) forming a two dimensional electron gas at the interface. In the energy band diagram $\Delta E_c$ and $\Delta E_v$ represent the conduction band and the valence band off-sets at the interface between the smaller gap semiconductor ($Eg_1$) and the larger gap semiconductor ($Eg_2$). $E_F$ is the Fermi level, $\Delta E_G$ is equal to $Eg_1-Eg_2$. The undoped GaAlAs layer (the so-called spacer) between the Si-doped GaAlAs and the GaAs reduces the coulombic interactions with the parent donors. The increase in mobility is then obtained at the expense of electron transfer. Large differences are observed in the electronic properties of different heterojunctions, depending on the growth technique and the manufacturer. Consequently, it is difficult to compare different heterojunctions since each of them is unique by nature.

In order to achieve a better understanding of the electronic properties of such a two dimensional electron gas, we chose to use hydrostatic pressure as an external parameter to change the energy diagram of the structure. The obvious advantage of this method is that the electronic properties can be modified under pressure without changing the intrinsic properties of the material. In this paper, we show that hydrostatic pressure techniques are particularly convenient in the case of GaAs/GaAlAs heterojunctions - because the deep impurity states that govern the activation processes in Si-doped GaAlAs are very sensitive to pressure. As a result, electron transfer in the quantum well can be controlled by pressure. In the first section, we give a short review of the effect of hydrostatic pressure on impurity states in a 3D semiconductor. The consequences of this effect on the energy diagram of the GaAs/GaAlAs heterojunctions is then discussed. The modification of the diagram is correlated with a deionization of Si-impurities when pressure increases - and as a result a decrease in the 2D electron gas density $n_s$ is observed. The possibility of controlling $n_s$ by the application of pressure enabled us to study various effects on the same heterojunctions. The following topics are discussed in the second part of this paper:

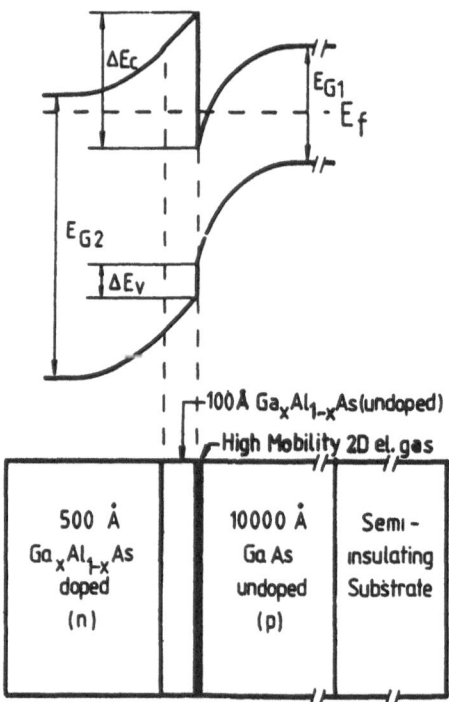

Fig. 1.  Basic structure of the GaAs/GaAlAs heterojunctions.

- the dependence of mobility on electron density. This problem is usually studied using several heterojunctions in which the width of the spacer layers is increased at the expense of electron transfer.

- the localisation effects, which are observable via the quantum Hall effect and also via the metal non metal transition in the ultra quantum limit on a dilute 2D electron gas.

This last point is of particular interest, since it demonstrates the existence of a new kind of magneto-donor, composed of donor atom and electron spatially separated from each other by the spacer.


I - PROPERTIES OF IMPURITY STATES UNDER HYDROSTATIC PRESSURE IN 3D AND 2D SYSTEMS

I.1  3D Case

It is well known that the electronic properties of semiconductors are highly dependent on the nature of the impurity states which give rise to the conduction electrons. The usual doping impurities introduce both shallow donor states and deep levels taking their predominant character from higher conduction band minima. In materials whose conduction band is located at the centre of the Brillouin zone, these deep levels are often resonant with the $\Gamma$ minimum of the conduction band, but they can also be located inside the forbidden band. In transport experiments, shallow and deep impurity states can be conveniently distinguished using hydrostatic pressure [2-3].

In contrast with shallow impurity states, which have a relatively small pressure coefficient, deep impurity states follow the shift in energy of the L and X minima when pressure is applied and their position relative to the $\Gamma$ minimum vary at a rate of the order of 10 meV/kbar. As a result, deep impurity states that are resonant with the $\Gamma$ minimum under zero pressure drop to the forbidden band when pressure is applied. Those that are located in the forbidden band move away from the $\Gamma$ minimum. When such levels are involved in the conduction processes, a strong decrease in the carrier concentration can be expected when pressure is applied. This is typically the case of Si-doped $Ga_{1-x}Al_xAs$ compound. Alloying has an effect that is similar to pressure, introducing deep levels into the forbidden gap when the Al content x > 0.2. In spite of uncertainties about the energies of the conduction minima, it can be said that the $\Gamma$-X cross-over occurs when x ~ 0.4. The two values of x correspond to the two limits between which the GaAlAs compound is usually grown when constructing GaAs/GaAlAs heterojunctions. Thus, it is obvious that, under these conditions, hydrostatic pressure can be used to pull the electrons back from the GaAs into the GaAlAs. It should be noted here that the Si level involved in this process has a metastable character : persistent photoconductivity experiments have clearly shown that this level is strongly coupled to the lattice.

I.2  2D Case

In the following we present a theoretical model that accounts for the pressure dependence of the 2DEG density. It corresponds to the case of $Ga_{.7}Al_{.3}As$/GaAs heterojunctions, the basic structure of which is given in Fig. 1.

The potential drop $V_{20}$ across the ionized part of the doped layer is related to $\Delta E_c$ by the following equation. (See Fig. 2.)

$$v_{20} = \Delta E_c - (E_F - E_{C2}) - (E_F - E_o) - E_o$$

$E_F - E_{C2}$ determines the position of the Fermi level far from the interface. $E_o$ is the position of the lowest electric subband in the quantum well. An approximate relation between $n_s$ and the charge stored in the accumulation layer of a semiconductor, has been derived under conditions of non degeneracy [4].

$$n_s = \sqrt{2\ \epsilon_2\ N\ v_{20}/e^2}$$

where N is the density of ionized donors, $\epsilon_2$ is the dielectric constant and e is the charge of electron.

This expression clearly shows that the decrease in $v_{20}$, obtained because of the shift of the Si-impurity level, when pressure is applied, leads to a decrease in $n_s$.

A more complete expression for $n_s$ can be derived for more realistic cases which assume that the doped GaAlAs layer is partially compensated ($N_{d2}$, $N_{a2}$) and that the donors are not completely ionized (this is particularly true under pressure, even at 300K). Considering the spacer as a perfect insulator and neglecting the variation of $\Delta E_c$ with pressure ($\Delta E_c$ is a fraction of $\Delta E_G$, whose variation is negligible because of the same symmetry in the conduction band minima of the two materials) we obtain (5):

$$n_s = \sqrt{2\epsilon_2\ N_{d2}\ v_{eff}/e^2}$$

$v_{eff}$ represents an effective band bending calculated in the triangular well approximation, which is given by the following expression:

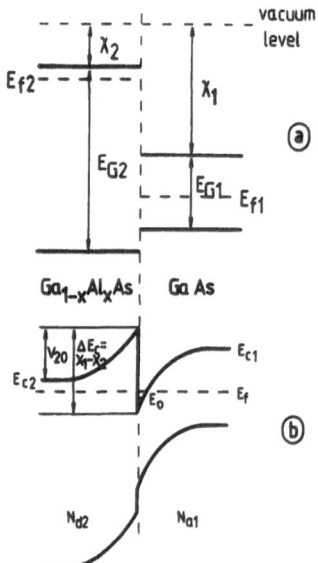

Fig. 2. (a) Energy band diagram before contact between
materials.
(b) Energy band diagram of the GaAs/GaAlAs
heterojunction.

$$v_{eff} = kT \left[ \ln \frac{1 + a}{a + \exp (v_{20}/kT)} + \frac{N_{a2}}{N_{d2}} \frac{v_{20}}{kT} \right.$$

$$\left. + 4 \frac{N_{C2}}{N_{d2}} \ln \frac{1 + \frac{1}{4} \exp \{-(E_F - E_{C2})/kT\}}{1 + \frac{1}{4} \exp \{-(E_F - E_{C2} + v_{20})/kT\}} \right]$$

where $a = 2 \exp \dfrac{(E_{d2} - E_F)}{kT}$

The previous expression enables us to take into account the experimental variations of $n_s$ versus pressure by considering that the pressure coefficient of the Si-impurity level is equal to about 11 meV/kbar and $E_{d2} \approx 60$ meV.

Typical results are given in Fig. 3 : they show that $n_s$ decreases linearly when the pressure increases; this variation is well supported by expression [2]. The value of $n_s$ can be determined experimentally by Hall effect measurements, but the Shubnikov de Haas experiments under hydrostatic pressure shown in Fig. 4 are also demonstrative. The shift of the oscillations towards lower magnetic fields when pressure increased was due to the decrease in $n_s$. It should be noted here that at temperatures lower than 200K $n_s$ did not vary when the sample was cooled. This is due to the metastable character of the Si-impurity level : the concentration of ionized impurities varies only in the high temperature range and becomes constant when the energy of the barrier is larger than kT.

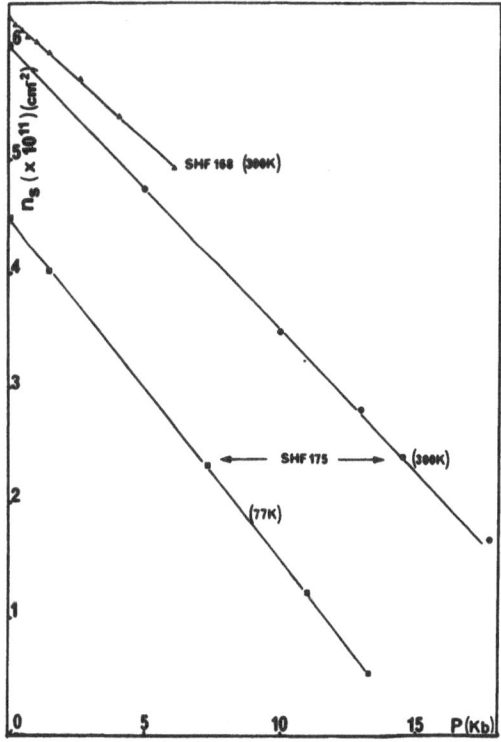

Fig. 3. Pressure dependence of $n_s$

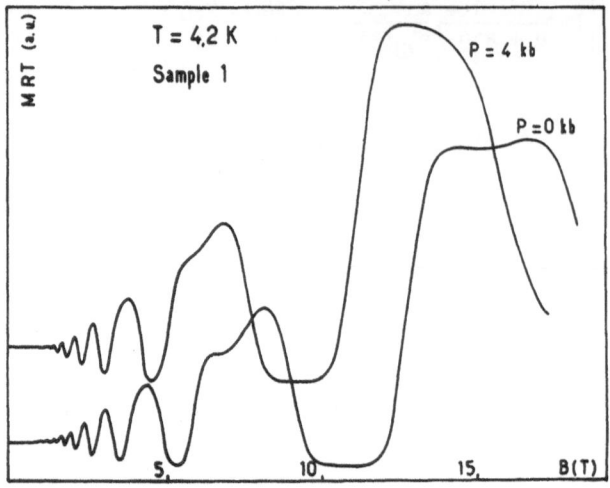

Fig. 4.   Transverse magnetoresistance $R_{xx}$ (MRT) of GaAlAs/GaAs
heterojunction (Table 1).


## II  -  PROPERTIES OF THE 2D ELECTRON GAS UNDER HYDROSTATIC PRESSURE

### II.1  Mobility versus Concentration

By cooling to low temperatures under appropriate pressure, the
density of the 2D electron gas can be changed.  It is then possible to
measure mobility on the same sample over a large concentration range.
Carrier concentrations of about $2.10^{10} cm^{-2}$ were obtained, which cannot
be achieved by changing only the growth parameters (doping level or
spacer thickness).

Fig. 5 represents typical curves, which are well suited to make
theoretical studies, insofar as the intrinsic parameters of the
heterojunctions are always the same.  To calculate the scattering by
Coulomb centres, we use a model proposed by Walukiewicz [6].  The
potential is calculated assuming a profile of ionized impurity
distribution which changes under pressure because of the change
occurring in the effective band bending (Fig. 6).

### II.2  Localisation effects under Hydrostatic Pressure

#### Quantum Hall effect

Quantum Hall plateaux observed in 2D electron gas occur because of
the existence of localised states (Fig. 7).  When the Fermi level is
located in localized states the resistance $R_{xx}$ vanishes and the Hall
resistance $R_{xy}$ keeps a constant value (Fig. 8).  For a given filling
factor $\nu$, the Hall resistance is quantized to $R_{xy} = h/\nu e^2$.

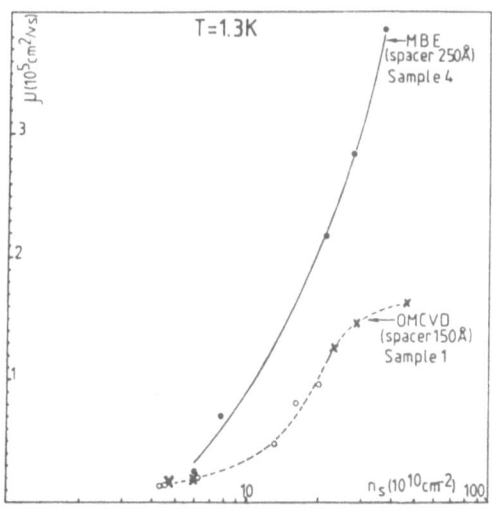

Fig. 5. Mobility versus $n_s$ for various samples.

As the density of localised states around the extended states increased, the plateaux became wider. It can be assumed that the extension of the localised states depends on the screening effect due to free electrons. Since the free carrier density can be changed in the same heterojunction, it is possible to observe the effect of screening on the width of the plateau. Keeping the same magnetic field, we measured the widths of the plateaux for different values of the filling factor. $\nu$ = 2, 4, 6 (Fig. 9).

The width of each plateau can be calculated if we assume that localized states are distributed over the interval between successive $\delta$-like extended states and that their density does not vary when pressure is applied. We found that this hypothesis leads to an increase by a factor 2 of the width of the plateau when $\nu$ goes from 6 to 2. This factor does not correspond to the experimental increase in the width of the pleateaux when $\nu$ changes from 6 to 2. As a result, it can be said that the density of localized states changes when the pressure changes. The increase in the widths of the plateaux must be associated with an increase in the density of localized states. This effect can be explained by the decrease in the screening effect.

Magnetic field induced metal non metal transition under
hydrostatic pressure in 2D systems

In order to reach the ultra quantum limit with available magnetic fields and to observe a metal non metal transition, it is necessary to use a 2D electron gas of sufficiently low density. In this case only the lowest Landau sub-level is occupied. This condition can be achieved by applying hydrostatic pressure. Several structures grown by MBE or MOCVD, were studied with spacer thicknesses varying between 60 Å and 250 Å. The sample characteristics at 4.2K with and without pressure are given in Table 1. The values of $n_s$ and mobility are the Hall values measured at B = 0.5T. In the high magnetic field range (up to 18T), $n_s$ is deduced from the $\sigma_{xy}$ component of the conductivity tensor. We have:

$$n_s = 1/e \ (R_H B^2/\rho_{xx}^2 + R_H^2 B^2)$$

where $R_H$ is the Hall coefficient and $\rho_{xx}$ is the transverse resistivity of the sample in the presence of the magnetic field B.

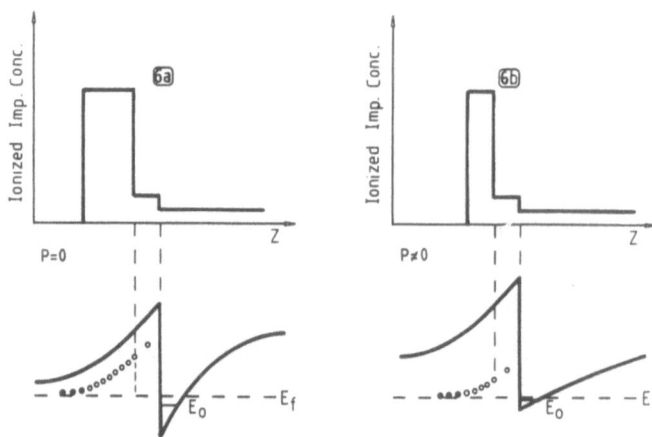

Fig. 6. Schematic representation of the ionized impurity distribution profile and the energy configuration for a heterojunction for (6a) P = 0 and (6b) P ≠ 0.

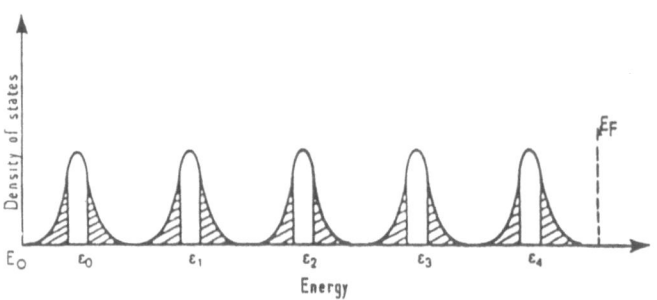

Fig. 7. Density of states versus energy.

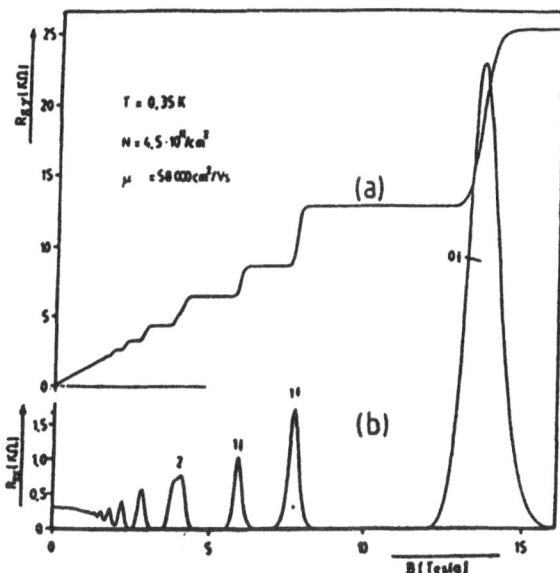

Fig. 8.   Magnetic field dependence of the Hall resistance
          $R_{xy} = R_H B$ and of the resistance $R_{xx}$ (Ref. 7).

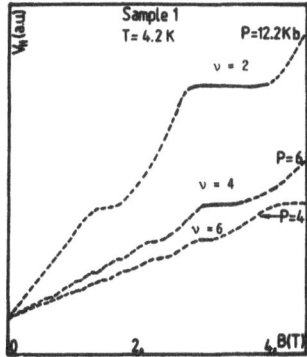

Fig. 9.   Width of the QHE plateau at different pressures.

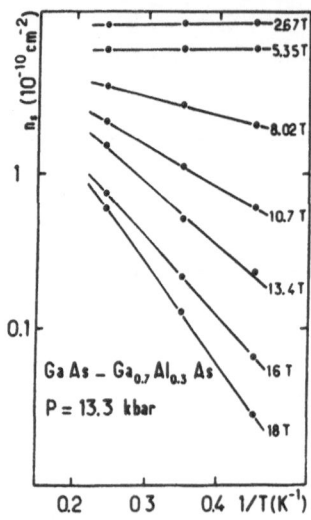

Fig. 10.   Temperature dependence of the surface
electron density for different magnetic
fields. Sample 1 under pressure of 13.3 kbar.


Typical results for $n_s$ with different magnetic fields and
temperatures are given in Fig. 10.   Above a critical value $B_c$, the $n_s$
electron density becomes activated.   This enables us [8] to determine a
critical value $n_{sc}$ corresponding to the transition between the metallic
and non metallic states (see Table 1).   The pressure applied to samples
1, 2 and 4 was chosen to obtain approximately the same critical density
$n_{sc} \sim 6.10^{10} cm^{-2}$.  A thermally activated density can be described as $n_s = n_0 \exp(-E_a/kT)$ where $E_a$ is the activation energy.

Fig. 11 shows that the magnetic field dependences of the activation
energies were distinctly different in different samples in spite of the
fact that the $n_{sc}$ were almost the same.   This suggests that the observed
localisation effect should not be ascribed to the Wigner condensation of
a dilute 2D gas, which, in the case of similar electron densities, would
lead to the same activation energy value.   On the other hand, if the
observed decrease in $n_s$ was due to trapping of the GaAs electrons in the
potential induced by the donors in GaAlAs, the $E_a$ (B) variation should
depend on the thickness of the spacer.   This is in fact what was
observed : the activation energy decreased with increasing spacer
thickness.

A metal non metal transition is usually associated in the Mott's
sense with an overlap of impurity wave functions.   In high magnetic
fields, surface electrons would be expected to move on an orbit with a
radius equal to $(\hbar/eB)^{1/2}$.   On the other hand, the average distance
between surface electrons at critical density was $n_{sc}^{-1/2}$.   Thus, the
overlap condition for Mott transition is given by $n_{sc}^{1/2} L_c \sim 0.5$ : we
find (Table 1) that the product $n_{sc}^{1/2} (\hbar/eB_c)^{1/2}$ was close to that
value.   Our results in GaAlAs/GaAs heterojunctions with spacer that were
not too wide, show that the Coulomb interaction between parent donors in
GaAlAs and electrons in GaAs cannot be neglected in investigations of 2D
electron gas.

## TABLE 1

Sample characteristics, critical magnetic fields and surface electron densities for metal-nonmetal transition in GaAs-Ga$_{1-x}$Al$_x$ heterostructures at different hydrostatic pressures. The last column gives the Mott-like criterion for the metal-nonmetal transition.

| Sample | x | Spacer thickness (Å) | P = 0 $n_s$ $10^{10}$ cm$^{-2}$ | T = 4.2K $\mu$ $10^4$ cm$^2$/Vs | Hydrostatic Pressure T = 4.2 K p Kbar | $n_s$ $10^{10}$ cm$^{-2}$ | $\mu$ $10^4$ cm$^2$Vs | $B_c$ (T) | $n_{sc}$ $10^{10}$ cm$^{-2}$ | Fig. 4 | $n_{sc}$ $L_c$ |
|---|---|---|---|---|---|---|---|---|---|---|---|
| 1 | 0.3 | 60 | 52 | 7.5 | 13.3 | 8.5 | 0.95 | 6 | 6 | | 0.26 |
| 2 | 0.25 | 150 | 24 | 12.9 | 8.8 | 6.3 | 3 | 4.8 | 6.4 | a | 0.3 |
| | | | | | 8.8 | 5.7 | 2.36 | 4.2 | 5.6 | b | 0.29 |
| | | | | | 8.8 | 7.8 | 1.55 | 3.5 | 6 | c | 0.34 |
| | | | | | 8.8 | 5.1 | 1.84 | 3.3 | 4.5 | d | 0.3 |
| 3 | 0.27 | 250 | 20.8 | 5.2 | 5.9 | 12.5 | 2.0 | 10 | 11 | | 0.27 |
| 4 | 0.3 | 250 | 35 | 41.9 | 13 | 6.5 | 6.82 | 8 | 6.5 | | 0.23 |

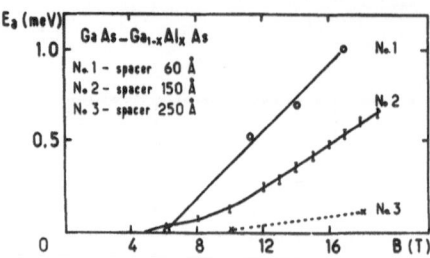

Fig. 11. Magnetic field dependence of the activation energy for several samples with different spacers.

ACKNOWLEDGEMENTS

The authors would like to thank Drs. J.P. André, P.M. Frijlink (LEP), F. Alexandre and J.M. Masson (CNET) for providing the samples. The authors acknowledge the SNCI-CNRS, Dr. J.M. Mercy, L. Konczewicz, E. Litwin-Staszewska, and R. Piotrzkowski for their participation in some experiments. This work has been supported by the MRT and the CNRS. The authors are grateful to Dr. G. Bastard and Pr W. Zawadzki for many valuable discussions.

REFERENCES

1. L. Esaki and R. Stu, IBM Res. Not. R.C. 2418 (1969).
2. W. Paul, Proc. IX Int. Conf. Physics of Semiconductors, Moscow, 1:16 (1968).
3. L. Konczewicz, E. Letwin-Staszewska and J. Porowski, Proc. 3rd Conf. on Narrow Gap Semiconductors, Warsaw, 211 (1977).
4. S.I. Cserveny, Int. J. Electronics, 25:65 (1968).
5. J.M. Mercy, C. Bousquet, J.L. Robert, A. Raymond, G. Gregoris, J. Beerens, J.C. Portal, P.M. Frijlink, P. Delescluse, J. Chevrier and N.T. Linh, Proc. of the Fifth Int. Conf. on Elec. properties of Two-Dimensional systems, Surface Science, ed. R.J. Nicholas, 142:298 (1984) (North-Holland, publ. Conf. Amsterdam).
6. W. Walukiewicz, H.E. Ruda, J. Lagowski and H.C. Gabos, Phys.Rev. B30, 8:4571 (1984).
7. K.V. Klitzing and G. Ebert, Solid State Sciences, ed. Bauer, Kuchar, Heinrich (Springer Verlag, Berlin), 53:242 (1985).
8. J.L. Robert, A. Raymond, L. Konczewicz, C. Bousquet, W. Zawadzki, F. Alexandre, J.M. Masson, R. Andre and J.M. Frijlink, Phys.Rev. B33, 8:5935 (1986).

# OPTICAL PROPERTIES OF InAs-GaSb SUPERLATTICES UNDER HYDROSTATIC PRESSURE

J.C. Maan

Max Planck Institut für Festkörperforschung
Hochfeld Magnetlabor
166X F-38042 Grenoble Cedex, France

ABSTRACT

At the interface between InAs and GaSb the GaSb valence band is 150 meV higher than the InAs conduction band. This fact leads for certain thicknesses of the layers of InAs and GaSb in superlattices to an electron subband at lower energy than a hole like subband. By means of hydrostatic pressure this arrangement can be inverted. The energy difference between these bands, measured by magneto-optical means, as a function of hydrostatic pressure, allows to determine the pressure dependence of the band offset. It is found that the InAs conduction band increases at a rate of 5.6meV/kbar with respect to the GaSb valence band, implying a pressure dependence of the valence band offset. Furthermore the results of the pressure dependence show a gradual transition from interband to intraband like character resulting from the band-mixing, which shows up clearly in this experiment.

## I INTRODUCTION

One of the most intriguing heterostructures which has been realized up to recently, is that based on thin layers of GaSb and InAs. This system was for the first time proposed and realized already several years ago [1,2] but still its fundamental physical properties are a source of many questions. Both InAs and GaSb bulk are standard direct small-gap semiconductors ($E_G$ = 410meV and 820meV respectively), however the peculiarity of the system is that the InAs conduction band edge is at a lower energy than the GaSb valence band edge. The actual value of this band offset between the InAs-CB and the GaSb-VB is experimentally determined to be 150meV [3]. This fact makes the electronic properties of InAs and GaSb heterostructures so intriguing, because it means that at the interface there is a continuum of electronic states, which is valence band like on the GaSb side and electron like at the InAs side of the interface. For a single interface the carriers spill over from the full valence to the empty conduction band states, which leads to charge separation and thereby to an electric dipole which bends the bands. However if one deals with a superlattice where the InAs and GaSb layer thicknesses are thin enough one can ignore this band bending and the electronic properties of this system are determined purely by the band-structure and not by the electrostatic effects. A more detailed description of the InAs-GaSb system can be found in refs 4 and 5. The band structure of this system has been thoroughly studied and will be discussed elsewhere by M. Altarelli [6] and G. Bastard [7]. Here experimental results of magnetooptical absorption measurements on an InAs-GaSb superlattice as a function of hydrostatic pressure will be discussed.

There are two, somewhat related, reasons, why the application of hydrostatic pressure to this system is interesting. The first reason is related to the question of the band offsets. It is well known experimentally how the energy gaps in InAs and GaSb depend on pressure (in fact they both increase with 10 and 14 meV/kbar respectively [8] ). However, it is not known how the conduction band in one material (InAs) shifts with respect to the valence band in the other (GaSb), or, what is the pressure dependence of the band line-up. This is an interesting experimental fact because in some sense the band offset between InAs and GaSb without pressure and that between InAs and GaSb at a high pressure is like comparing two different samples with different lattice constants and thereby different band structure. Therefore the pressure dependence of the band offsets can constitute a test for band-line-up theories. The second reason is that since this band offset is pressure dependent one can study the change of the electronic structure with pressure. In particular, the staggered band-line-up leads in a superlattice to hole-like (with an in-plane dispersion relation curving downward) and electron like (with an in-plane dispersion relation curving upward) subbands which are very close in energy. It is known that theoretically [9] the interaction between these bands leads to strong non-parabolicities of the in plane dispersion. This non-parabolicity depends strongly on the energy of the subbands with respect to each other and thereby on hydrostatic pressure.

## II  INTERBAND MAGNETO-OPTICAL PROPERTIES OF InAs-GaSb SUPERLATTICES

The band structure of InAs-GaSb superlattices can qualitatively be understood by considering the system as two interwoven Kronig-Penney potentials, one for the electrons (the discontinuity in the conduction bands) and one for the holes (the discontinuities in the valence bands,

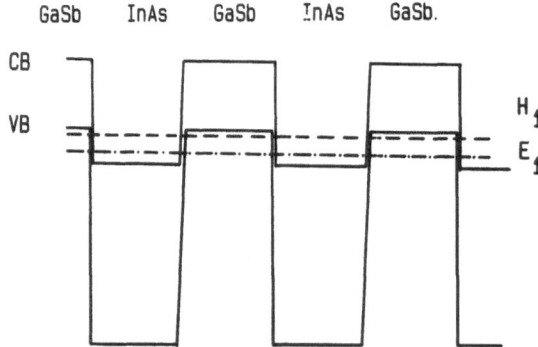

Fig. 1.  Spatial band-edge variation of an InAs-GaSb superlattice.

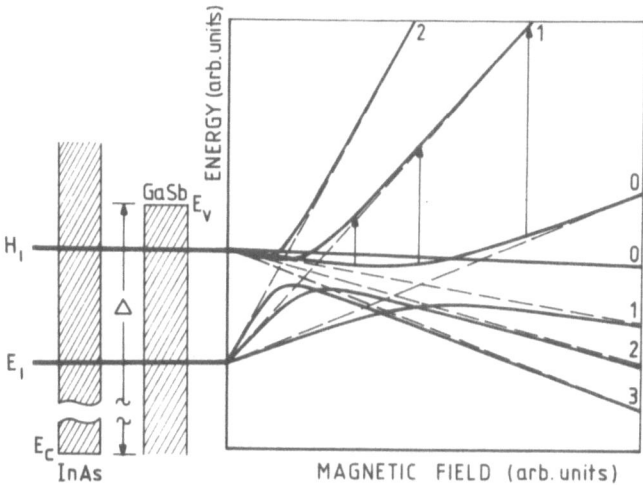

Fig. 2. The relative position of the hole ($H_1$) subband and
the electron subband ($E_1$) with respect to the InAs-CB
and GaSb-VB edges, respectively. In the right part
of the figure the hole-like and the electron-like
Landau levels are shown schematically in the presence
(solid) and in the absence (dashed) of coupling
between them.

see Fig. 1). The electrons are mainly confined in the InAs layers and
the holes in the GaSb layers. For thin layers two dimensional subbands
are formed as a consequence of the quantum size effect and the band edge
of the electrons (the energy shift of the subband with respect to the
InAs CB edge) increases, and similarly the band edge energy for the
holes (the shift of the hole subband with respect to the GaSb VB)
decreases. If the layers are very thin the hole like subband will be at
lower energy than the electron like and the superlattice is like a
normal semiconductor. If the layer thickness increases at some point
the hole like subband will become higher in energy than the electron
like subband, and in this case the system is called semimetal like.

The sample studied here was a superlattice with many successive
layers of 12 nm of InAs and 8 nm of GaSb and for these layer thicknesses
the electron like subband $E_1$ is about 40meV lower in energy than the
hole-like $H_1$. (This is the same sample which has been studied earlier
[10,11] at zero pressure.) In addition these layer thicknesses are
sufficiently thin to allow the band bending due to charge transfer to be
neglected. The schematic subband structure of the sample is shown in
Fig. 2. Also shown in the figure are the Landau levels of the electrons
and the holes from these subbands if a magnetic field is applied
perpendicular to the layers. Neglecting at present the coupling between
the layers the Landau levels constitute a set of linearly field
dependent equidistant energy levels which are given by:

$$E_N = (N + 1/2)\hbar eB/m^*_{h,e} \tag{1}$$

where N is the Landau level index and $m^*_{h,e}$ the (negative) hole or the electron mass respectively (shown as the dashed lines in fig. 2). With the usual $\Delta N = 0$ selection rule the resonance condition for interband absorption of a photon with energy $\hbar\omega$ is given by:

$$\hbar\omega - (E_1 - H_1) = (N + 1/2)\hbar eB.(m_e^{-1} - m_h^{-1}) \qquad (2)$$

From this equation one can immediately see that for a given N one can observe a transition with a linear magnetic field versus energy dependence which at zero magnetic field extrapolates to $-(E_1 - H_1)$. In fact this behaviour is that of interband absorption in a magnetic field of a semiconductor with a negative energy gap. Therefore one can measure quite accurately $E_1 - H_1$ with such an experiment. As mentioned before the energy of the electronic subband is essentially that of the InAs band-edge minus the confinement energy of the electron in the well. Similarly the hole subband energy is essentially that of the GaSb band-edge plus the confinement energy of the holes. Therefore $E_1 - H_1$ is determined by $\Delta$ (the band offset between the InAs-CB and the GaSb-VB) reduced by the respective confinement energy, and a measurement of this quantity is a sensitive probe for measuring $\Delta$. As was mentioned in the introduction, since $\Delta$ will generally be dependent on hydrostatic pressure one can determine the pressure dependence of this quantity in this way.

The previous description has ignored any coupling between the InAs conduction band and the GaSb valence band. However this neglection is only valid at zero wavevector, or similarly, at zero magnetic field, as has been shown theoretically [9]. At finite magnetic field there is an interaction between the InAs conduction band and the GaSb valence band and this leads to an anticrossing between the hole-like and electron like levels, as shown schematically in Fig. 2. It is obvious that the magnetic field at which, for a certain Landau level, this anticrossing occurs depends strongly on $E_1 - H_1$. As this quantity depends on hydrostatic pressure one can study also this anticrossing behaviour as a function of $E_1 - H_1$. Such a study has not only an academic interest, because this anticrossing behaviour has qualitative consequences. No anticrossing implies that the energy versus wavevector dispersion relations, for wavevectors in the plane of the layers, of the holes and the electrons cross each other. In this case there would be no gap in the density of states, which implies that such an InAs-GaSb superlattice would be a semimetal (coexistance of holes and electrons at the same energy). Due to the anticrossing there will appear a gap in the density of states and for an intrinsic sample the Fermi energy will be in the middle of this gap, and the system is still a semiconductor with a very small gap although there can still exist hole like states ($E$ vs $k$ relation bending up at $k = 0$) at a higher energy than electron like states ($E$ vs $k$ relation bending up at $k = 0$). Or, similarly, with hole Landau levels decreasing in energy with magnetic field at a higher energy than electron Landau levels which increase with field.

III THE EFFECT OF HYDROSTATIC PRESSURE ON InAs-CB and GaSb-VB BAND OFFSET

Both experimentally and theoretically the problem of the line-up of the bands at the interface of two semiconductors is poorly understood. Clearly since the band offset at zero pressure is badly known, its pressure dependence will be even less so. Still it can be interesting to see whether at least there is an agreement between the trend predicted by several theories or models for the band offset, and an experimental observation. A widely used model to estimate the valence band offsets is Harrison Atomic Orbital theory [12,13] (HAO). In this theory the energy of the valence band maximum with respect to the vacuum is given as:

$$E_v = \frac{E_p^{\,c}+E_p^{\,a}}{2} - \left[\frac{2.16h^2}{md^2} + \frac{(E_p^{\,c}- E_p^{\,a})^2}{4}\right]^{1/2} \tag{3}$$

where $E_p^{\,c}$ and $E_p^{\,a}$ are the atomic term values of the cation (c) and the anion (a), which are tabulated [11], and d is the lattice constant. The atomic term energies are not pressure dependent therefore the only effect of hydrostatic pressure is on the lattice constant d. The valence band offset can be calculated as the difference of the valence band energies calculated using eq. 3 for the two different materials. This theory predicts that the GaSb valence band is 100 meV higher than the InAs CB, which is a remarkably good agreement with the experimental 150meV. From the compressibility of InAs and GaSb [8] one can calculate the change in the lattice constant as a consequence of the hydrostatic pressure. This way one finds that the valence band edges relative to each other remain within 1 meV at the same energy between zero and 10 kbar. The InAs bandgap increased with a pressure coefficient of

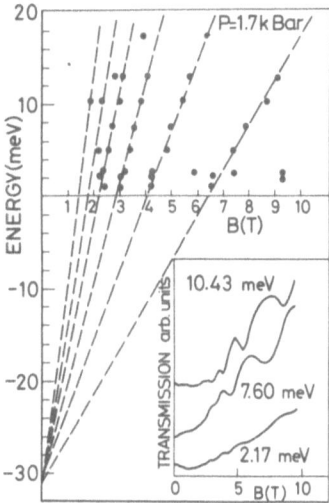

Fig. 3. Observed transition energy between Landau levels of the subbands of a InAs-GaSb superlattice as a function of the magnetic field at 1.7 kbar. The inset shows the experimental spectra. The dashed lines are calculated with eq. 2. (no coupling)

10 meV/kbar. Therefore this way one estimates that the energy difference $\Delta$ between the InAs CB edge and the GaSb VB edge will decrease with 10 meV/kbar. With this pressure dependence the overlap between the InAs-CB and the GaSb-VB will be zero at 15 kbar. In the case of the superlattice studied here one can therefore shift $E_1$ higher than $H_1$ with pressure, implying a pressure induced semimetal to semiconductor like transition.

## IV EXPERIMENTAL RESULTS

In the experiment the transmission of Far Infrared Radiation (FIR) at fixed radiation frequency is measured as a function of the magnetic field. The sample was mounted in a CuBe pressure bomb with piston using mineral oil as the pressure medium. A 6mm thick sapphire window was used to pass the FIR radiation and the pressure was measured with a InSb pressure gauge. The radiation was detected with a 470 Ohm Allen and Bradley carbon resistor used as a bolometer which was mounted directly behind the sample which itself was glued with silver paint onto the sapphire window thus minimizing the leakage of radiation. The ensemble was He gas cooled in a tube immersed in liquid Helium.

In Fig. 3 some represenative transmission curves are shown together with a plot of magnetic field position of the transmission minima for different radiation energies. The dashed lines in the figure represent a calculated fan chart of interband transitions as given by eq. 2. One can clearly observe the interband Landau level transitions of a negative gap semiconductor, as discussed before. The extrapolation point at zero magnetic field gives the value of $E_1-H_1$ determined in this way. In Fig. 4 the pressure dependence of this extrapolation point obtained in this fashion is shown and is approximately linearly decreasing at a rate of 4 meV/kbar. By identifying this rate of decrease with that of $\Delta$ then this pressure dependence is much weaker than that estimated from the Harrison line up theory as described before (10 meV/kbar). However, before coming back to this question we will continue our analysis of the experimental results.

We now focus our attention on behaviour of the last, high fields, interband transition of Fig. 2 as a function of pressure, shown in Fig. 5. The dashed lines in this figure are a linear fit to the experimental points. The slopes of this line as a function of pressure are shown in Fig. 6. One observes that it decreases roughly linearly from 5.3 meV/kbar at zero pressure to 3.3 meV/kbar at 10.7 kbar. In the simple description of non-interacting Landau levels as given before this transition corresponds to the N = 1 interband transition. In that case its slope is given by $(1+1/2)$ $e\hbar(1/m^*_e-1/m^*_h)$. The relevant hole mass here is a combination of the light and the heavy hole mass of GaSb but can be considered in any case much heavier than the electron mass. The latter mass is the mass of the InAs-CB at an energy $E_{conf}$ above the InAs band edge. Taking into account the non-parabolicity of InAs;

$$m^*_e\ (E) = m^*_{0,e}\ (1 + 2E_{conf}/E_G) \tag{4}$$

with $E_G$ the InAs bandgap and $m^*_{0,e}$ the band edge mass, this slope is calculated as $1.5 \times \hbar\omega = 1.5 \times 3.5 = 5.3$ meV/T which is very close to the observed slope at zero pressure, but quite different from that at 10.7 kbar. To explain the slope at 10.7 kbar within the simple model, it has to be assumed that the mass increases substantially with hydrostatic pressure. The pressure dependence of the band edge mass in InAs can be calculated from the expression which relates the band edge

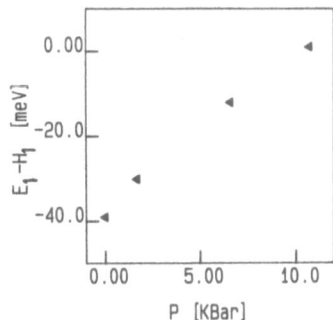

Fig. 4.  $E_1$-$H_1$ as a function of pressure as
determined from the extrapolation
of the transition energy vs. field
dependence to zero magnetic field.

mass to the energy gap, $E_G$, the energy between the valence band edge and
the split-off valence band $\Delta$.   (N.B. this is the usual name for  the
split-off  valence band gap,  not to be confused with the CB-VB  offset)
and the interband matrix element P, and which is given by:

$$\frac{m}{m^*_e} = 1 + \frac{2P^2}{3} \left(\frac{2}{E_G} + \frac{1}{E_G + \Delta}\right)$$  (5)

P  and  $\Delta$ may  be  assumed to have  a  small  pressure  dependence,  and

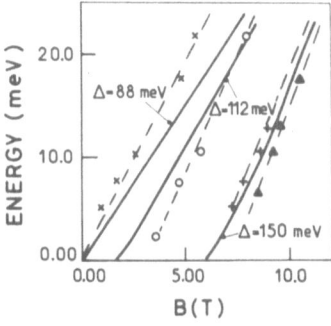

Fig. 5.  Pressure dependence of the high field
transition at O($\blacktriangle$), 1.7(+), 6.6(O) and
10.7 kbar(x) in an InAs-GaSb super-
lattice.  The drawn lines are the
transitions calculated theoretically
with the full 6-band model with coupled
bands.  The dashed lines are straight
lines drawn through the experimental points.

331

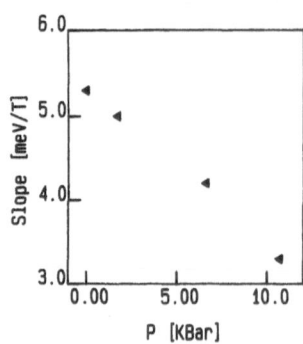

Fig. 6.  The slope of the last high field
transition as a function of pressure,
obtained from the results in Fig. 5.

therefore the entire pressure dependence of the band edge mass is
determined by the change in $E_G$ (10 meV/kbar).  This way one finds that
the band edge mass increases from 0.023 m to 0.028 m (m is the free
electron mass) between zero and 10 kbar, which is an increase of 20%.
This increase is not sufficient to explain the decrease in slope
observed experimentally which was twice as much.  Furthermore the mass
in the subband at an energy $E_{conf}$ will increase even less than the band-
edge mass because of a reduction in the non-parabolicity.  This can be
seen from eq. 4.  Since on one hand $E_G$ increases as a direct consequence
of the pressure and on the other hand the confinement energy (which is
inversely proportional to the band-edge mass) is reduced by the increase
of the band-edge mass.

At this stage it is clear that we have to admit that the simple
model using uncoupled Landau levels for electron and holes fails to
describe the experimental results correctly, because it can never
explain the gradual change in the slope of the transition we have been
discussing.  Therefore the preceding analysis of the data is only
qualitatively valid, and describes what in essence is observed, i.e. a
pressure induced transition from a semimetal like to a semiconductor
like superlattice.  It is interesting to note that this experiment is
the first that shows qualitatively the importance of the interaction
between electron and hole subbands of different materials. Historically
the experiments in InAs-GaSb superlattices have been analyzed with non-
interacting bands for the motion in the plane of the layers.  This type
of analysis was sufficiently precise to describe the results but was
criticized strongly from the theoretical side.  The theory therefore
carried the burden to explain experimental results with a more
complicated, although more accurate [14], band structure, while a
simpler, although in principle wrong, bandstructure seemed already
sufficient.

V   ANALYSIS

To illustrate the effect of the anticrossing we have calculated the
Landau levels of two interacting bands.  Following previous theoretical
work [9], the Hamiltonian in this case is given by:

$$H = \begin{vmatrix} E_1 - \dfrac{\hbar eB}{m^*_e}(a^+a^- + 1/2) & Q(\dfrac{2eB}{h})^{1/2} \\ \\ Q(\dfrac{2eB}{h})^{1/2} & H_1 - \dfrac{\hbar eB}{m^*_h}(a^+a^- + 1/2) \end{vmatrix} \qquad (6)$$

with $a^+$ and $a^-$ the creation and annihilation operators, B the magnetic field, Q the interband matrix element, and $E_1$ and $H_1$ the energy of the hole and the electron like subband edges at zero magnetic field. In Fig. 7a, b and c we show the resulting energy levels of eq. 6, for different values for $E_1 - H_1$. We have chosen the masses, Q and the energies close to those corresponding to the experiment. The N = 1 transition which is the one we have been discussing before is indicated by the arrows. This model calculation explains directly the change of slope of this transition. As $E_1$ is much lower than $H_1$ the initial state of the transition has a hole like behaviour (decreasing slightly in energy because of the much heavier hole mass) and the final state an electron like (increasing in energy with a field dependence of $1.5\hbar\omega_{c,e}$, $\omega_{c,e}$ the cyclotron frequency of the electrons), and the transition between these levels will behave like an interband Landau level

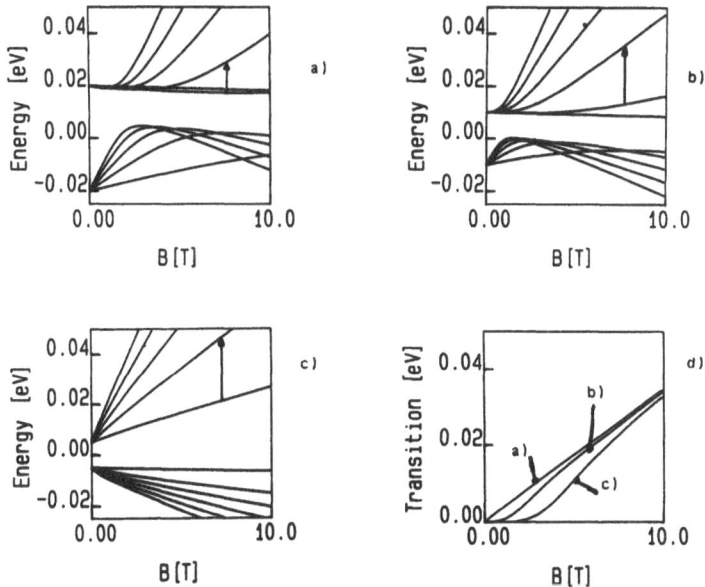

Fig. 7.  Landau levels for coupled bands calculated from the two band model (eq. 6), for different energy separations ($\Delta E$) between conduction band like and valence band like states. (a) $\Delta E$ = -0.04 eV, (b) $\Delta E$ = -0.02 eV, and (c) $\Delta E$ = 0.005 eV, showing the gradual change from interband to intraband like character of the transition indicated by the arrow. The magnetic field dependence of this transition is shown in (d).

transition with a slope of about $1.5\hbar\omega_{c,e}$ (that is neglecting the weak field dependence of the initial state). As $E_1$ is higher than $H_1$ the initial state has become almost entirely electron like with a slope $0.5\hbar\omega_{c,e}$ and the final state is still electron like with as before a slope of $1.5\hbar\omega_{c,e}$, and the transition between the two states is plainly cyclotron resonance with $\hbar\omega_{c,e}$ as a magnetic field dependence. Fig. 7d shows the corresponding transition energy as a function of field dependence for the three different cases shown in 7 a, b and c, and illustrates the gradual change of the transition from interband Landau level like to intraband cyclotron resonance. This behaviour is exactly what we observe for the transition shown in Fig. 5 and the slope of which changes from 5.3 meV/T which corresponds to $1.5\hbar\omega_{c,e}$ at zero pressure to 3.3 meV/T, which is $\hbar\omega_c$, at 10.7 kbar which is only 10% less than 2/3 of the slope at zero pressure, as should be expected. That the slope is less than 2/3 can easily be understood because of the increase of the InAs band edge mass with pressure as was calculated before.

Having understood the experimental results qualitatively we can analyze them as correctly as possible with the calculated Landau levels using the full six band model, which is described elsewhere [6]. In the calculation we have included the effect of hydrostatic pressure on the band structure of the bulk materials through its effect on the energy gaps. This way the masses are adjusted automatically in a manner as described before in eq. 5. The offset between the InAs-CB and the GaSb-VB is varied in such a way as to obtain the best agreement with the experimental results. The comparison between the experiments and the theory is shown by the drawn lines in Fig. 5, and the agreement can be considered satisfying. The pressure dependence for $\Delta$ obtained this way is 5.8 meV/kbar and is approximately linear. This rate is more than that determined from the simpler previous analysis shown in Fig. 4. The main reason for this difference is the fact that as a function of pressure the confinement energy of the electrons is reduced as a consequence of the increase of the band edge mass. What is experimentally observed is $E_1-H_1$ as a function of pressure which equals $\Delta$ minus the confinement energy of the electrons minus the confinement energy of the holes. Since the confinement energy of the electron decreases with pressure, $\Delta$ has to increase more in order to explain the experimentally observed $E_1-H_1$. This statement is illustrated in Fig. 8 where we show the energies of the $E_1$ and $H_1$ subbands with respect to the GaSb-VB as a function of the increase in the InAs bandgap (which increases proportionally to the pressure) but keeping the VB-edges at the same energy. One sees that $H_1$ and $E_1$ are basically independent of the relative position of the band edges ($\Delta$) with respect to each other, but are determined mainly by the position of the subband with respect to the relevant band edge, (which a posteriori justifies the simplified description of this type of superlattice as two independent interwoven Kronig-Penney potentials) $E_1$ decreases slightly (20%) with respect to the position of the InAs-CB edge with pressure. This is essentially an effect of the increase of the InAs band-edge mass, and a corresponding reduction of the confinement energy.

In section III we estimated the pressure dependence of $\Delta$ on the basis of the HAO method to be 10 meV/kbar. This number was derived from the result that the valence band offset was pressure independent and that therefore $\Delta$ varied with pressure as the InAs gap. Experimentally we find a much weaker pressure dependence, and this implies that the valence band offset must be pressure dependent. Concretely it has to increase with a rate of $\sim$ 4 meV/kbar to explain our experimental result. This result is in direct contradiction with the predictions of the HAO method. As mentioned before if a model of band line-up has to be successful in predicting the band offsets of different materials with a

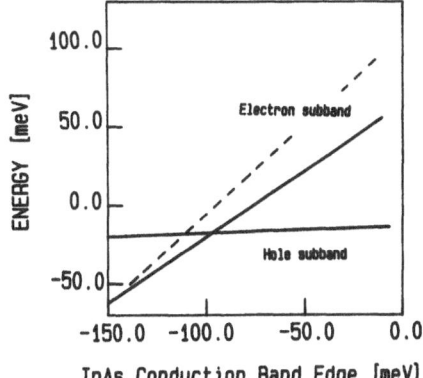

Fig. 8. Calculated relative positions of $H_1$ and $E_1$ as
a function of the position of the InAs CB edge
as measured with respect to the GaSb-VB. In
the calculation the valence band offsets are
kept constant and the InAs bandgap is
increased. A shift of 10 meV in the InAs band-
edge corresponds to 1 kbar, this way. The
dashed line shows the variation of the band
edge itself.

different band structure, it necessarily also has to predict the correct
dependence on hydrostatic pressure for a particular heterojunction. In
this sense experimental results on the pressure dependence of the band
line-up constitute a severe test for band line-up theories. The HAO
method is one of the pioneering theories, but there exist many other
models; the electron affinity rule [15], the theory by Frensley and
Kroemer [16], ab initio band structure calculations [17], the heuristic
model in which deep impurity levels are lined up [18] and the line-up of
the charge neutrality level [19,20]. It goes beyond the scope of this
paper to review the merits of these models. In many cases it is not
easy to extract a prediction of the pressure dependence out of these
theories, and therefore a comparison with our results is difficult. All
that can be said is that the latter two models, which relate the energy
level that has to be lined up at the·interface (deep impurities and
charge neutrality level) directly to the relevant bulk bands (mostly to
the valence bands) suggest strongly that the valence band edge will in
general be pressure dependent with respect to this level, in concordance
with the experiments.


VI CONCLUSIONS

Optical experiments as a function of hydrostatic pressure in
heterostructures are interesting because they permit a determination of
the pressure dependence of the band line-up, and a study of the
consequences on the band structure as caused by this different band
offset. In the case of InAs and GaSb this type of experiment has shown
for the first time qualitatively the effect of band mixing between two
different materials, because with the pressure one has a handle to
switch these effects on and off. Furthermore it was shown that the

line-up of the valence bands changes with pressure, which to my knowledge has not been seen before and which constitutes a critical test of line-up theories. Experimentally this technique, which has shown its merits in the study of bulk materials, up till now has found only limited application in heterostructures. It is hoped that our results will stimulate further work in this direction.


## ACKNOWLEDGEMENTS

The work described here is the result of the collaboration of many people. In particular L.L. Chang and L. Esaki who conceived the InAs-GaSb system and have shown many of its interesting possibilities the first time, Michiel Claessen for doing the experiments, and Massimo Altarelli for clarifying the band structure problems. Finally we wish to thank Gerard Martinez for many useful discussions about all sorts of problems related to hydrostatic pressure.

## REFERENCES

1. G.A. Sai-Halasz, R. Tsu, L. Esaki, Appl.Phys.Lett. 30:651 (1977).
2. H. Sakaki, L.L. Chang, G.A. Sai-Halasz, C.A. Chang, L. Esaki, Solid State Commun. 26:589 (1978).
3. G.A. Sai-Halasz, L.L. Chang, J.M. Welter, C.A. Chang, L. Esaki, Solid State Commun. 27:935 (1978).
4. L.L. Chang, L. Esaki, Surface Sci. 98:70 (1980).
5. J.C. Maan, "Infrared and Millimetre waves", edited by K.J. Button (Academic Press, New York), Vol. 8, Ch. 9 (1982).
6. M. Altarelli and references therein, this volume.
7. G. Bastard and references therein, this volume.
8. G. Martinez, "Handbook of semiconductors", edited by M. Balkanski (North Holland, Amsterdam), Vol. 2, p. 132 (1980).
9. A. Fasolino, M. Altarelli, Surface Sci. 142:322 (1984).
10. Y. Guldner, J.P. Vieren, P. Voisin, M. Voos, L.L. Chang, L. Esaki, Phys.Rev.Lett. 45:1719 (1981).
11. J.C. Maan, Y. Guldner, J.P. Vieren, P. Voisin, M. Voos, L.L. Chang, L. Esaki, Solid State Commun. 39:683 (1981).
12. W.A. Harrison, J.Vac.Sci.Technol. 14:1016 (1977).
13. W.A. Harrison, "Electronic Structure and the Properties of Solids", (W.H. Freeman and Cy., San Francisco), 1980.
14. M. Altarelli, J. of Luminescence 30:472 (1985).
15. R.L. Anderson, Solid State Electron. 5:341 (1962).
16. W.R. Frensley, H. Kroemer, Phys.Rev. B16:2642 (1977).
17. M.L. Cohen, Adv.Electron.Phys. 51:1 (1980).
18. J.M. Langer, H. Heinrich, Phys.Rev.Lett. 55:1414 (1985).
19. F. Flores, C. Tejedor, J.Phys. C12:731 (1979).
20. J. Tersoff, Phys.Rev. B30:4874 (1984).

# MAGNETOTRANSPORT MEASUREMENTS UNDER HYDROSTATIC PRESSURE IN

# TWO-DIMENSIONAL ELECTRON AND ELECTRON-HOLE SYSTEMS

G. Gregoris*, J. Beerens+, L. Dmowski, S. Ben Amor
and J. C. Portal

INSA avenue de Rangueil, 31077 Toulouse Cedex
and SNCI-CNRS, 166X, 38042 Grenoble, France

## Abstract

We discuss the application of the hydrostatic pressure technique for
the investigation of transport properties in two-dimensional systems at
high magnetic fields.  The experimental results cover various systems like
two-dimensional electron gas systems (GaAs-AlGaAs and GaInAs-AlInAs
heterojunctions) and a two-dimensional electron-hole system (a
GaSb-InAs-GaSb double heterostructure).  In both cases, the carrier
density decreases with increasing pressure and this gives a unique
opportunity to study transport phenomena as a function of either the
carrier density or the degree of electron-hole compensation without having
to change the sample.

## INTRODUCTION

Because it acts on the band structure, hydrostatic pressure has been
used for a long time in optical and transport measurements to determine
band parameters of III-V semiconductors (Martinez 1980, Spain this
workshop).  Hydrostatic pressure has also shown its usefulness in the
investigation of the transport properties of two-dimensional (2D) systems
(Robert et al 1984 and this workshop).  This method was first applied to
GaAs-AlGaAs heterojunctions (Mercy et al 1984) and the 2D carrier density
was found to decrease linearly with increasing pressure.  This behaviour
was demonstrated to be related to the deepening of the Si-donor level in
the gap of the AlGaAs doped layer and it gives an interesting experimental
method for the study of 2D electron localization (Mercy et al 1985).

In this paper we put special emphasis on the investigation of some of
the transport properties which depend on the carrier density and we show
how successful hydrostatic pressure is in this field.  The experimental
results cover essentially two different kinds of heterostructures which
are modulation doped heterojunctions with a 2D electron gas (GaAs-AlGaAs
and GaInAs-AlInAs) and a type II system with spatially separated
2D electron and hole gases (a GaSb-InAs-GaSb double heterostructure).  In

---

*   present address:  National Research Council, Division of Physics,
Ottawa, Canada  K1A OR6

+present address:  C.N.E.T, 196 av H. Ravera, 92 220 Bagneux, France

both cases the 2D carrier density decreases with increasing pressure and this has either an extrinsic cause (the behaviour of the donor level under pressure), or an intrinsic cause (a pressure-induced change in the band discontinuity at the interface).

In the first part we present the basic ideas about the pressure-control of the 2D carrier density in a modulation doped heterostructure. We report magnetotransport measurements in the GaInAs-AlInAs system over a wide range of temperature (from 300K to 4.2K) and we pay special attention to the effect of the parallel conduction. We finally report the first observation of quantum Hall effect in this system, which was only observable at high pressure when parallel conduction was reduced. In the second part we discuss the magnetophonon resonance effect in both heterojunctions. We report measurements of effective mass and non-parabolicity and we put emphasis on the study of the strength of the electron-LO phonon coupling measured via the amplitude of the resonances. In both cases, the carrier density is an important parameter because of non-parabolicity effects and because free carriers screen the electron-phonon interaction. In the last part, we touch upon our investigation of the electron-hole system, the GaSb-InAs-GaSb heterostructure. This system is partially compensated (the density of electrons is greater than the density of holes) and hydrostatic pressure induces a decrease of both carrier concentrations which lead to a semimetal-semiconductor transition.

EXPERIMENTAL

We use a liquid-medium pressure cell with a 6mm inside bore diameter (a 4mm useful diameter for the sample) to produce hydrostatic pressure (P) as high as 18 kbar, from 300K to 4.2K. Pressure is measured through the calibrated four-point resistance of a heavily doped InSb:Te bar placed near the sample. This gauge provides the value of the pressure with an uncertainty of less than 100bars at any temperature (Konczykowski et al 1978). Pressure is always applied at room temperature and the cell can then be slowly cooled down. All the samples studied were MBE grown and Hall-bridge shaped using current lithography and contacting procedures.

HYDROSTATIC PRESSURE ON A MODULATION-DOPED HETEROSTRUCTURE

1. Theoretical Background

Although experimental results presented in this section concern only the GaInAs-AlInAs system, the analysis we give is qualitatively relevant for the GaAs-AlGaAs system (Beerens et al 1986).

Under hydrostatic pressure the band gaps of most of III-V semiconductors increase at specific rates for each of the $\Gamma$, X and L points of the Brillouin zone. The $\Gamma$X and $\Gamma$L separations increase at typical rates of respectively 14±2 and 7±2meV/kbar (Martinez 1980). In other words, a donor level $E_D$, possibly related to a X or L minimum of the conduction band, will move deeper into the gap relative to the $\Gamma$ minimum, thereby increasing the activation energy $\varepsilon_D = E_\Gamma - E_D$ as pressure is applied. Therefore, in the case where the absolute minimum of the conduction band is at the $\Gamma$ point the separation $E_\Gamma - E_F$ is expected to increase as the Fermi level follows the donor level, and, for a modulation-doped heterojunction, the free carrier density in the doped layer as well as the charge transfer to the well will then decrease (Fig. 1).

In order to estimate this effect we have to calculate the conduction band diagram across the whole heterostructure, from the surface to the quantum well. For that purpose we can proceed in the following way: the Poisson equation is numerically integrated in the doped layer, the spacer

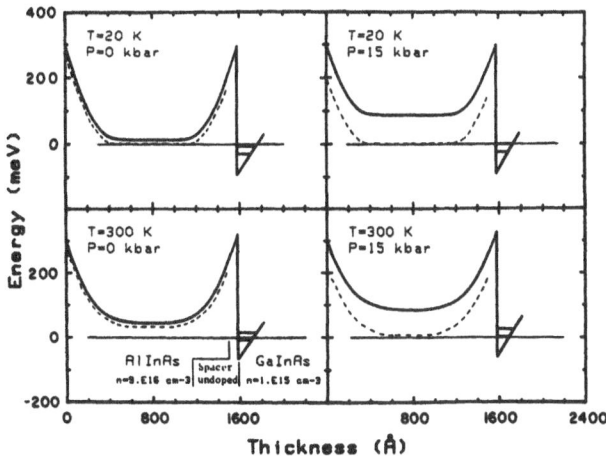

Fig. 1: Calculated energy band diagram of the GaInAs-AlInAs
heterojunction as a function of pressure and temperature. The dashed line
represents the Si-donor level. Zero energy is taken at the Fermi level.


layer is assumed to be fully compensated (constant electric field) and the
quantum well is treated in the triangular-well approximation (Grégoris et
al 1986). Poisson's equation reads:

$$\frac{\partial^2 E_c}{\partial z^2} = \frac{q^2}{\varepsilon^2} \left[ N_D^+ - N_A - n(z) \right] . \tag{1}$$

$E_c$ is the conduction band energy, $N_D^+$ and $N_A^- = N_A$ are the ionized donor
and acceptor densities, and $n(z)$ is the free electron density. z is the
distance from the surface along the growth axis. $N_D^+$ is given by the
usual thermal statistics and the 2D carrier density in the well is
obtained from:

$$N_s = \sum_i \int_{E_i}^{\infty} \frac{m^*}{\pi \hbar^2} f(E) \, dE \tag{2}$$

where $f(E)$ is the Fermi distribution and $E_i$ is the energy of the electric
sub-bands. The two first quantized levels (i=0,1) have been considered in
our calculations.

In this model both sides of the interface are treated separately and
a solution is found by requiring that the displacement vector be
continuous at the interface, i.e. the sheet depletion charge at the
barrier side and the sheet accumulation charge in the well must be equal.
From these calculations we get the conduction band diagram (Fig. 1), the
free carrier density in the doped layer and the 2D carrier density in the
well. Therefore this model enables one to account for high temperature
measurements when both the quantum well and the doped layer conduct
(parallel conduction).

In a multilayered structure, parallel conduction can strongly affect
magnetotransport measurements and lead to mistaken characterization of the
2D electron gas by the classical Hall effect. However the association of
high magnetic field and hydrostatic pressure provides a way to
discriminate the two components of parallel conduction. There are two
interesting magnetic field regimes where the Hall mobility $\mu_H$ and the Hall

carrier density $N_H$ do not depend on B and depend only on the carrier densities $N_i$ and mobilities $\mu_i$ of each of the conductive layers (Kane et al 1985, Grégoris et al 1986).

At low magnetic field ($\mu_i B \ll 1$):   At high magnetic field ($\mu_i B \gg 1$):

$$N_H = \frac{1}{eR_H} = \frac{(N_s\mu_s + N_b\mu_b)^2}{N_s\mu_s^2 + N_b\mu_b^2} \quad (3) \qquad N_H^h = \frac{1}{eR_H^h} = N_s + N_b \quad (5)$$

$$\mu_H = \frac{R_H}{\rho_o} = \frac{N_s\mu_s^2 + N_b\mu_b^2}{N_s\mu_s + N_b\mu_b} \quad (4) \qquad \mu_H^h = \frac{R_H^h}{\rho_{xx}^h} = \frac{N_s + N_b}{\dfrac{N_s}{\mu_s} + \dfrac{N_b}{\mu_b}} \quad (6)$$

The indexes s and b respectively stand for the 2D electron gas and the bulk. $\rho_o$ is the resistivity at B=0 and $R_H$ is the Hall constant. $\rho_{xx}$ is the transverse component of the resistivity and h stands for high B. $N_H^h$ gives the total carrier density in the system. The measurement of $N_H$, $\mu_H$, $N_H^h$ and $\mu_H^h$ give access to the four unknowns $N_i$ and $\mu_i$ of a two-layer conduction process.

In practice, the main difficulty is to reach the high field limit. When it is not possible to have $\mu B \gg 1$ for all layers, one can use independent measurements of $\mu_b$ and $N_s$ to estimate the remaining parameters from the low field Hall measurements. In certain cases, $\rho_{xy}$ does seem to reach the high field limit, while $\rho_{xx}$ does not (Grégoris et al 1986) and eqs. (3), (4) and (5) can be used provided $N_s$ is known. We then found it useful to define $\mu^*$ as:

$$\mu^* = \frac{R_H^h}{\rho_o} = \frac{N_s\mu_s + N_b\mu_b}{N_s + N_b} \qquad (7)$$

This "high field mobility" has the advantage of having a simpler physical meaning than $\mu_H^h$ (eq. 6): it is a weighted mean value of the mobilities in every layer, and signifies how much conduction is dominated by one of the layers.

## 2. Experimental Results In GaInAs-AlInAs

From Shubnikov-de Haas (SdH) and Hall measurements at 4.2K we observe a linear decrease of $N_s$ with increasing P (Fig. 2). Two sub-bands, $E_0$ and $E_1$ are occupied in this system at low temperature and their densities, $N_0$ and $N_1$, are measured from the corresponding series of SdH oscillations. The Hall density $N_H$ agrees fairly well with the sum $N_s = N_0 + N_1$. The rate of decrease in $N_s$ is 1% kbar$^{-1}$ in this system while a rate of 6% kbar$^{-1}$ was observed in GaAs-AlGaAs (Mercy et al 1984). The model presented above accounts for these results with an increase of the activation energy $\varepsilon_D$ of the Si donor level of 5±1meV/kbar in AlInAs and 11±1meV/kbar in AlGaAs respectively. These rates are comparable to that of $\Gamma L$ or $\Gamma X$ separations and this suggests an association of the donor state with a satellite minimum. But, although some authors claim that Si forms X- or L- like states in AlGaAs (see e.g. Henning et al 1984), no analogous work has been done in AlInAs to prove the existence of such a donor state.

At higher temperatures (T>150K) free carriers are thermally excited into the conduction of the doped layer, causing strong parallel conduction. The Hall curves $\rho_{xy}$(B) are not linear and show clear evidence of conduction by two types of carriers (Grégoris et al 1986).

340

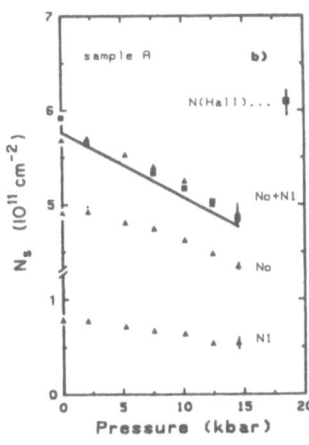

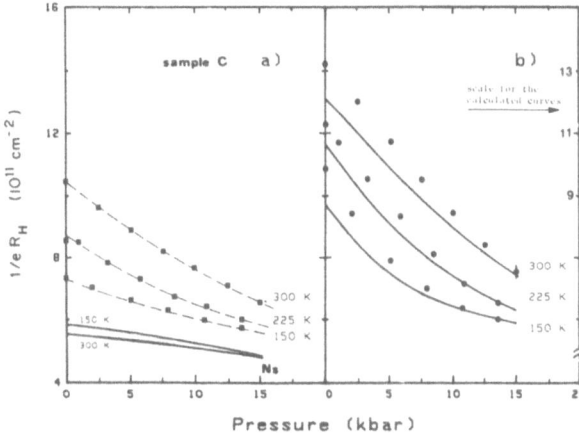

Fig. 2: Carrier concentration from (2D) Shubnikov de Haas and low field Hall effect at 4.2K ($N_s \simeq N_0 + N_1 \simeq N_H$). The full line is the calculated $N_s(P)$.

Fig. 3: Sheet carrier concentration $N_H$ from low field a) and $N_H^h$ from high field b) Hall effect at high T vs pressure. Full lines are calculated concentrations.

Low and high field Hall densities are plotted in Fig. 3. The high field density $N_H^h$ shows a strong exponential decrease with increasing P, typical of bulk conduction. The low field density $N_H$, less sensitive to low mobility carriers, gets closer to $N_s$. As pressure is increased $N_H^h$ and $N_H$ reach a comparable value, close to $N_s$, which suggests that parallel conduction is progressively vanishing. Calculations of the total carrier density (2D+bulk) qualitatively agree with $N_H^h$ (Fig. 3), giving a good description of its exponential decrease with P. A slight discrepancy in the absolute value comes from conduction in the thick GaInAs buffer layer (n-type residual) which is not included in the model.

The mobility of the 2D electron gas, extracted from the data, is

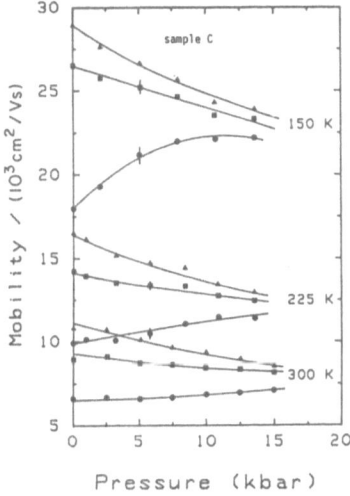

Fig. 4: Pressure dependence of the electron mobility from Hall measurements at low field (squares) and high field (dots) at high T. Triangles are the mobility of the 2D gas. Lines are guides for the eye.

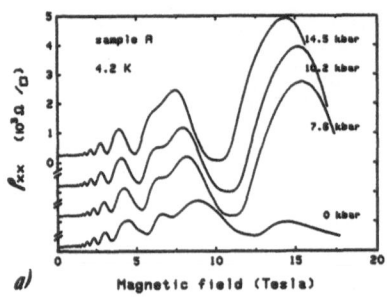 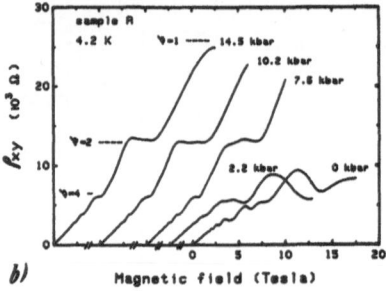

Fig. 5: Shubnikov-de Haas a) and Hall b) recordings at high magnetic field and for several pressures at 4.2K.

shown in Fig. 4 and compared with the classical $\mu_H$ and with $\mu^*$. The values $\mu_s$, $\mu_H$ and $\mu^*$ are very different at ambient pressure when parallel conduction is maximum. When P increases $\mu_s$, $\mu_H$ and $\mu^*$ get closer to each other since $N_b$ tends to zero. Unlike $\mu_H$, $\mu_s$ shows the expected $1/m^*(P)$-like decrease.

## 3. Quantum Transport

The Quantum Hall effect (QHE) has never been observed at ambient pressure in GaInAs-AlInAs: $\rho_{xx}(B)$ shows positive magnetoresistance and $\rho_{xy}$ gives no sign of plateau formation (Fig. 5). At pressures greater than 7 kbar, plateaus appear owing to a drastic reduction in parallel conduction.

We have seen (Fig. 2) that Hall and SdH give the same carrier density at 4.2K. This is rigorously true only at high pressure while we observe $N_H > H_1 + N_0$ at ambient pressure. A study of the high field Hall effect below 150K demonstrates the existence of a large density of free carriers in AlInAs ($\simeq 5 \cdot 10^{11} \mathrm{cm}^{-2}$) that is still present at 4.2K (Grégoris et al 1986). Since the ratio $N_s \mu_s / N_b \mu_b$ is much greater than unity at 4.2K ($\mu_s \simeq 80\ 000 \mathrm{cm}^2 V^{-1} s^{-1}$, $\mu_b \simeq 600 \mathrm{cm}^2 V^{-1} s^{-1}$), the classical low field Hall effect gives precise values for $N_s$ and $\mu_s$, but quantum effects are strongly affected by parallel conduction. At high pressure conductivity becomes negligible in AlInAs and consequently $\rho_{xx}(B)$ shows no more positive magnetoresistance. Quantum plateaus develop in $\rho_{xy}(B)$ together with zeros in $\rho_{xx}(B)$ at values of B corresponding to integer Landau level filling factors.

MAGNETOPHONON RESONANCE UNDER HYDROSTATIC PRESSURE

## 1. The Magnetophonon Resonance

The magnetophonon resonance (MPR) effect manifests itself as an oscillatory behavior of the magnetoresistance caused by resonant interaction of longitudinal optical (LO) phonons and electrons. Phonon absorption and emission are favoured when two Landau levels are separated by the energy of the LO-phonon. Therefore a maximum develops in $\rho_{xx}(B)$ at the resonance condition:

$$\omega_{LO} = N\omega_c = N\frac{eB_N}{m^*} \qquad (8)$$

where $\hbar\omega_{LO}$ and $\hbar\omega_c$ are the phonon and cyclotron energies, $m^*$ is the polaron effective mass and N is the resonance index. (See e.g. Nicolas 1985). Stradling and Wood (1968) first proposed the following empirical

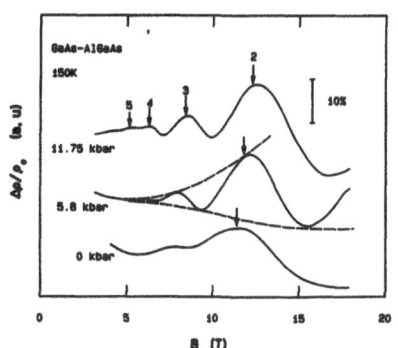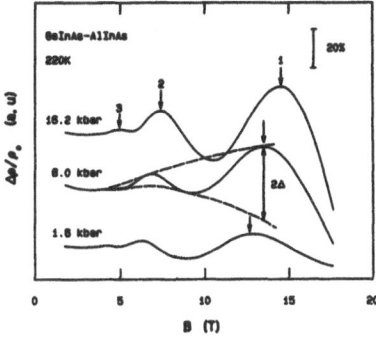

Fig. 6: Magnetophonon resonance oscillations in GaAs-AlGaAs and GaInAs-AlInAs for several pressures. The broken line shows the exponential damping of the oscillations, $\Delta$ is the amplitude and arrows point the resonances. At 0 kbar the fundamental field ($NB_N$) is found at 22.1T and 12.5T for GaAs-AlGaAs and GaInAs-AlInAs respectively .

rule for the amplitude of the oscillating part of $\rho_{xx}(B)$

$$\rho_{osc}/\rho_o \propto \exp - \gamma \frac{\omega_{LO}}{\omega_c} \cos 2\pi \frac{\omega_{LO}}{\omega_c} \tag{9}$$

which accounted for results in GaAs. The MPR effect has been widely used to investigate electronic properties of semiconductors and has received new attention with the development of heterostructures (Nicholas 1985). The observation of MPR requires high mobility and low temperature to satisfy the quantization conditions $\mu B > 1$ and $kT \ll \hbar\omega_c$, however high temperatures are also necessary to generate phonons. A compromise temperature is reached around 100-150K, but MPR can still be observed at 300K. The resonance condition enables one to deduce either $m^*$ or $\omega_{LO}$, and the amplitude of the oscillations gives one information about the strength of the electron-phonon coupling (through $\rho_{osc}/\rho_o$) and about the scattering processes (through the damping factor $\gamma$).

The amplitude of MPR oscillations usually represents a few percent of the $\rho_{xx}$ value, and, therefore most of the time, an amplification technique (second derivative or compensation of the monotonous part of the magnetoresistance) must be used to allow measurements to be made. We used the latter since we wanted to develop a quantitative study of the oscillations amplitude which would be affected by the band pass filter in the derivative technique. Typical recordings are shown for GaAs-AlGaAs and GaInAs-AlInAs for various pressures in Fig. 6. We will discuss here general aspects of these results. A detailed analysis will be the subject of a subsequent publication (Grégoris et al to be published).

## 2. Effective Mass and Non-parabolicity

For a non-parabolic conduction band, the effective mass of electrons is a function of their kinetic energy. In 2D systems, high kinetic energies are brought by electric quantization (electric sub-band $E_i$) and high carrier density (high Fermi energy $E_F$), and this brings about a corresponding increase of non-parabolicity. From an expression of the Landau levels based on kp theory (Palik et al 1961) and including statistical occupation of the Landau levels we can correct $m^*$ for non-parabolicity by using

$$\frac{1}{m^*} = \frac{1}{m^*_o} \left[ 1 + \frac{2K_2}{E_g} \sum_L W_L \langle E \rangle_L \right] \tag{10}$$

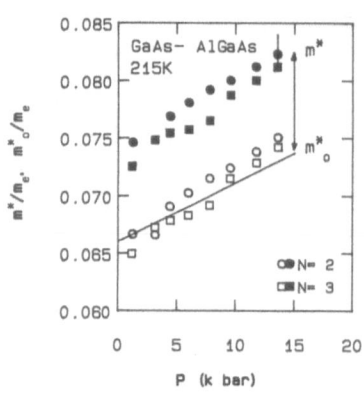
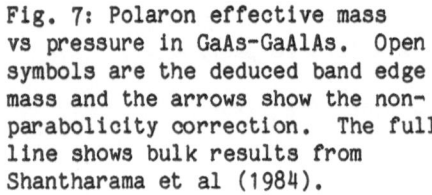
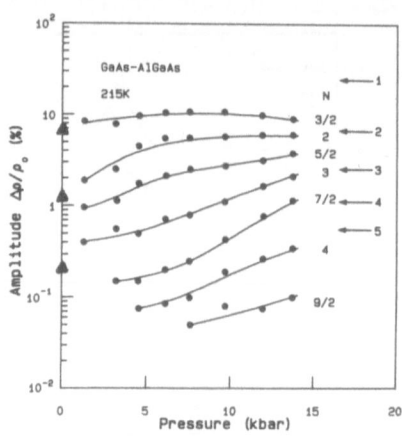

Fig. 7: Polaron effective mass vs pressure in GaAs-GaAlAs. Open symbols are the deduced band edge mass and the arrows show the non-parabolicity correction. The full line shows bulk results from Shantharama et al (1984).

Fig. 8: Amplitude $\Delta$ of the MPR oscillations with increasing P (decreasing $N_s$). Arrows indicate bulk results and triangles are 2D results from Kido et al (1982). Full lines are guides for the eye.

where $\langle E \rangle_L = (E_L + E_{L+N})/2$ is the mean energy at the transition L between Landau levels $E_L$ and $E_{L+N}$, and $W_L$ accounts for the occupancy of these levels. $K_2 (<0)$ is the non-parabolic coefficient, $m*_o$ is the band-edge effective mass and Eg is the gap. Equation (10) can be extended to take into account transitions in the two first sub-bands. We have considered the interface-induced shift of the LO-phonon energy towards the TO energy as observed in GaInAs-based heterojunctions ($\omega_{LO} \simeq 222$ cm$^{-1}$) (Nicholas et al 1985) and in GaAs-AlGaAs heterojunctions ($\omega_{LO} \simeq 282$ cm$^{-1}$) (Brummell et al to be published). A considerably enlarged non-parabolicity coefficient ($K_2 \simeq -1.5$) compared to kp predictions ($K_2 \simeq -0.85$) gave a reasonable agreement with ambient pressure band-edge masses measured in bulk GaAs and GaInAs. Such enlarged $K_2$ coefficients agree with previous cyclotron resonance measurements in GaInAs (Sarkar et al 1985), GaAs (Hopkins and Nicholas, to be published) and GaAs-AlGaAs (Hopkins et al to be published).

Experimental masses m* amd $m_o$* are plotted vs pressure in Fig. 7 (GaAs-AlGaAs only). As expected from kp theory, $m_o$* increases linearly with P, in proportion to the gap increase. Results are directly compared to measurements in bulk GaAs (Shantharama et al 1984) and in bulk GaInAs (Shantharama et al 1985) and a reasonable agreement is found. The deviation (m*-$m_o$*) /$m_o$* gives an experimental estimate of non-parabolicity in the related 2D systems. Non-parabolicity reaches 10% and 30% in the GaAs-AlGaAs and GaInAs-AlInAs heterojunctions respectively . It is smaller in the former case since GaAs has a larger gap. Slight discrepancies are however observed between bulk and heterojunction results for $m_o$*. The slope d$m_o$*/dP is too large in GaAs-AlGaAs and it is too small in GaInAs-AlInAs compared to bulk results. To clarify these discrepancies one will have to consider the four following remarks. First, kp theory does not fully agree with measurements of $m_o$*(P) under pressure in bulk GaAs and GaInAs (Shantharama et al 1984 and 1985). Second, the interface-induced shift of the phonon energy has not been fully interpreted. Third, the measurements of the 2D carrier density and the calculation of the sub-band energies in the well are poorly controlled at high temperatures. Finally it is noteworthy that we have not considered any polaron correction of m*. However, we observe an enhancement of the polaronic coupling with increasing P in GaAs-AlGaAs (discussed in section 3) which is likely to be the origin of the slightly larger slope $dm_o*/dP$ obtained in the heterojunction compared to that in bulk GaAs.

## 3.    Polaronic Coupling and Screening Effects

There is a direct relationship between the strength of the electron-LO phonon coupling and the amplitude of the oscillations. Theory predicts enhanced polaronic coupling for ideal 2D systems compared to bulk situations but it also mentions that this is reduced below bulk values when finite extension of the quasi-2D electron gas and screening effects are included (Das Sarma 1983), in agreement with experiment (Englert et al 1982, Brummel et al 1983 and Brummel et al to be published).

The measured amplitudes of the resonances are plotted in Fig. 8 (GaAs-AlGaAs only) on a semi-log scale as a function of P, i.e. as a function of decreasing $N_s$, for all the maxima N and minima N+½ that could be revealed.

In the GaAs-AlGaAs system, all the resonances are enhanced with increasing P while the 2D carrier density is simultaneously reduced by a factor of 3 at room temperature (from 3 to 1 $10^{11}$ cm$^{-2}$, $E_F < E_0$). As $N_s$ becomes very low the electron slowly experiences a transition from quasi-degenerate to non-degenerate statistics with a corresponding decrease of the screening efficiency. At high pressure, the amplitudes tend to reach the values observed at ambient pressure in bulk GaAs (non degenerate) for comparable high temperature mobility (Kido and Miura 1983).

In GaInAs-AlInAs, the fundamental resonance $\omega_c = \omega_{LO}$ is strongly enhanced as P is increased while the amplitude of the other resonances remain almost constant. Moreover, this N=1 resonance is abnormally small at low pressure compared to the expected exponential damping of the oscillations (eq. 9). Again we attribute the enhancement of the polaronic coupling to a decrease of screening. But this time the high degeneracy of the gas (5.8 $10^{11}$ cm$^{-2}$) causes the lowest Landau level to be heavily populated. Our calculations show that the Fermi level lies almost at the energy of this first Landau level (within a few meV). This might bring an influence on the screening, since it is known that screening depends on the density of states within kT of the Fermi energy (Ando and Murayama 1985).

The theory of Lassnig and Zawadzki (1984) about the MPR did not enable us to fit the curves $\rho_{osc}/\rho_o(B,P)$, even after including the formalism of the 2D screening in the Thomas-Fermi approximation for a quasi-2D system (Sigg et al 1985). We believe a k-dependent screening approach must be used to account for these results.

## A PRESSURED-INDUCED SEMIMETAL-SEMICONDUCTOR TRANSITION IN A GaSb-InAs-GaSb DOUBLE HETEROSTRUCTURE

The GaSb-InAs system has given rise to a great deal of interest since its peculiar band alignment leads to the coexistence of spatially separated 2D electron and hole gases (Esaki 1985). In this structure, the top edge of the valence band of GaSb is at higher energy than the bottom of the conduction band of InAs. The structure is undoped and the charge transfer to the well is mainly determined by the band discontinuity Δ and the thickness L of the InAs layer. The contribution of holes to conduction was quantitatively demonstrated for a 150A InAs layer imbedded into two GaSb layers (Mendez et al 1985). In our sample the electron and hole carrier densities, $N_e$ and $N_h$, deviate from the ideal condition ($N_e = N_h$) with $N_e \simeq 4 N_h$ (Mendez et al 1985). This could reflect the presence of

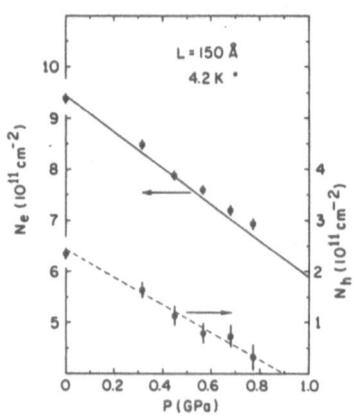

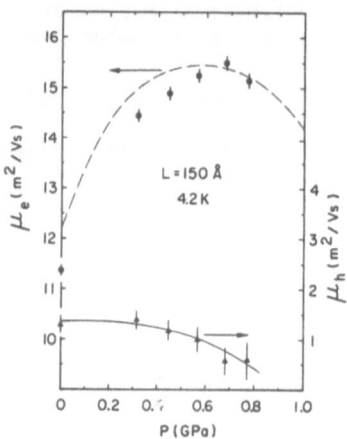

Fig. 9: Electron and hole concentration vs P. The full line is
the SdH carrier density. A semimetal-semiconductor transition
is observed at 8.5 kbar.

Fig. 10: Electron and hole
mobilities vs P. The broken
line is the low field Hall
mobility.

interface donor states. Low magnetic field Hall curves $\rho_{xy}(B)$ show
positive curvature, typical of mixed conduction by electrons and holes.
Using a best-fit of the experimental data to the classical expressions for
$\rho_{xx}(B)$ and $\rho_{xy}(B)$ permitted us to estimate electron and hole densities, $N_e$
and $N_h$, and mobilities, $\mu_e$ and $\mu_h$, at every pressure (Figs. 9 and 10).
The electron and hole densities decrease linearly with increasing
pressure, both at the same rate. The hole density vanishes around 8.5
kbar and a pressure-induced semimetal-semiconductor transition is observed
(Beerens et al submitted for publication). The electron mobility first
increases with pressure and starts to decrease as soon as $N_h$ becomes
negligible. No increase of $\mu_e$ has been observed in similar structures
with thinner InAs layers where no holes are present (Beerens et al to be
published). These results suggest scattering processes between electrons
and holes across the interface. Notice that $N_e$ and $\mu_e$ are close to the
classical Hall values $N_H$ and $\mu_H$ since the hole mobility is much smaller
than the electron mobility. Also $N_e$ obtained through the best-fit
procedure agrees with the density deduced from the Shubnikov-de Haas
oscillations.

The pressure-induced decrease of the carrier densities is attributed
to a reduction of the band discontinuity Δ. This is due to the fact that
the conduction band edge (which goes up in energy as P increases) is more
sensitive to pressure than the valence band edge. A comparative study
(Bereens et al to be published) between a semimetallic and a semiconductor
sample (different InAs thicknessess) enabled us to estimate the interface
donor density (of the order of $10^{12} cm^{-2} eV^{-1}$) and dΔ/dP ($\simeq$ -7meV/kbar), the
latter being in agreement with far-infrared magneto-absorption
measurements under pressure in InAs-GaSb superlattices (Maan this
workshop).

CONCLUSION

Hydrostatic pressure has been successfully applied to high magnetic
fields to investigate transport properties in semiconductor

heterostructures. Its success comes in essence from the fact that it enables one to control the carrier density in a given sample.

Hydrostatic pressure permitted us to develop a quantitative study of parallel conduction in a GaInAs-AlInAs heterojunction and made possible the first observation of the quantum Hall effect in this structure, due to the fact that parallel conduction becomes negligible over 7 kbar. Magnetophonon resonance measurements under pressure in GaAs-AlGaAs and GaInAs-AlInAs heterojunctions provided unique experimental data about the strength of the polaronic coupling in two dimensions as a function of the carrier density, i.e. of the 2D screening effect which is not yet well described by present theories. We do believe that these results will encourage further theoretical investigations in this field. Finally, by applying pressure on a GaSb-InAs-GaSb semimetallic system, we have observed a semimetal-semiconductor transition at 8.5 kbar. This gives interesting experimental grounds for the investigation of mixed transport phenomena as a function of the electron-hole compensation ratio.

## ACKNOWLEDGEMENTS

We would like to thank many of our coworkers who have provided the samples, in particular, D.L. Sivco and A.Y. Cho of Bell Labs., E.E. Mendez, L.L. Chang and L. Esaki of IBM and F. Alexandre of CNET. This work has been sponsored in part by the NATO and the Conseil Regional Midi-Pyrénées.

## REFERENCES

Ando, T. and Murayama, Y., 1985, J. Phys. Soc. Japan 54, 1519.

Beerens, J., Grégoris, G., Portal, J.C., Alexandre, F. and Aubin, M., Proc. of the 18th ICPS, Stockholm, Sweden 1986, to be published.

Beerens, J., Grégoris, G., Ben Amor, S., Portal, J.C., Mendez, E.E., Chang, L.L. and Esaki, L., submitted to Phys. Rev. B15.

Beerens, J., Grégoris, G., Portal, J.C., Mendez, E.E., Chang, L.L., and Esaki, L., to be published.

Brummell, M.A., Nicholas, R.J., Portal, J.C., Cheng. K.Y. and Cho, A.Y., 1983 J. Phys. C16, L579.

Brummell, M.A., Hopkins, M.A., Nicholas, R.J., Harris, J.J. and Foxon. C.T, to be published.

Das Sarma, S., 1983, Phys. Rev. B27, 2590.

Englert, T., Tsui, D.C., Portal, J.C., Beerens, J. and Gossard, A., 1982, Solid State Comm. 44, 1301.

Esaki, L., 1985, in "Molecular Beam Epitaxy and Heterostructures", L.L. Chang and K. Ploog ed., Martinus Nijhoff Publ., Dordrecht.

Grégoris, G., Beerens, J., Ben Amor, S., Dmowski, L., Portal, J.C., Sivco, D.L. and Cho, A.Y., 1986, accepted for publ. to J. of Phys. C.

Grégoris, G., Beerens, J., Dmowski, L., Ben Amor, S., Portal, J.C., Alexandre, F., Sivco, D.L. and Cho, A.Y., to be published.

Henning, J.C.M., Ansems, J.P.M., de Nijs, A.G.M., 1984, J. Phys. C17, L915.

Hopkins, M.A. and Nicholas, R.J. to be published.

Hopkins, M.A., Nicholas, R.J., Brummell, M.A., Harris, J.J. and Foxon, C.T., to be published.

Kane, M.J., Apsley, N., Anderson, D.A., Taylor, L.L., Kerr, T. 1985, J. Phys. C18, 5629.

Kido, G., Miura, N., Ohno, H. and Sakaki, H., 1982, J. Phys. Soc. Jap. 51, 2168.

Kido, G. and Miura, N., 1983, J. Phys. Soc. Jap. 52, 1734.

Konczykowski, M., Baj, M., Szafarkiewicz, E., Konczewicz, L., Porowski, S., 1978, Proc. Int. Conf. High Pressure and Low Temperature Physics, Cleveland, Ohio 1977 (Plenum Press, New York), p. 523.

Lassnig, R. and Zawadzki, W. 1984, Surf. Sci. 142, 361.

Martinez, G. 1980, in "Handbook of Semiconductors", vol. 2, M. Balkanski Ed. (North Holland Publ., Amsterdam), p. 181.

Mendez, E.E., Esaki, L. and Chang, L.L. 1985, Phys. Rev. Lett. 55, 2216.

Mercy, J.M., Bousquet, C., Robert, J.L., Raymond, A. Grégoris, G., Beerens, J., Portal, J.C., Frijlink, P.M., Delescluse, P., Chevrier, J. and Linh, N.T., Surf.

Mercy, J.M., Bousquet, C., Robert, J.L., Raymond, A., Grégoris, G., Beerens, J., Portal, J.C. and Frijlink, P.M., 1985, Proc. 17th Int. Conf. Physics of Semiconductors, San Francisco, 1984, J.D. Chadi and W.A. Harrison Ed. (Springer-Verlag, New York), p. 1099.

Nicholas, R.J., 1985, Prog. Quant. Electr. 10, 1.

Nicholas, R.J., Brunel, L.C., Huant, S., Karrai, K., Portal, J.C., Brummell, M.A., Razeghi, M., Cheng, K.Y. and Cho, A.Y., 1985, Phys. Rev. Lett. 55, 883.

Palik, E.D., Picus, G.S., Teitler, S., Wallis, R.F., 1961, Phys. Rev. B122, 475.

Robert, J.L., Mercy, J.M., Bousquet, C., Raymond, A., Portal, J.C., Grégoris, G., Beerens, J., 1984, in "Two-Dimensional Systems, Heterostructures and Superlattices, G. Bauer, F. Kuchar and H. Heinrich Ed. (Springer-Verlag, Berlin), p. 252.

Sarkar, C.K., Nicholas, R.J., Portal, J.C., Razeghi, M., Chevrier, J. and Massies, J., 1985, J. Phys. C18, 2267.

Shantharama, L.G., Adams, A.R., Ahmad, C.N. and Nicholas, R.J., 1984, J. Phys. C17, 4429.

Shantharama, L.G., Nicholas, R.J., Adams, A.R. and Sarkar, C.K., 1985, J. Phys. C18, L443.

Sigg, H., Wyden, P., Perenboom, J.A.A.J., 1985, Phys. Rev. B31, 5253.

Stradling, R.A. and Wood, R.A., 1968, J. Phys. C1, 1711.

# PARTICIPANTS

Dr Gerhard Abstreiter
Physik Dept der Technischen
Universität Munchen
8046 GARCHING b. MUNCHEN
Federal Republic of Germany

Dr Massimo Altarelli
Max-Planck-Institut
Hochfeld Magnetlabor
166 X
38042 GRENOBLE CEDEX
France

Dr Gerald Bastard
Laboratoire de Physique ENS
24 rue Lhomond
75005 PARIS CEDEX 05
France

Professor Gunther Bauer
Institut für Physik
Monatuniversität Leoben
Franz-Josef Strasse 18
A-8700 LEOBEN
Austria

Dr Recai Ellialtioglu
Marmara Gebze Research Centre
P.O. Box 74
GEBZE-KOCAELI,
Turkey

Dr Annalisa Fasolino
SISSA Strada Costiera 11
34100 TRIESTE
Italy

Professor Jean Pierre Faurie
Dept of Physics
University of Illinois at Chicago
P.O. Box 4348 / Chicago
ILLINOIS 60580
USA

Professor J K Furdyna
Dept of Physics
Purdue University
West Lafayette
INDIANA 47907
USA

Dr Vasco Pires S. Gama
Dept Quimica
ICEN / LNET I
Estrada Nacional 10
2686 SACAVEM
Portugal

Dr Dolores Golmayo
Centro Nacional de Microelectronica
Serrano 144
28006 MADRID
Spain

Dr Luisa Gonzalez
Centro Nacional de Microelectronica
Serrano 144
28006 MADRID
Spain

Professor Erich Gornik
Institut für Experimentalphysik
Technikerstr. 15
A-6020 INNSBRUCK
Austria

Dr Guy Gregoris
Microphysics M23A-147
National Research Council
OTTOWA   K1A   OR6
Canada


Dr Philip Klipstein
Blackett Laboratory
Imperial College
Prince Consort Road
LONDON   SW7   2BZ


Professor Frederik Koch
Physik Department
Technische Universität Munchen
8046   GARCHING b. MUNCHEN
Federal Republic of Germany


Professor Friedemar Kuchar
Ludwig Boltzmann Institut
Kopernikusgasse 15
A-1060   WIEN 1
Austria


Professor Thomas C. McGill
T J Watson Sr Lab. of App.Physics
California Inst. of Tech. 128-95
Pasadena
CALIFORNIA   91125
USA


Dr Jankees C. Maan
Max-Planck-Institut
Hochefeld Magnetlabor
166 X
38042   GRENOBLE CEDEX
France


Dr J Y  Marzin
C N E T
196 rue de Paris
92220   BAGNEUX
France


Professor Ulrich Merkt
Institut für Angewandte Physik
Universität Hamburg
Jungiusstrasse 11
2000   HAMBURG  36
Federal Republic of Germany

Dr Francisco Mezeguer
Dept de Fisica
Universidad Autonoma de Madrid
Ciudad Universitaria
Canto Blanco
28049   MADRID 34
Spain


Dr Alan Miller
R S R E
St. Andrews Road
Great Malvern
WORCS. WR14   3PS


Dr David A B  Miller
AT & T Bell Laboratories
Crawford Corner Road
Holmdel
NJ   07733
USA


Dr Maria-Helena Nazare
Dept de Fisica
Universidade de Aveiro
AVEIRO
Portugal


Professor Carl Pidgeon
Dept of Physics
Heriot-Watt University
Riccarton
EDINBURGH   EH14   4AS


Professor Jean-Claude Portal
SNCI-CNRS
B P 166
38042   GRENOBLE CEDEX
France


Dr Manijeh Razeghi
Thomson-CSF
Laboratoire Central de Recherche
B P 10
91401   ORSAY
France


Professor Brian K. Ridley
Dept of Physics
University of Essex
Wivenhoe Park
COLCHESTER  CO4   3SQ

Professor J.L. Robert
Groupe d'Etudes de Semicon.
U S T L
Place E Bataillon
34060  MONTPELLIER CEDEX
France


Professor Wilson Sibbett
Dept of Physics
University of St. Andrews
North Haugh
ST. ANDREWS
Fife  KY16  9SS


Dr John Singleton
Clarendon Laboratory
Parks Road
OXFORD  OX1  3PU


Richard Sizmann
Physik Department
Technische Universität München
8046  GARCHING b. MUNCHEN
Federal Republic of Germany


Dr Maurice Skolnick
R S R E
St. Andrews Road
Great Malvern
WORCS.  WR14  3PS


Dr Clivia M. Sotomayor-Torres
Dept of Physics
University of St. Andrews
North Haugh
ST. ANDREWS
Fife  KY16  9SS


Professor Ian Spain
Dept of Physics
Colorado State University
Fort Collins
COLORADO  80523
USA


Professor R. Anthony Stradling
Blackett Laboratory
Imperial College
Prince Consort Road
LONDON  SW7  2BZ


Dr Seigo Tarucha
Max-Planck-Institut for Solid State
    Physics
Heisenbergstrasse 1
7000  STUTTGART  80
Federal Republic of Germany


Dr Paul Voisin
Laboratoire de Physique ENS
24 rue Lhomond
75231  PARIS CEDEX  05
France


Professor Claude Weisbuch
Thomson- CSF
Laboratoire Central de Recherche
B P 10
91401  ORSAY
France


Dr Colin Whitehouse
R S R E
St. Andrews Road
Great Malvern
WORCS. WR14  3PS


Professor H.H. Wieder
Dept of Elect.Engineering & Comp.Sc.
University of California at San Diego
Mail Code C-014 La Jolla
CALIFORNIA  92093
USA


Professor Wlodek Zawadski
Institut für Experimentalphysik
Technikerstr. 15
A-6020  INNSBRUCK
Austria

Participants in the NATO ARW on Optical Properties of Narrow Gap Low Dimensional Structures

# AUTHOR INDEX

Quantum Hall effect, 15-16, 32-34, 138, 193, 318, 342

Q.W. lasers (see lasers)


Raman scattering, 269
  CdMnTe/CdMnTe SLS, 143-144
  GaSb/AlGaSb, 55-56, 60
  HgTe/CdTe SLS, 29
  single particle excitation, 271
  collective excitation, 143, 272-273
  depolarisation shift, 271
  electric field induced, 275
  lineshapes, 273
  CARS, 119, 124
  Resonant, 34, 92, 270, 273

RHEED, 111

Rutherford backscattering, 55-56, 58


Semimetal-semiconductor transition, 31, 87
  pressure induced, 330, 338, 345-347

SIMS of III-Vs, 39, 41-42

Skipping orbits, 195-202

sp-d interaction, 135-139, 143

Spin density fluctuations, 273

Spin-flip
  resonances, 124
  transitions, 124, 131

Spin-orbit interaction, 192, 207, 221

Spin-split bands, 187-188, 192

Spin splitting in DMS, 135, 137, 139, 141

Spin sheets, 144

Spin superlattice, 141

Strain
  biaxial, 55, 68, 91, 93, 104, 119-120
  misfit, 85, 91, 100
  surface degradation with, 111

Strained layers

critical thickness, 93, 100, 109, 276
heterostructures, 39-40
superlattice, 26, 68, 101-102, 106, 108-110, 228
elastic energy, 100 112

Subband occupancy, 196

Subband splitting, 118, 192, 271
  strain induced, 139

Superlattice
  open gap/zero gap (type III), 73, 139
  subband width, 18, 20
  type I, 31, 106, 118, 121, 124
  type I', 118, 122
  type II, 85, 106-107, 122, 337
  type III, 25, 30-31, 139
  type III-type I transition, 25-28, 33-34
  zero gap/zero gap, 139


Surface migration, 64


Transmission electron microscopy, 55-56, 61-64, 112-113

Tunnelling, 240
  resonant, 142-143, 205

Vibronic excitations, 275

Virtual crystal approximation, 222

Weber function, 205, 207-209, 212

X-ray diffraction
  GaAs/InP, 41
  GaSb/AlSb, 57-58, 91
  GaSb/AlGaSb, 55-56, 64, 68
  HgTe/ZnTe, 33
  HgMnTe/CdTe SLS, 28
  InAs/GaAs/InP, 40-41
  InP/GaAs, 41
  InGaAs/GaAs, 102-103

X-ray interference in PbTe/PbSnTe, 119-120

X-ray photoemission spectroscopy, 30, 34
  CdTe/HgTe, 29
  HgTe/CdTe, 29